Nature of
EARTH MATERIALS

ANTHONY C. TENNISSEN

Department of Geology
Lamar University

Prentice-Hall, Inc., *Englewood Cliffs, New Jersey*

Library of Congress Cataloging in Publication Data

TENNISSEN, ANTHONY C.
  Nature of earth materials.

  Bibliography: p. 413
  1. Mineralogy. 2. Petrology. 3. Mineral
industries. I. Title.
QE364.T46   552   73–7565
ISBN 0–13–610501–7

© 1974 by Prentice-Hall, Inc., Englewood Cliffs, New Jersey

10 9 8 7 6 5 4 3 2 1

Printed in the United States of America

Prentice-Hall International, Inc., *London*
Prentice-Hall of Australia, Pty. Ltd., *Sydney*
Prentice-Hall of Canada, Ltd., *Toronto*
Prentice-Hall of India Private Limited, *New Delhi*
Prentice-Hall of Japan, Inc., *Tokyo*

Dedicated to Virginia

Contents

Preface

This book is concerned with *earth materials*, chiefly minerals and rocks, but also petroleum and water. Particular emphasis is placed on the nature of these materials; however, the origin of minerals and rocks is also briefly discussed. Commercial uses of metals, minerals, and rocks are brought out in the latter part of the book. The book is intended primarily as a text for a college course of Earth Materials, but it should have general appeal to geologists and rockhounds as well.

A fact which strongly influenced the writing of this book is that, unfortunately, a book covering the full topics of Earth Materials for college instruction is unavailable. In most colleges and universities offering a course of Earth Materials as part of an Earth Science curriculum, students are required to have completed at least an introductory course in Physical Geology. Such students are not science majors, and hence the context material has been kept fairly uncomplicated.

The book is arranged in four parts, dealing respectively with (1) atoms, ions, and crystals, (2) minerals and their properties, (3) rocks, their origins and characteristics, and (4) industrial use of minerals and rocks.

The first part attempts to show the nature of atoms, ions, and molecules, and how they are bonded together to make crystals. Crystal growth is discussed, crystal geometry is explained, and the classification of crystals on the basis of symmetry is presented. The text material is augmented by numerous sketches of atoms, ions, and crystal forms. The discussion is not an advanced treatment of crystallography, but is intended to introduce the subject for non-science students.

The second part deals with minerals. About 114 of the most common minerals are described. A definition of a mineral is presented, along with a discussion of its physical properties and certain chemical aspects. An attempt has been made to present a fairly concise description of the field of mineralogy as an area of study, and the text is augmented by photographs that reveal the physical nature of minerals. Again, no advanced treatment of mineralogy is presented, but basic mineralogical material is given.

The third part is concerned with the origin and nature of igneous, sedimentary, and metamorphic rocks. Compositions and environments of formation are briefly explained, and classifications are presented. Photographs of a great number of rock types assist in evaluating the nature of the rocks. However, the text material is not a full treatment of rock study. It is hoped that this section of the book will appeal to students who are interested in rocks and who have toyed with the idea of a career in geology.

The fourth part is designed for those who wish to obtain information on the utilization of ores, metals, industrial minerals and rocks, and even petroleum. The nature and origins of mineral deposits are discussed in uncomplicated terms. In addition, selected commercial ores, minerals, and rocks are discussed in some detail. Properties which make these materials valuable and sources of supply are brought out. The materials discussed are by no means all of the commercial minerals and rocks, but they represent the more common and important ones. The discussion of any particular ore, metal, mineral, or rock is necessarily brief, but it is hoped that these discussions will arouse the curiosity of many students to pursue the topics further.

The book is informative and easily understood. It is hoped that it can be used as a readable, instructive text, and also in part as a reference on minerals and rocks. The book will hardly serve for advanced study, but it may be inspirational to a few students.

Appreciation is given to Professor William H. Matthews III for generous discussion and suggestions regarding the manuscript. Other colleagues at Lamar University who made noteworthy suggestions are Drs. Robert R. Wheeler, James B. Stevens, and William R. Pampe. Thanks are due to Dr. H. E. Eveland, Department Head, for absorbing incidental costs with departmental funds.

Margaret Stevens drew all the black and white line illustrations, and special thanks are extended to her. Unless otherwise indicated, photographs were made by Jackie F. James, Director of Photographic Services, Lamar University. Typing of the manuscript was done by Linda Mack, and I wish to extend my gratitude to her. Specimens of turquoise and beryl were loaned for photographic purposes by Eugene P. Crowell. All other mineral and rock specimens are from the Geology Department collection, Lamar University.

Anthony C. Tennissen

Chapter One

Introduction

The science of Earth Materials deals primarily with the inorganic solid materials of the Earth's crust, i.e., minerals and rocks, but it does include the liquids water and oil. It does not include organic material such as living trees, plants, flowers, or animals. However, certain materials such as coal, which contain fossilized trees and plants, and also some material derived from the decomposition of trees, plants, and various organisms, are also included. Actually, "Earth Materials" as a field of study includes minerals and mineral materials of several types, rocks of three main types, and also the commercial uses and applications of these solid materials.

The terms *mineral* and *rock* are both rather flexible, each meaning something different to different people. To some people, mineral might mean any nonliving substance and thus it has an extremely broad connotation. On the other hand, a pharmacist may call a vitamin a mineral and be perfectly correct in his much smaller sphere of usage. A prospector may call anything he recognizes as valuable a mineral, and he, too, would be perfectly correct in his somewhat more restricted meaning of the term. However, a mineralogist usually refers to something as a mineral only if it possesses certain physical and chemical properties, and very likely he would not include the substances that a pharmacist and a prospector might call mineral. How the term mineral has come to mean different things to different people will not be discussed here, but it is important to know that it is one of the flexible English words.

Perhaps, though, the term mineral should be restricted to mean what a mineralogist says it means. After all, a mineralogist has been trained to recognize a large number of minerals. Since it is generally agreed that nearly all so-called minerals scientifically belong in the realm of mineralogy, a mineralogist is in a much better position to evaluate what is mineral and what is not. When a mineralogist is in training in college, he learns that, in order for a substance to be a true mineral, it must have certain specific properties, both physical and chemical, which are used to make the evaluation. In fact, there are six properties which are evaluated, and they can be seen in the following mineralogical definition: A mineral is a (1) *naturally* occurring (2) *inorganic* (3) *solid* (4) *crystalline* substance with a (5) *definite chemical composition* and (6) *characteristic crystal structure*. The only one of these properties which is very easily determined is whether a substance is *solid* or not. It is a pretty simple matter to say that a substance is solid, and not liquid or gas. These are states of matter that even a child can differentiate. The remaining five properties are not so easy to detect. For example, it would be pretty difficult to say that a particular solid substance, in some cases, had been formed by *natural* rather than synthetic processes. More explicitly, a synthetic quartz crystal has the same physical and chemical properties and looks just like a natural quartz crystal. Thus it would be rather difficult to say which one of them was a natural substance. It would be necessary to have additional explanatory information. And to prove that a certain substance had been formed by *inorganic* and not organic processes also would be difficult because a few substances may be formed by organic as well as by inorganic processes. Again, more information would be required to explain the circumstances.

In a similar manner, to say that a substance has a definite chemical composition would require some sort of chemical analysis to show which elements are present and their proportions. In college, a mineralogist learns that the composition of some minerals may vary a little by the substitution of one atom or ion for another atom or ion *within certain limits*. However, such substitution and resultant variation does not bias the definition.

To make things even more complicated, the atoms (ions) of which the substance is composed must be arranged in some systematic repetitive pattern in the crystals. We all know that atoms are very small and that we cannot see them. In order to determine that the atoms are arranged in a systematic pattern in a substance requires some very sophisticated and expensive equipment such as X-ray diffraction or neutron diffraction devices. In addition, considerable time is required to make a crystallographic study which would determine the systematic repeat-pattern of the atomic arrangement and also provide details of the distances between various atoms or ions, the positions of the atoms or ions, and the sizes of the constituent atoms or ions.

So you see, already the introductory material in this chapter has a very complicated but scientific connotation. It is probably more proper then to allow the mineralogist to define a mineral in the mineralogical sense and agree that exceptions do exist. It is also true that even the mineralogical sense is not always clear. For example, is oil (and natural gas) a mineral in the mineralogical sense? Not really, because oil is a liquid (and natural gas is a gas), not a solid as specified in the definition. Furthermore, oil forms mainly by organic processes of decomposition and regeneration of new substances.

Oil and gas may not be minerals in the mineralogical and scientific sense of conforming to the six restrictive properties. But ask any oil company lawyer if oil and gas are minerals—the answer will undoubtedly be yes. And so another modification of the definition comes into play, or rather an exception is permitted. The same problem is encountered concerning ground water being extracted from below the surface in many areas for ranch, city, farm, and individual uses, and this again is an exception to the hard and fast scientific definition.

To many people—even scientists—the definition of a mineral given above seems perhaps too restrictive to be very practical; perhaps it should be more flexible. The definition seems to be very suitable for most earth substances, but it should also include those other materials which normally fall into the sphere of earth materials. However, before we consider "mineral" any further with respect to the term's flexibility, let us look briefly at the term "rock."

In the scientific (geologic-petrologic) sphere of usage, a rock is considered as an aggregation of minerals and/or of mineral-like substances. But this is not perfectly accurate. Rarely, a rock will be composed of crystals of only one mineral, and only occasionally will the crystals of only two minerals be involved. Most rocks are composed of crystals of two or more (often four or five) different minerals.

Since a rock is usually composed of minerals, then it too must be a naturally occurring inorganic solid substance, usually made up of crystals of minerals or mineral-like substances, but whose chemical composition may vary within broad limits. Even so, there are great numbers of exceptions to this definition, just as there are numerous exceptions to the definition of mineral. Take the case of a volcanic rock known as *obsidian*. Obsidian is composed almost exclusively of volcanic glass—there are *no crystals* of any mineral because the material cooled so rapidly from a very hot condition that no crystals could form. The glass formed as the volcanic lava was rapidly cooled. Yet obsidian is considered a *rock* in the geological sciences even though it is *not* composed of an aggregation of different minerals. It is, however, composed of a mineral-like substance. Obsidian is more like a liquid than a solid as far as the random arrangement of the constituent atoms (ions) is concerned. Furthermore, some rocks, such as felsite, are composed of crystals which are so small that it is impossible to determine which minerals are present if we use the common means of identification (hand lens, scratch test, or streak test). To determine specifically just which minerals are present might require optical study, X-ray study, or various chemical analyses before a rock name can be properly applied.

Another example is coal. Is coal a rock? By petrologic (rock) definition, to be called a rock, coal should be made of natural inorganic crystalline material. But it isn't. Coal is composed of organic material produced by decomposition of vegetation along with resistant portions of plants; it also includes regenerated organic solids as decomposition proceeded to various stages of completion. The composition is highly variable, and the coal actually is more like a liquid (like supercooled obsidian) in its random atomic arrangement than most other so-called rocks. Nevertheless, in the field of petrology, coal is generally considered a rock because it is composed of mineral-like materials. But this again is an exception.

Going a step further, there are some who say that, because the composition of oil is variable like the composition of coal, perhaps oil might better be classed as a rock rather than as a mineral as it usually is. Even though it is liquid, oil possesses the other physical properties of coal except for its state of being. Why not call oil a *rock*, then, the same as coal?

A rock may vary somewhat in composition from place to place, according to the proportion of the various minerals present, within very broad limits, without having its name changed, even when the crystals are large and easily identified. Not so for a mineral; the composition may change only within very narrow limits. If changes are too large, the mineral name gets changed. It is quite easy to see that, because of such a great variety of materials in the Earth's crust, there will be exceptions to any definitions which may be devised for minerals and for rocks. So perhaps it is best to include those materials as minerals or rocks which are considered as such by the mineralogists and petrologists. It is not really essential that we adhere so closely to those hard and fast definitions of minerals and rocks; we must make some exceptions.

Chapter Two

Atoms, Ions, and Crystals

A substance composed of atomic groups which are alike is considered a pure substance. Some pure substances *cannot* be decomposed into simpler substances because the atomic groups contain only one kind of atom. These substances composed of only one type of atom are the *elements*, such as iron, aluminum, copper, cobalt, and lead (see Table 2-1 for a list of common ele-

Table 2-1: COMMON COMPOUNDS

Common Elements in Combination	*Compound Name and Formula*
Carbon (C) + oxygen (O)	Carbon dioxide (CO_2)
Calcium (Ca) + chlorine (Cl)	Calcium chloride ($CaCl_2$)
Copper (Cu) + sulfur (S)	Copper sulfide (Cu_2S)
Iron (Fe) + oxygen (O)	Iron oxide (Fe_2O_3)
Silicon (Si) + oxygen (O)	Silicon dioxide (SiO_2)
Calcium (Ca) + carbon (C) + oxygen (O)	Calcium carbonate ($CaCO_3$)
Sodium (Na) + chlorine (Cl)	Sodium chloride (NaCl)

ments). But there are other pure substances which *can* be decomposed into two or more simpler substances because the atomic groups consist of more than one kind of atom. These are called *compounds*, such as iron oxide (containing iron and oxygen), copper sulfide (copper and sulfur), and aluminum fluoride (aluminum and fluorine).

Most materials of the earth are compounds and are made up of more than one kind of atom (or ion) which are not preferentially associated with one another, that is, there are no *molecules*. On the other hand, a few materials exist which do consist of atoms or ions preferentially arranged with one another and thus are made up of molecules.

What is the nature of these constituent particles of matter? How can we best picture the construction or makeup of atoms, ions, and molecules? Why are there various kinds of chemical elements? We all know that atoms, ions, and molecules are extremely small particles and that they cannot be seen as individuals. In order to reveal their nature, very sophisticated equipment is used to obtain information about them. Data so accumulated over the past two or three decades by nuclear scientists have given us a very clear picture of atoms, ions, and molecules, as well as of the nature of binding forces attracting and holding them together.

Structure of Atoms, Ions, and Molecules

Nuclear science reveals that an atom is made up of an extremely small, positively charged nucleus which is surrounded by one or more shells (or layers) of negatively charged electrons, the whole acting as a sphere. The radius of an atom is usually measured in Angstrom units (1 Angstrom = 1/100,000,000 cm and is symbolized Å). Atoms of different elements have somewhat different sizes, although still in the realm of 1/100,000,000 cm. For example, the radius of the iron atom is 1.27/100,000,000 cm (1.27 Å); and the radius of the oxygen atom is 0.61/100,000,000 cm (0.61 Å).

Each atom consists of a tiny nucleus containing protons, neutrons, and other particles. The protons bear positive (+) electrical charges, and the nucleus itself is located at or very near the center of the space available for the whole atom. The number of positive charges in the nucleus is called the *atomic number*. Surrounding the nucleus are negatively (−) charged particles called electrons. The number of electrons corresponds to the number of protons in the nucleus, as shown in Fig. 2-1.

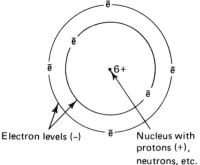

Electron levels (−)   Nucleus with
                             protons (+),
                             neutrons, etc.   *Fig. 2-1:* Structure of an atom.

The reason there are various kinds of chemical elements is that atoms of different elements consist of different numbers of protons with corresponding electrons. In fact, there are 104 different numerical arrangements of protons and electrons, and hence there are 104 different chemical elements so far discovered. The electrons revolve around the nucleus in concentric or nearly concentric orbit, traveling at very high velocities approaching the speed of light. The positively charged nucleus exerts an attractive force on the negative electrons which tends to pull the electrons into the nucleus. But this force is balanced by the centrifugal force which results from the circular motion of the electrons at different levels around the nucleus and tends to move the electrons away from the nucleus. Consequently, a definite quantity of energy is associated with an electron while it is revolving in its orbit. In an orbit far away from the nucleus, the energy associated with an electron is greater than that associated with an electron closer to the nucleus. If an atom becomes "excited" (absorbs more energy) by heat, radiation, or other means, one or more electrons in an inner orbit may move up to a higher energy level or may leave the atom altogether.

It is also known that atoms do not stand absolutely still; rather, they vibrate in their positions around a point. Because the electrons travel so rapidly as they revolve around the nucleus, there is a great deal of energy involved in holding the atoms together. This electrical energy of attraction of nucleus for electrons (and vice versa) is very strong. But it is not so strong, in cases of atoms of many elements, that one or more electrons in the outer energy levels (orbits) may not move away from the nucleus and its sphere of influence as a result of heat or radiation. When this happens, the atom becomes ionized. Neither is the attraction so weak that one or more electrons from another atom may not be added to the sphere of influence of the nucleus to which it does not properly belong, in the cases of many other elements. The atom becomes ionized in this way. This phenomenon of ionization, where atoms of some elements lose one or more electrons and atoms of other elements gain one or more electrons, is extremely common. In fact, it is nearly impossible for any nucleus to hold, with its attractive forces, exactly the number of electrons it is supposed to hold. Because of a great many other environmental factors which cause changes in the amounts of energy in atoms, electrons are lost easily by some atoms and gained easily by others.

When there are exactly the same number of electrons as there are protons, the atom has no residual unbalanced electrical charge because all charges are satisfied. But when electrons are lost, all of the remaining positive charges are not satisfied, and the atom is then a *positive ion* called a *cation* (Fig. 2-2). If

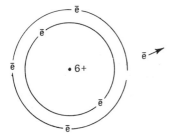

Fig. 2-2: Cation, since one electron (ē) is lost.

an atom gains electrons, it accepts extra negative charges which are not satisfied because there are not enough positive charges. The atom is then a *negative ion* called an *anion* (Fig. 2-3). Furthermore, a *positive ion* is very

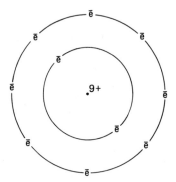

Fig. 2-3: Anion, since one electron (ē) is gained.

capable of being attracted by a *negative ion*, and vice versa. It is this type of electrical attraction of oppositely charged ions which permits matter to be held firmly together to form solid compounds, including minerals, of varying complexities.

Sizes of Atoms and Ions

When an atom loses an electron, there is a loss of an energy level around the nucleus. There is not only a loss in the amount of energy itself, but also a reduction in the sizes of the remaining electron levels. For example, the size (radius) of the iron atom is 1.27 Å, but when iron loses two electrons, the effective size becomes 0.74 Å. Similarly, when an atom gains an electron, there is a gain of an energy level around the nucleus, not only a gain in the amount of energy itself, but also an increase in the size of the whole ion. The radius of the chlorine atom is 0.99 Å. But when it gains an electron, its effective size increases to 1.81 Å (see Table 2-2). Consequently, the effective radius of an ion depends partly on the nature of the element itself, but more importantly, on the state of ionization and the manner in which it is linked to adjacent ions.

Electrostatic Bonding Among Minerals

Atoms are attracted to each other by electromagnetic forces. These interatomic forces are responsible for the coherence and stability of the atom groups. X-ray crystal structure analysis, developed in the early part of the 20th century, revealed the nature of these interatomic forces. It detected the atomic configuration of structures, thereby opening the way to a physical analysis of the chemical forces.

Table 2-2: ATOMIC RADII AND IONIC RADII OF COMMON ELEMENTS
(in Angstroms)

Atomic No.	Element	Symbol	Atomic Radius	Ionic Radius	
1	Hydrogen	H	0.53	H^+	
3	Lithium	Li	1.56	Li^+	0.68
4	Beryllium	Be	1.12	Be^{2+}	0.35
5	Boron	B	1.39	B^{3+}	0.23
6	Carbon	C	0.77	C^{4+}	0.16
7	Nitrogen	N	0.53	N^{5+}	0.13
8	Oxygen	O	0.61	O^{2-}	1.40
9	Fluorine	F	0.68	F^-	1.36
11	Sodium	Na	1.91	Na^+	0.97
12	Magnesium	Mg	1.60	Mg^{2+}	0.66
13	Aluminum	Al	1.43	Al^{3+}	0.51
14	Silicon	Si	1.17	Si^{4+}	0.42
15	Phosphorus	P	1.08	P^{5+}	0.35
16	Sulfur	S	1.06	S^{6+}	0.30
				S^{2-}	1.85
17	Chlorine	Cl	0.97	Cl^-	1.81
19	Potassium	K	2.38	K^+	1.33
20	Calcium	Ca	1.96	Ca^{2+}	0.99
22	Titanium	Ti	1.46	Ti^{4+}	0.68
23	Vanadium	V	1.35	V^{3+}	0.74
24	Chromium	Cr	1.28	Cr^{3+}	0.63
25	Manganese	Mn	1.25	Mn^{2+}	0.80
26	Iron	Fe	1.27	Fe^{2+}	0.74
27	Cobalt	Co	1.25	Co^{2+}	0.72
28	Nickel	Ni	1.24	Ni^{2+}	0.69
29	Copper	Cu	1.27	Cu^{2+}	0.72
				Cu^{1+}	0.96
30	Zinc	Zn	1.37	Zn^{2+}	0.74
33	Arsenic	As	1.25	As^{5+}	0.46
38	Strontium	Sr	2.14	Sr^{2+}	1.12
40	Zirconium	Zr	1.59	Zr^{4+}	0.79
42	Molybdenum	Mo	1.40	Mo^{6+}	0.62
47	Silver	Ag	1.44	Ag^{1+}	1.26
50	Tin	Sn	1.58	Sn^{4+}	0.71
51	Antimony	Sb	1.61	Sb^{5+}	0.62
				Sb^{3+}	0.76
56	Barium	Ba	2.24	Ba^{2+}	1.34
73	Tantalum	Ta	1.47	Ta^{5+}	0.68
74	Tungsten	W	1.41	W^{6+}	0.62
78	Platinum	Pt	1.38	Pt^{2+}	0.80
79	Gold	Au	1.43	Au^{1+}	1.37
80	Mercury	Hg	1.55	Hg^{2+}	1.10
82	Lead	Pb	1.74	Pb^{4+}	0.84
				Pb^{2+}	1.20
83	Bismuth	Bi	1.82	Bi^{3+}	0.96
90	Thorium	Th	1.80	Th^{4+}	1.02
92	Uranium	U	1.50	U^{4+}	0.97

A review of crystal structures, analyzed by X-ray methods, of a great many types of solids during the past several decades emphasized the very important role of the interatomic forces which determine the structural arrangement of the atoms. These interatomic forces are conveniently divided into the following four types:

(1) ionic,
(2) covalent,
(3) metallic, and
(4) Van der Waals.

The distinction between these four bond types is not absolute. In many crystals, the bond between atoms or ions may possess an intermediate character and may show properties of two or more types. In addition, two or more different types of bonding may function in the same crystal between different atoms.

The Ionic Bond

The ionic bond is the simplest of the four types of interatomic bonding. This type of bond occurs as a result of the force of electrostatic attraction between oppositely charged ions, i.e., between cations and anions.

A simple example of a mineral with ionic bonding is halite (NaCl). Sodium is an element which, by losing one electron, becomes a positively charged ion—a cation. On the other hand, chlorine is an element which, by adding an additional electron, becomes a negatively charged ion—an anion. This type of adjustment, of course, requires the presence of an atom that can give up the additional electron and still maintain a stable configuration. Sodium can do this. In halite, the sodium ions and chlorine ions occupy alternating positions throughout the crystal. An apparent feature of this structure is that there is no trace of any molecule NaCl. Rather, each sodium ion (Na) is symmetrically surrounded by six chlorine (Cl) neighbors. It is common practice to say in this arrangement that each sodium ion is in sixfold coordination with chlorine; similarly, each chlorine is symmetrically surrounded by six sodium ions and thus is in sixfold coordination with sodium. The sodium, now positively charged, and the chlorine, now negatively charged, can combine easily to form into a structure of oppositely charged ions bonded by electrostatic attraction (Fig. 2-4). Each ion is surrounded by ions of opposite charge; there is no pairing of individual positive ions (Na) with negative ions (Cl) to give discrete molecules. Every sodium ion is equally attracted to six sodiums. In an ionic crystal, the number of neighbors to which an ion may be bound is determined primarily by geometrical considerations and relative sizes of the ions.

Most of the inorganic compounds are bonded by ionic types of bonding.

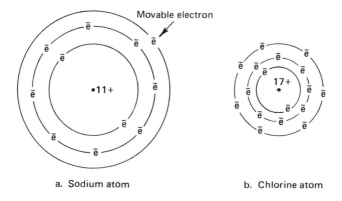

a. Sodium atom   b. Chlorine atom

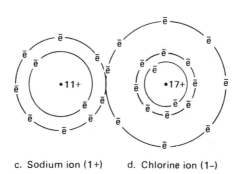

c. Sodium ion (1+)   d. Chlorine ion (1−)

Fig. 2-4: The ionic bond.

The minerals, with the exception of sulfides and native elements, are ionic compounds.

The Covalent Bond

The most stable configuration for an atom is one in which the outer shell of electrons is completely filled. A few atoms, such as hydrogen require only two electrons to achieve this stable arrangement. However, a great many other atoms require eight electrons to achieve this state. The covalent bond, which depends on atoms filling their outer orbits, is important in the formation of such molecules as shown by fluorine (F_2). Each fluorine atom has only seven electrons in the outer level, but would be more stable if it had eight electrons in this level. The two atoms in each of these molecules actually attain a more stable configuration by *sharing* of two electrons in a special way, which can be shown diagramatically in Fig. 2-5 where $\bar{e}$'s represent electrons in the electron shells. If the shared electrons, one from each atom, are considered as

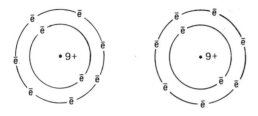

a. Two fluorine atoms, each with 7 outer electrons

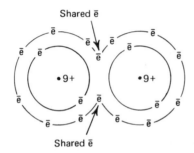

b. Two fluorine atoms, each sharing an electron
with the other to form a molecule F_2

Fig. 2-5: The covalent bond.

belonging to both atoms, each atom has effectively obtained an extra electron in its outer shell to make eight. The number of such bonds by which a given atom can be linked to other atoms is limited and is determined by quite different factors than in ionic bonding. In the molecules of fluorine, each atom is able to achieve a stable configuration by sharing one of its electrons with the other atom (Fig. 2-5). But once this state has been achieved, no further covalent bonds can be formed because those two F atoms depend on one another. Thus, the symbol F_2 is applicable, rather than F_3 or F_4. In forming covalent compounds, a single covalent bond allows an atom to increase by one the number of electrons in its outer orbit. Actually, the covalency of most ions is determined by the extent to which its outermost orbit falls short of being filled by electrons, that being eight. For example, an oxygen ion is two short and is divalent; the nitrogen ion is three short and is trivalent; and the carbon ion is four short and is quadrivalent.

Covalent bonding is relatively common in organic compounds, but as a rule, it is rare but not absent among minerals. A good mineralogical example is diamond, in which each carbon atom is attached to four other carbon atoms, each sharing one electron with the central atom, and the whole crystal being a giant molecule.

The Metallic Bond

This type of bond is responsible for the very strong cohesion of a metal. Metals are different from other types of solid matter by several physical properties, including high electrical and thermal conductivity, as well as optical opacity. Metals are elements whose atoms easily detach outer electrons; the atoms thus are all positively charged ions. The geometrical arrangement of atoms in a metal is determined by the packing of the positively charged ions. However, the detached electrons are not lost completely from the ions but are considered as being dispersed among the ions and as possessing rather high mobility (Fig. 2-6), which accounts for the good electrical and

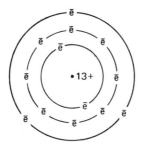

a. Aluminum atom

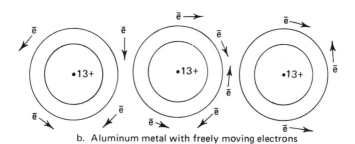

b. Aluminum metal with freely moving electrons

Fig. 2-6: The metallic bond.

thermal conductivity of metals. The electrons do not leave the structure, but are continually mobile throughout the structure. The bonds from any one atom are considered to be spherically distributed, and the bonds are capable of acting on as many neighbors as can be packed around any atom. The structure or atomic arrangement is determined by geometrical considerations only. Most metallic elements exhibit close-packed atomic structures which represent the most compact arrangement possible. This close packing accounts for high densities and ductilities of metals, properties which are not present in nonmetallic solids.

The Van der Waals Bond

The Van der Waals bond is generally considered to be a residual weak bond between all atoms and ions in all solids. It is weak compared to the ionic, covalent, and metallic bonds which usually mask the effect of the Van der Waals in any structure. The Van der Waals bond is similar to the metallic bond because it can attach an atom to an indefinite number of neighbors (Evans, p. 111).

In certain molecular structures with discrete molecules, the molecules themselves are bonded internally by strong ionic or covalent bonds, but the molecules are often bonded to one another by weak Van der Waals bonds. A number of minerals exhibit this type of residual bonding between nonspherical molecules, which of necessity cannot be close-packed. For example, in the sheet structures of graphite and mica, strong bonds hold the individual sheets intact as molecules, but the sheet molecules are held together (to each other) only weakly by Van der Waals bonding. Consequently, it is very easy to rub apart the graphite molecule as a smudge on the fingers or to separate the mica sheets by placing the fingernail between layers.

Crystallography

Introduction

Crystals have been known to man for a long time, long before Christ. Egyptians mining turquoise in the Sinai area were well aware of the geometrical perfection of those gemstones. In the 4th century B.C., Theophrastus published a *Book of Stones* in which garnet crystals are described. Later, about 64 B.C., Strabo compared quartz to ice, and the term *crystal*, an Anglicized Greek word meaning ice, came into use when quartz was discussed. The term, crystal, later was applied to all natural polyhedral solids. Pliny's *Natural History* describes forms and faces of crystals, but in 1669, Steno announced *the constancy of interfacial angles* as a fundamental law of crystallography. In 1611 Kepler suggested that internal atomic building blocks resulted in crystal regularity. But in 1784, Hauy suggested the concept of crystal structure by showing that calcite was made up of small polyhedral units, each having a characteristic shape. He also brought forth a second law of crystallography—the *law of simple rational indices*—a simple mathematical relation between faces on a crystal.

Early in the 20th century, X-ray diffraction by crystals was discovered by Max von Laue at the University of Munich. Many crystal studies were conducted by Laue and his associates, and the science of crystallography began to move rapidly forward. Bragg, in 1914, analyzed sodium chloride, and analyses of other compounds soon followed. Since then, great strides in crystallography have been made by X-ray methods. The application of X-rays to crystal studies was probably the most important advance in the field of crystallography.

Formation of Crystals

Crystals of minerals are found wherever the atoms or ions were able to come together, attracted to one another by electrostatic forces, in correct proportions and under proper conditions to initiate a solid. When enough of the atoms or ions take on a regular ordered arrangement, a crystal begins to take shape. A solid substance of this type may be initiated from a solution, from a melt, or from a vapor under certain temperature and pressure conditions. When solid material begins to form from a solution which is cooled or evaporated, or from a vapor which is cooled quickly, the atoms or ions develop into many minute beginnings of crystals. As crystallization continues along with cooling or evaporation, the minute crystal nuclei continue to grow and develop a characteristic geometrical shape. Growth will continue and crystals will become enlarged until they begin to impinge on one another as growth space is reduced. Then the geometrical shape begins to be interfered with, and as crystal growth proceeds, the solid mass will consist of crystals without their characteristic geometrical shapes.

Well-developed crystals of minerals are apt to develop only in places where growth of good geometrical shapes is not impeded. Even though great differences exist in the conditions, rates of growth, and supplies of material in crystal development, poorly shaped crystals and well-shaped crystals of the same substance have the same orderly internal arrangement of atomic or ionic constituents. Both are crystalline.

Definition of a Crystal

During growth, early organizations of the constituent ions of a mineral will continue until the source of material is exhausted or until some other factor causes crystals to stop growing. Early beginnings (called crystallites) will be atomically arranged according to the nature of the atoms or ions available and to the general chemical environment, and each mineral has its own characteristic manner of arrangement. As more material is added to early crystallites, the new material will continue to become attached to them, by electrical attractions, in such a way that the groups of new material will repeat the pattern of the earlier groups. Regardless of how large the crystallites and succeeding material may become, there will always be a whole series of repeat-patterns produced by the added material, in three dimensions throughout the crystals. Growth may not be uniform in all three directions in all crystals, but the repeat-patterns will always continue, no matter how rapidly or how slowly growth occurs in any direction, because the repeat-pattern is controlled by the nature of the atoms available and the environment.

When the supply of material is limited, the crystal will display flat surfaces or *faces* on the outside which are parallel to surfaces in the interior repeat-patterns. The outside faces will be formed best when nothing has interfered with their development and particularly where free space exists. However, in many cases, good face development is hindered by the crowding together of crystals during growth, a large supply of constituent atoms, and other factors. The poor development of exterior faces does not take away from the fact that the interior is well arranged.

In terms of crystal growth and the resulting repeat-pattern, a crystal may be defined as a homogeneous solid with three-dimensional internal atomic order, or ionic repeat-pattern, which under favorable conditions may be expressed externally as smooth plane faces (Hurlbut, p. 3), or under unfavorable conditions may not be expressed externally.

Usually, though, *crystal* suggests such familiar things as sapphire, gypsum, and topaz because we tend to picture mentally a solid bounded by a series of plane faces. Crystalline solids with poorly developed faces (as a result of growth interference) are just as much crystals as those which have well-developed faces. Thus, the terms *euhedral* (good development of faces), *subhedral* (moderately well to poorly developed faces), and *anhedral* (very poorly developed faces or no faces at all) are appropriate for describing degrees of development of exterior faces on crystalline solids.

The external shapes of crystals and all the physical properties of a particular substance depend heavily on the internal structure. When the other physical properties of a mineral crystal are known, the external shape becomes less important in identification. Once a thorough understanding of orderly arrangements in well-formed crystals is obtained, poorly organized structures can be recognized. In fact, we have now come to recognize varying degrees of atomic orderliness in many substances which rarely form euhedral shapes. Modern crystallography is no longer only the science of crystals, but rather it concerns the crystalline state. For teaching purposes, crystallography usually deals mainly with the geometry of well-shaped crystals. However, it must be borne in mind by the reader and student that such a narrow area of study only lays the foundations for the more encompassing field of modern crystallography revolving around the crystalline state.

Descriptions of Crystals

Shapes, Sizes, and Symmetry

Mineral collectors usually regard their well-crystallized specimens as most valuable. In fact, those crystals with the best development of faces are the most highly prized, even though the actual mineral has no economic or commercial value. This is because specimens with good development of faces are more rare than specimens of the same material with poor face development. The reason for this is that rather special conditions of growth must prevail for the development of good exterior faces, and such special conditions are not often encountered in nature.

A crystal is simply a stacking together of the crystallites, and the shape of the accumulating crystallites ultimately will be reflected in the faces of the final crystal. Faces on crystals do not develop at random, but they can develop only in certain directions, depending on several factors such as the number of atoms in planar directions and strength of attractive forces in certain directions. The fact that faces are poorly developed or not developed at all does not mean that the substance is not a crystal. It indicates that the development of crystal faces was interfered with when that particular crystal stopped growing.

All faces on a crystal are related to the same fundamental atomic structure of the constituent ions; thus they are related to each other according to fixed rules. Since faces are smooth planes on the exterior of a crystal, the edge between adjacent faces actually is an angular corner and has a measurable value in angular degrees. Even in two crystals of the same mineral, the angles between similar sets of faces are equal, regardless of the origins of the crystals or how misshapen they are (Fig. 2-7). This angular constancy of angles between faces on crystals of the same substance was discovered by Nicolaus Steno in 1669 and is called Steno's *law of constant interfacial angles.*

Since crystals grow by the addition of small atomic groups or units, a crystal is therefore composed of a regular stacking of these much smaller groups. Each such unit is called the *unit cell* and actually is the smallest volume which contains the true properties of the crystal. These properties include chemical, physical, and geometrical aspects. Each unit cell contains enough of the constituent atoms so that its formula can be determined, its atomic pattern can be ascertained, and the main crystallographic directions and their relative ratios to one another can be determined.

Crystals of different minerals normally exhibit quite different shapes. Even those euhedral crystals that exhibit well-developed faces may assume different shapes among the different minerals. Among anhedral crystals, i.e., those in which crystal faces are poorly developed or not developed at all, there is a tremendous variation in possible shapes. Although great variation exists, all of the basic shapes of unit cells or larger euhedral crystals can be related to six crystal systems. These broad groups of crystalline shapes are based on the fundamental geometry of the crystal. The six crystal systems are: isometric, hexagonal, tetragonal, orthorhombic, monoclinic, and triclinic. The various crystal systems are further subdivided into *classes* according to their symmetry. Symmetry is usually defined as the regularity of arrangement

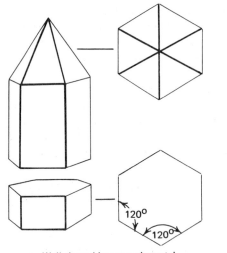
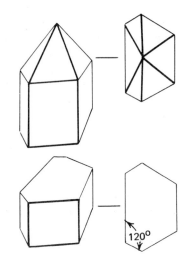
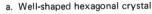

a. Well-shaped hexagonal crystal

b. Poorly shaped hexagonal crystal

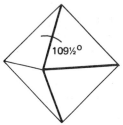
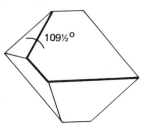

c. Well-shaped octahedron

d. Poorly shaped octahedron

Fig. 2-7: Constancy of interfacial angles.

of faces on a crystal. There is a fairly strong tendency for faces to be arranged so that edges between many faces are parallel. In addition, faces of the same size and shape frequently occur in parallel pairs on opposite sides of a point in the crystal. Also, faces may be so arranged that a crystal is bilaterally symmetrical, if viewed from several sides, and may appear to have a dividing plane which cuts the crystal in two. Regularity of faces may also yield a line through the crystal such that, if the crystal is rotated 360° around the line, the crystal will achieve a similar position two or more times. Thus, symmetry is based on repetition of faces, and the repetition may be relative to a point, to a plane, or to a line. In this respect, a crystal may have a *center of symmetry*, a *plane of symmetry*, or an *axis of symmetry*. These are known as elements of symmetry.

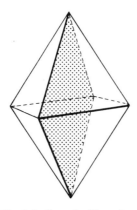

Fig. 2-8: Crystal with a plane of symmetry.

These elements of symmetry may be visualized with the aid of a few sketches. For example, a *plane of symmetry* is really an imaginary plane which divides the crystal into two halves which are mirror images of each other. This means that each face or edge on one half appears to be reflected across the plane to the other half, as if viewed in a mirror (Fig. 2-8). A *center of symmetry* is merely an imaginary point inside the crystal, situated so that each face on the crystal has a similar face parallel to it on the opposite side of the crystal (Fig. 2-9). An *axis of symmetry* is an imaginary line through the crystal. If rotated around this line, the crystal arrives at corresponding positions two or more times. Depending upon the regularity of faces on the crystal, correspondence may occur two, three, four, or six times during a 360° rotation (Fig. 2-10).

In many cases, crystals do not grow symmetrically, and the true geometric

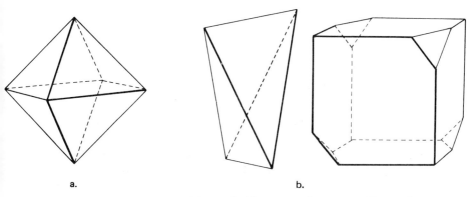

a.   b.

Fig. 2-9: Center of symmetry: (a) crystal with a center of symmetry; (b) crystals with no center of symmetry.

19

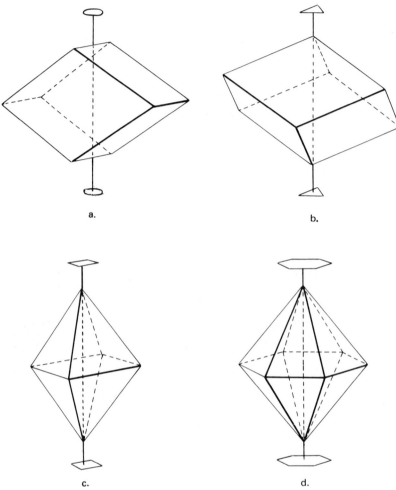

Fig. 2-10: Crystals showing axes of symmetry: (a) 2-fold; (b) 3-fold; (c) 4-fold; (d) 6-fold.

symmetry is hard to detect. In fact, it frequently happens that symmetry described for a particular mineral is obtained from X-ray data of the atomic unit cell and not from the external crystal shape.

Faces, Forms, Habits, and Twins

Faces. A well-developed crystal can never have less than four faces, but it may have considerably more than four. Crystal growth may produce a geometrically regular crystal or a geometrically irregular crystal, depending on just how and where material was added during growth. A geometrically regular crystal is one in which similar faces of the same size and shape are

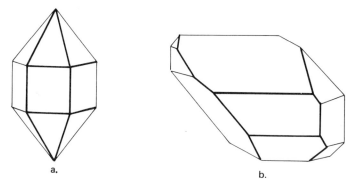

a.   b.

Fig. 2-11: Geometrical regularity of crystals: (a) regular; (b) irregular; (c) regular; (d) irregular.

located at correspondingly similar positions on the crystal. When similarly positioned faces show differential development in size and shape, then a geometrically irregular crystal is formed. However, interfacial angles between adjacent faces will always be constant for a given substance, regardless of whether the crystal is regular or irregular (Fig. 2-11).

The positions of the faces on a crystal exist in a systematic way, and the faces are referred to three main directions, called *axes*, through the crystal. These crystallographic axes (or axial directions) are imaginary lines through the crystal and are oriented from front to back, from right to left, and from top to bottom. These axes are usually designated *a*, *b*, and *c*, and each has a positive and a negative end. They indicate directions and are not specific lines which emerge at specific points. These axial directions are normally oriented parallel to the edges of the crystal formed by the intersection of the main faces. In most crystals, the axes are at right angles to each other (Fig. 2-12).

Each face on a crystal has *parameters*. We define the parameters of a crystal face by indicating the position of the face relative to the crystallographic

a   b

Fig. 2-12: Axial angles: (a) all 90°; (b) one greater than 90°.

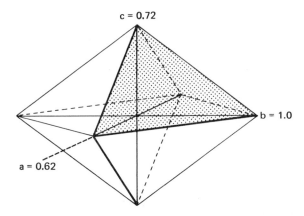

Fig. 2-13: Shaded face has intercepts 1a, 1b, 1c.

axes. To obtain the parameters of a particular face, it is necessary to determine whether or not that face intersects all of the crystallographic axes and at what relative distances. For example, aragonite has axial ratios of 0.62: 1: 0.72. If a crystal face intersects the *a*, *b*, and *c* axes at these relative distances, then the parameters are 1 on *a*, 1 on *b*, and 1 on *c* since these are *unit distances* on the axes. These intercepts then are written 1*a*, 1*b*, and 1*c* (Fig. 2-13). A face that cuts the two horizontal axes at distances proportional to these unit lengths and cuts the vertical axis at a distance which is twice its relative unit length will have intercepts of 1*a*, 1*b*, and 2*c* (Fig. 2-14). Again these are relative values and not absolute.

The most widely used method to express the parameters or intercepts of a crystal face on the axes is by means of *indices*. Although others have been used in the past, the system of *Miller indices* is fairly simple and is widely used today. The Miller indices of a face consist of a series of whole numbers which are derived from the parameters of that face. To obtain the Miller indices from the intercepts, three steps are followed: (1) invert the intercepts, (2) find the least common denominator, and (3) delete the denominators. The numerators are now the Miller indices.

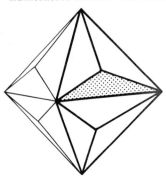

Fig. 2-14: Shaded face has intercepts 1a, 1b, 2c.

As an example, let's use the intercepts as given in the preceding section on parameters, i.e., a face with intercepts 1*a*, 1*b*, 2*c*. Steps:

(1) inverting: 1/1, 1/1, 1/2.

(2) least common denominator is 2:   2/2, 2/2, 1/2.

(3) delete denominators: 2/*x*, 2/*x*, 1/*x*.

The numerators are now the Miller indices 221. These refer to the *a*, *b*, and *c* axes, respectively.

The Miller indices thus are numbers which indicate the position of a crystal face because these numbers are derived from the axial relationships. Each face on a crystal cuts one or more axes, and the numbers in the Miller indices merely indicate relative distances on the crystallographic axes where the face intersects and the amount of slope of the face.

Forms. In crystallography, the term *form* refers to more than the general outward appearance of a mineral crystal. A form consists of a group of faces on a crystal which are alike in all respects, including position with respect to the axes, size and shape, and atomic composition. Individual faces on a crystal are described by the use of the Miller indices, as 113. Since each crystallographic axis has a positive end and a negative end, the Miller indices for a particular face may be written 11$\bar{3}$, meaning that the face cuts the negative end of the *c* axis and the positive ends of the *a* and *b* axes. A well-developed crystal may have a series of eight faces similarly disposed to the axes and of the same size and shape, and they have Miller indices as follows:

(1) 113;   (5) $\bar{1}$13;

(2) 11$\bar{3}$;   (6) $\bar{1}$1$\bar{3}$;

(3) 1$\bar{1}$3;   (7) $\bar{1}\bar{1}$3;

(4) 1$\bar{1}\bar{3}$;   (8) $\bar{1}\bar{1}\bar{3}$.

These eight faces comprise a *form*. The number of faces in any form is determined by the symmetry of the crystal system displayed. When we describe a single face by use of the Miller indices, the indices may or may not be placed in parentheses, thus 113 or (113). However, to designate this *form* for this crystal, which this set of eight comprises, the Miller indices with all *positive* numbers are placed in brackets, thus [113] or {113}. This symbol of the form automatically includes all eight similar faces with positive and negative indices. A crystal may have more than one form shown by the faces.

When one form such as a cube encloses space, it is called a *closed form* (Fig. 2-15). A form which does not enclose space by itself is called an *open form* (Fig. 2-16).

Because it does not enclose space, one *open form* is not sufficient to produce a crystal. Thus, it cannot exist by itself or as an actual crystal. At least two open forms are required, or one open form may be in combination with

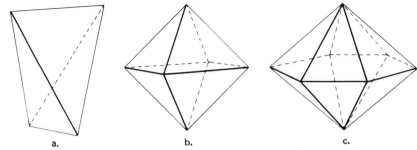

Fig. 2-15: Closed forms: (a) Tetragonal disphenoid; (b) Tetragonal dipyramid; (c) Hexagonal dipyramid.

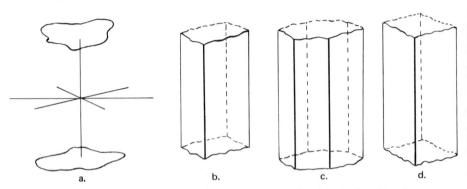

Fig. 2-16: Open forms: (a) Hexagonal basal pinacoid; (b) Orthorhombic prism; (c) Hexagonal prism; (d) Tetragonal prism.

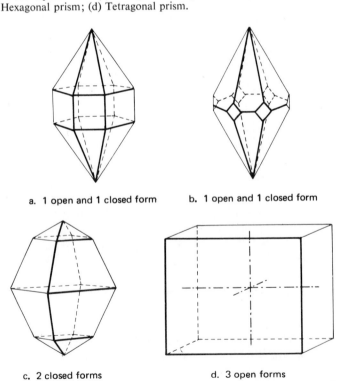

a. 1 open and 1 closed form   b. 1 open and 1 closed form

c. 2 closed forms   d. 3 open forms

Fig. 2-17: Open and closed forms in combination.

one or more closed forms to make a crystal. There are forty-eight different types of crystal forms which can be identified by angular relations of the faces (Fig. 2-17).

Habits. The *habit* of a crystal refers to the characteristic shape which the mineral assumes most of the time. The habit is determined by the crystal structure and thus includes the sizes and shapes of faces (the forms) as well as their positions. The characteristic shape is often quite important in identifying the mineral (see Physical Properties of Minerals in Chapter 3).

There are several important factors which affect the habit and growth of a crystal. Among the more important are the nature of the solution in which crystal growth is taking place, the temperature at which the crystallization processes are occurring, and the presence of impurities in the solution. Other factors such as relative humidity, pressure, and acidity or alkalinity may also partly determine the final habit of the crystal.

Twins. So far, our discussion of crystallography has considered only ideally developed crystals; however, well-formed symmetrical crystals are the exception. Minerals are more commonly produced, under natural conditions, either in groups of crystals which are not symmetrically shaped or are intergrown with one another. Groups of crystals (crystal aggregates) occur accidentally and simply represent individual crystals which grew at the same time and in the same neighborhood.

Quite often there are intergrowths of crystals called *twins*. These may develop according to conditions of atomic structure at the time crystal growth was taking place. Twins develop very early during rapid growth when ions become slightly offset from their normal positions when added to the growing crystal. Thus a somewhat different growing direction is followed. All atomic positions do not tolerate slightly offset growth, but usually there are certain ions in the crystal structure near which some slight offset is tolerated. This results in twins growing only in certain directions. Each portion of a twin crystal has its own proper orientation, and each orientation is related to the crystallography of each portion.

Various types of twins are encountered among minerals. For example, a *contact twin* is developed when one crystal seems to be cut in half, and the front half rotated to twin position. *Repeated twins* occur when a series of individual crystals seem to alternate in position and each has alternate orientation. Repeated twinning is also called *lamellar* or *polysynthetic twinning*. *Cyclic twins* occur when crystals twin on more than one side at the same time. Another type, called *penetration twin*, is developed when two crystals appear to be pushed into each other (Fig. 2-18).

Some minerals twin in several distinct ways, while others twin only in one way. There are very few minerals which never twin; consequently twin crystals are quite common, rather than rare.

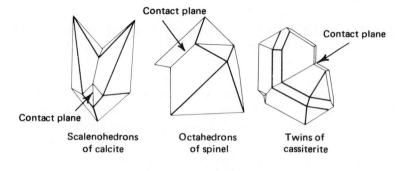

Scalenohedrons
of calcite

Octahedrons
of spinel

Twins of
cassiterite

a. Contact types

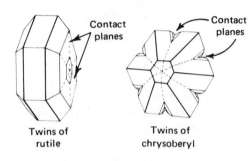

Twins of
rutile

Twins of
chrysoberyl

b. Cyclic types

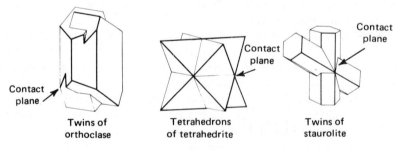

Twins of
orthoclase

Tetrahedrons
of tetrahedrite

Twins of
staurolite

c. Penetration types

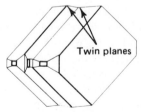

Twins of albite

d. Repeated type

Fig. 2-18: Types of twinning.

The Crystal Systems

All crystals can be classified into six crystal systems, according to the symmetry (regularity of faces) exhibited by the crystals. The various elements of symmetry were defined and explained earlier in this chapter. The elements of symmetry are:

(1) center of symmetry, shown by parallel similar faces on opposite sides of the crystal;

(2) plane of symmetry, which divides the crystal into two halves which are mirror images;

(3) axis of symmetry, an imaginary line through the crystal around which the crystal may be rotated in such a way that it repeats itself in space two or more times during a complete rotation.

Isometric System

The name *isometric* means *equal measure* in the three main directions. In this system, all three crystallographic axes are of equal length and are disposed at right angles to each other. The axes are designated as a_1 (front to back), a_2 (right to left), a_3 (top to bottom). Since the axes are of equal length, the crystals of the isometric system are equidimensional (Fig. 2-19).

There are several classes within the isometric system, according to the symmetry of the crystals. The class with the greatest number of symmetry elements (twenty-three), the hexoctahedral class, contains nine planes, six axes of twofold, four axes of threefold, three axes of fourfold, and one center. The other classes have fewer symmetry elements.

Common isometric forms and their Miller indices are presented in Table 2-3. These crystal forms are illustrated in Fig. 2-20.

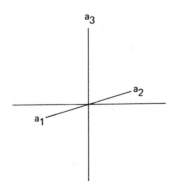

Fig. 2-19: Isometric axes.

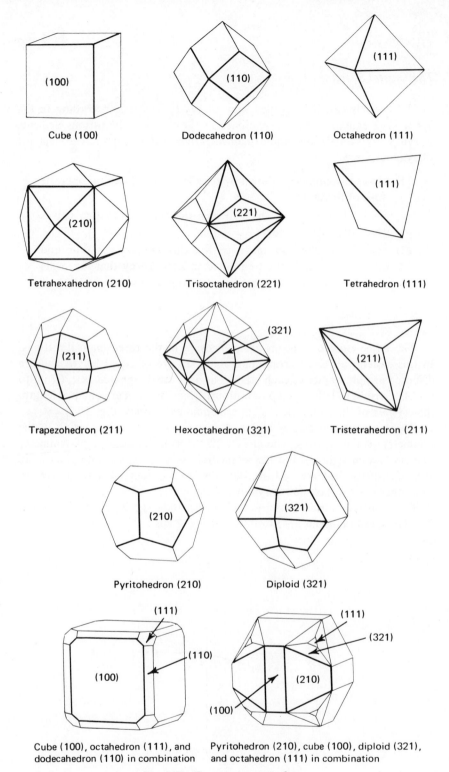

Cube (100)

Dodecahedron (110)

Octahedron (111)

Tetrahexahedron (210)

Trisoctahedron (221)

Tetrahedron (111)

Trapezohedron (211)

Hexoctahedron (321)

Tristetrahedron (211)

Pyritohedron (210)

Diploid (321)

Cube (100), octahedron (111), and
dodecahedron (110) in combination

Pyritohedron (210), cube (100), diploid (321),
and octahedron (111) in combination

Fig. 2-20: Common isometric forms.

Table 2-3: COMMON ISOMETRIC FORMS

Form	No. of Faces	Miller Indices
Cube	6	{100}
Dodecahedron	12	{110}
Octahedron	8	{111}
Tetrahexahedron	24	{210}
Trisoctahedron	24	{221}
Tetrahedron	4	{111}
Trapezohedron	24	{211}
Hexoctahedron	48	{321}
Tristetrahedron	12	{211}
Pyritohedron	12	{210}
Diploid	24	{321}

Hexagonal System

Crystals in this system are six-sided, or at least display a hexagonal (six-sided) outline when viewed parallel with the vertical axis. In this system, there are four crystallographic axes: three horizontal, intersecting at 120°, and a fourth vertical. The horizontal axes, designated a_1 (left front to right rear), a_2 (right to left), a_3 (left rear to right front), are of equal length. The vertical axis, designated c, is longer than the a axes in some minerals and shorter than the a axes in other minerals (Fig. 2-21).

There are several classes within this system, according to the symmetry of the crystals. The class with the greatest number of symmetry elements (fifteen), the dihexagonal-dipyramidal class, displays seven planes, six axes of twofold, one axis of sixfold, and a center. The other classes in this system have fewer symmetry elements.

Common hexagonal forms and their Miller indices are given in Table 2-4. Hexagonal forms are illustrated in Fig. 2-22.

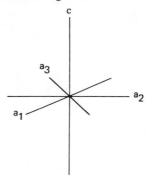

Fig. 2-21: Hexagonal axes.

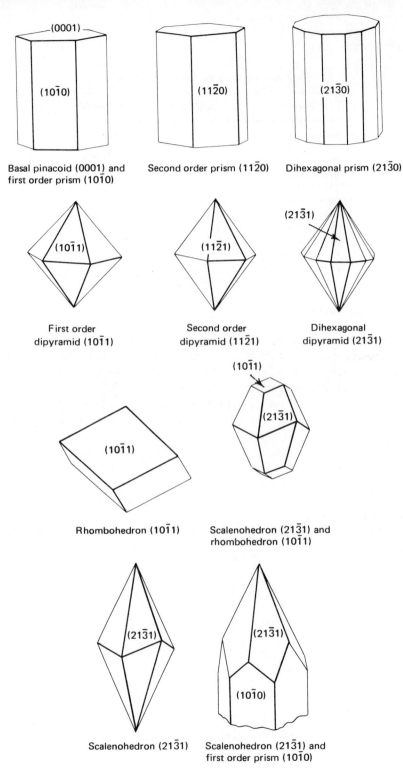

Basal pinacoid (0001) and first order prism ($10\bar{1}0$)

Second order prism ($11\bar{2}0$)

Dihexagonal prism ($21\bar{3}0$)

First order dipyramid ($10\bar{1}1$)

Second order dipyramid ($11\bar{2}1$)

Dihexagonal dipyramid ($21\bar{3}1$)

Rhombohedron ($10\bar{1}1$)

Scalenohedron ($21\bar{3}1$) and rhombohedron ($10\bar{1}1$)

Scalenohedron ($21\bar{3}1$)

Scalenohedron ($21\bar{3}1$) and first order prism ($10\bar{1}0$)

Fig. 2-22: Common hexagonal forms.

Table 2-4: COMMON HEXAGONAL FORMS

Form	No. of Faces	Miller Indices
Basal pinacoid	2	$\{0001\}$
First-order prism	6	$\{10\bar{1}0\}$
Second-order prism	6	$\{11\bar{2}0\}$
Dihexagonal prism	12	$\{21\bar{3}0\}$
First-order dipyramid	12	$\{10\bar{1}1\}$
Second-order dipyramid	12	$\{11\bar{2}2\}$
Dihexagonal dipyramid	24	$\{21\bar{3}1\}$
Rhombohedron	6	$\{10\bar{1}1\}$
Scalenohedron	12	$\{21\bar{3}1\}$

Tetragonal System

Crystals in the tetragonal system generally display a four-sided outline when viewed parallel with the c axis, hence, the name *tetragonal* (four-sided). In this system, there are three crystallographic axes: two horizontal and one vertical, mutually at right angles. The two horizontal axes are of equal length and are designated a_1 (front to back) and a_2 (right to left). The third axis is designated as c (top to bottom) (Fig. 2-23). In some minerals, the c axis is longer than the a axes; in others, the c axis is shorter than the a axes.

Several classes exist in this system, according to the symmetry displayed. The class with the largest number of symmetry elements (eleven), the ditetragonal-dipyramidal class, displays five planes, one axis of fourfold, four axes of twofold, and a center. The remaining classes exhibit fewer elements of symmetry.

The common tetragonal forms and their Miller indices are presented in Table 2-5. Tetragonal forms are illustrated in Fig. 2-24.

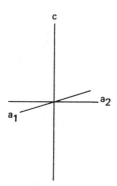

Fig. 2-23: Tetragonal axes.

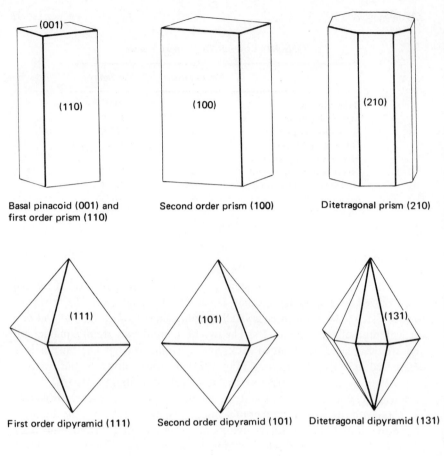

Basal pinacoid (001) and
first order prism (110)

Second order prism (100)

Ditetragonal prism (210)

First order dipyramid (111)

Second order dipyramid (101)

Ditetragonal dipyramid (131)

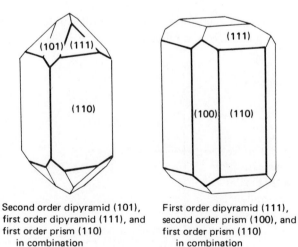

Second order dipyramid (101),
first order dipyramid (111), and
first order prism (110)
in combination

First order dipyramid (111),
second order prism (100), and
first order prism (110)
in combination

Fig. 2-24: Common tetragonal forms.

Table 2-5: COMMON TETRAGONAL FORMS

Form	No. of Faces	Miller Indices
Basal pinacoid	2	{001}
First-order prism	4	{110}
Second-order prism	4	{100}
Ditetragonal prism	8	{120}
First-order dipyramid	8	{111}
Second-order dipyramid	8	{101}
Ditetragonal dipyramid	16	{131}

Orthorhombic System

The crystals often display a rectangular outline or a diamond-like outline when viewed parallel to *c*. In orthorhombic crystals there are three unequal crystallographic axes: two horizontal and a third vertical, all three mutually at right angles. Since the three axes are unequal, they are designated as *a* (front to back), *b* (right to left), and *c* (top to bottom). Because the axes have different lengths, there is a problem of choosing which one is the *c*, which is the *b*, and which is the *a* axis. For orientation purposes, any of the three axes can be chosen as the vertical *c* axis. Then the longer of the other two becomes the *b* axis, and the shorter one becomes the *a* axis (Fig. 2-25).

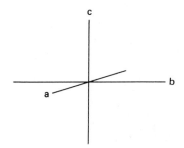

Fig. 2-25: Orthorhombic axes.

Several classes exist in this system, according to the symmetry present. The class with the largest number of elements of symmetry (seven), the rhombic-dipyramidal class, contains three planes, three axes of twofold, and a center. The remaining classes exhibit fewer elements of symmetry.

Common forms in the orthorhombic system and their Miller indices are given in Table 2-6 and are illustrated in Fig. 2-26.

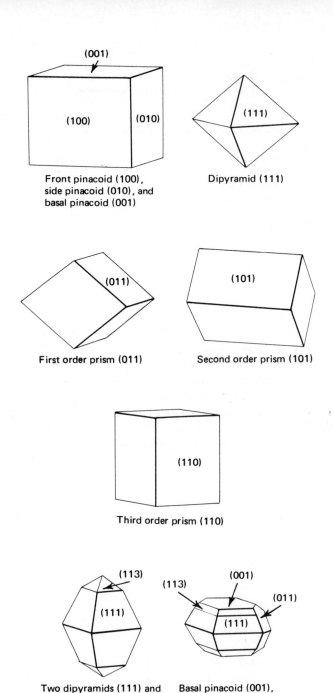

Front pinacoid (100),
side pinacoid (010), and
basal pinacoid (001)

Dipyramid (111)

First order prism (011)

Second order prism (101)

Third order prism (110)

Two dipyramids (111) and
(113) in combination

Basal pinacoid (001),
first order prism (011),
and dipyramids (111) and
(113) in combination

Fig. 2-26: Common orthorhombic forms.

Table 2-6: COMMON ORTHORHOMBIC FORMS

Form	No. of Faces	Miller Indices
Basal pinacoid	2	{001}
Side pinacoid	2	{010}
Front pinacoid	2	{100}
First-order prism	4	{011}
Second-order prism	4	{101}
Third-order prism	4	{110}
Dipyramid	8	{111}

Monoclinic System

In this system, there are three crystallographic axes: one horizontal, one vertical, and one neither horizontal nor vertical, but rather tilted at an angle. Thus the name *monoclinic* (one inclination) is applied. The horizontal axis is designated *b* (right to left), the vertical axis is *c* (top to bottom), and the *a* axis is tilted down to the front. The three axes are all of unequal length. The *b* and *c* axes are disposed at right angles, and the *a* and *b* axes are also disposed at right angles, but *a* is tilted with respect to *c* at some angle greater than 90°, an angle designated as β (Fig. 2-27).

Several symmetry classes are present in this system. The class containing the largest number of elements of symmetry (three) is the prismatic class, with one plane, one axis of twofold, and a center. The other classes in the system have fewer symmetry elements.

Common monoclinic forms and their Miller indices are presented in Table 2-7 and illustrated in Fig. 2-28.

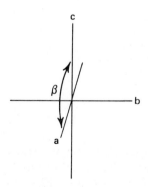

Fig. 2-27: Monoclinic axes.

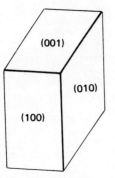
(001)

(010)

(100)

Basal pinacoid (001)
Side pinacoid (010)
Front pinacoid (100)

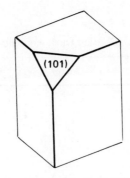
(101)

Second order pinacoid (101)

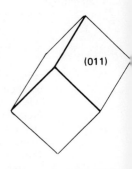
(011)

First order prism (011)

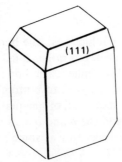
(111)

Fourth order prism (111)

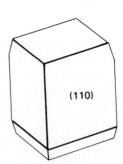
(110)

Third order prism (110)

Fig. 2-28: Common monoclinic forms.

Table 2-7: COMMON MONOCLINIC FORMS

Form	No. of Faces	Miller Indices
Basal pinacoid	2	{001}
Side pinacoid	2	{010}
Front pinacoid	2	{100}
First-order prism	4	{011}
Second-order pinacoid	2	{101}
Third-order prism	4	{110}
Fourth-order prism	4	{111}

Triclinic System

Crystals in this system have three unequal crystallographic axes: one vertical, one tilted down in front, and one tilted to the side, so that there are no right angle relationships between them. The term *triclinic* means three inclinations. The axis tilted down toward the front is designated *a*; the axis

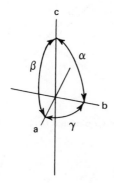

Fig. 2-29: Triclinic axes.

tilted to one side (right or left) is the *b* axis; the vertical axis (top to bottom) is designated *c*. Since the axes are not perpendicular to one another at all (Fig. 2-29), the angles between are designated as follows:

α between *b* and *c*

β between *a* and *c*

γ between *a* and *b*

There are only two classes in this system according to symmetry displayed. The pinacoidal class has only a center of symmetry, while the remaining class has no element of symmetry whatsoever.

Common triclinic forms and their Miller indices are given in Table 2-8. These forms are illustrated in Fig. 2-30.

Table 2-8: COMMON TRICLINIC FORMS

Form	No. of Faces	Miller Indices
Front pinacoid	2	{100}
Side pinacoid	2	{010}
Basal pinacoid	2	{001}
First-order pinacoid	2	{011}
Second-order pinacoid	2	{101}
Third-order pinacoid	2	{110}
Fourth-order pinacoid	2	{111}

X-Ray Crystallography

X-ray diffraction methods of mineral identification and atomic geometry determinations are used frequently in modern laboratories. X-ray examination of a very small (less than 0.5 mm across) single crystal of a mineral or any crystalline substance will reveal the positions of the constituent atoms and

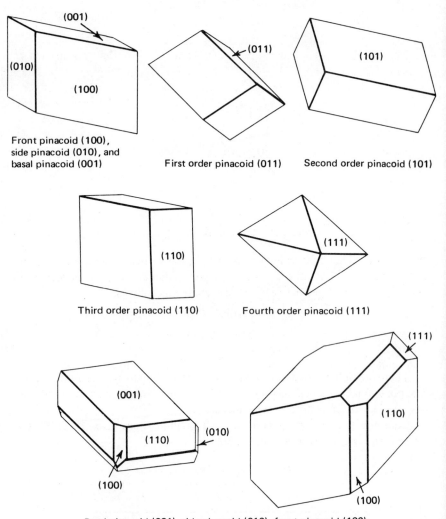

Front pinacoid (100),
side pinacoid (010), and
basal pinacoid (001)

First order pinacoid (011)   Second order pinacoid (101)

Third order pinacoid (110)   Fourth order pinacoid (111)

Basal pinacoid (001), side pinacoid (010), front pinacoid (100),
third order pinacoid (110), and fourth order pinacoid (111)
in combination

Fig. 2-30: Common triclinic forms.

the interatomic distances in the unit cell. Since any crystal is made up of multiple numbers of unit cells forming the repeat-pattern, the crystal shape and classification can be determined. Such studies are usually completed when a new mineral is named or a new chemical compound is synthesized. Detailed studies of this type require rather sophisticated equipment and a number of computer programs.

For identification purposes, another method of X-ray study is employed. A small amount of powdered substance (a few milligrams) is used, and consequently the method is known as the *powder method*. The powdered mineral may be placed either in a cylindrical camera or in a diffractometer. The same principles are applicable in both instruments. The sample is irradiated with a small beam of monochromatic X-rays. Reflections from atomic planes are recorded on film in the cylindrical camera or on graph paper in the diffracto-meter.

A reflection occurs from a set of atomic planes only at a critical angle θ, called the *Bragg angle*. The Bragg angle is established according to the inter-atomic distance, *d*, between the planes by a mathematical formula called the Bragg Equation:

$$n\lambda = 2d \sin \theta$$

This formula states that, in order for a reflection to occur at some angle θ, the interatomic distance *d* must be a certain value which keeps the X-ray wavelengths, λ, in phase.

In the cylindrical camera, cones of diffracted X-ray beams are produced from the various atomic planes in the specimen, and these various cones of beams are recorded on a strip film. The cones show up as arcs (or lines), since only portions of the cones are obtained on the narrow film. Actually, the arcs are made up of a large number of small diffracted beams reflected from atomic planes in the numerous crystalline fragments which are lying at random in the specimen. Since the specimen rotates while being irradiated, a solid, uniform line is produced.

A reflection is produced only when an atomic plane is at a definite angle to the X-ray beam. Since there are a large number of fragments with this atomic plane so positioned with respect to the beam, each contributes to the reflection. This critical angle is called θ (theta). The reflection angle is equal in size to the angle of incidence; so the reflected beam makes an angle of 2θ with the primary beam (Fig. 2-31).

In the diffractometer method of studying a powdered substance, a Geiger counter, a proportional counter, or a scintillation counter is used instead of X-ray photographic film to record reflections. The powdered mineral is

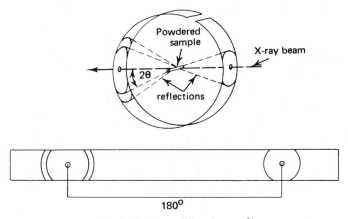

Fig. 2-31: X-ray diffraction on film.

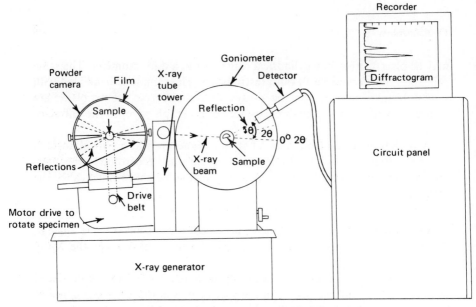

Fig. 2-32: Diagram of powder camera and diffractometer.

mounted on a flat surface and placed in the X-ray beam. Beginning at 0°, the Geiger counter or other sensitive counter revolves uniformly around the specimen. While the sample rotates through the angle θ, the counter revolves through the angle 2θ. This is necessary so that a relationship between X-ray source, sample, and counting tube can be maintained for the recording of all reflections, and none are blocked by the specimen holder (Fig. 2-32). With this gear arrangement between specimen rotation and Geiger counter motion, the counting tube picks up each reflection separately, even though all reflections from all atomic planes will be taking place simultaneously, just as in the powder camera.

The Geiger counter (or other counter) is connected through electronic circuits which amplify the reflections. In turn, the electronic circuits are connected to a recorder. The recorder contains a uniformly moving paper chart. During operation, the mineral specimen, the counting tube, and the paper drive of the recorder are all started simultaneously, beginning at 0°. The recorder continuously records the intensity of radiation picked up by the counter. Any reflections from the sample are recorded as peaks on the chart after being amplified through the electronic circuits.

As stated earlier, X-ray reflections occur only at certain angles of θ. If a particular set of atomic planes has a *d*-spacing so that a reflection occurs at $\theta = 15°$, no peak for this reflection would occur until the Geiger counter had traveled from 0° through $2\theta = 30°$. The reflected beam would then cause the tube to conduct a pulse, which would be amplified to produce a deflection on the recording chart paper. With continued scanning by the counting tube, each reflection becomes recorded as a peak. Scanning should continue until the entire pattern is obtained. In general, the heights of the peaks are proportional to the intensities of the reflections which produced them. A complete diffractogram consists of a set of peaks of varying heights at specific angular positions. The chart paper is divided into tenths of inches. For routine work, the counting tube travels 1° per minute, while the chart paper moves at a constant rate of 0.5 inch per minute. Thus, 0.5 inch on the chart is equivalent to 2θ of 1° of counter travel. Peak positions (in angles) are read off the paper as precisely as possible (to 0.05 or less). The corresponding d-spacings are then determined by the Bragg equation, $n\lambda = 2d \sin \theta$. Conversion tables for solving this equation for d are available.

Since any crystalline substance has a number of different atomic planes, each plane of atoms requires a certain angle of incidence. Thus, there is a reflected beam from each plane. All reflections from all atomic planes will take place simultaneously, and a different angle 2θ is measurable from each reflection.

The complete X-ray powder photograph consists of a set of arcs on the film. Each line has a specific angular position and darkness (intensity). Since each crystalline substance produces its own characteristic set of reflections, its complete X-ray pattern is more or less unique and can be used with great reliability in identification of the substance. The lines on the film actually are reflections as they occurred in the camera. These angles are measured as accurately as possible in terms of 2θ, and the corresponding d-spacings are obtained from a set of conversion tables. The darkness of the lines is generally proportional to the intensities of the reflections.

In attempting to identify an unknown mineral, a list of d-spacings is obtained using either the powder camera or the diffractometer method. Since each crystalline substance produces its own powder pattern, with respect to both reflection positions and intensities, the pattern may be compared to patterns of other minerals until a match is found. The matching must be accurate in d-spacings and intensities of all reflections.

The matching may be accomplished by the use of the *X-ray Powder Data File* (cards) published by the Joint Committee on Powder Diffraction Standards. On each card are listed the d-spacings of several thousand crystalline substances, including minerals. The d-spacings of the most prominent (strongest) reflections of the unknown mineral are used to seek out a corresponding series of d's in the card file. The card file is arranged in order of the d-spacings of the most prominent reflections, thereby providing as much help as possible. In addition, the strongest d-spacings are cross-indexed. The intensity of the strongest one is equated to 100, and the remaining prominent ones are scaled down accordingly. Using several prominent d-spacings, the search of the card file may yield several possible matches. Then the weaker reflections must be compared for more detailed matching for decisive identification (Fig. 2-33).

Although the diffractometer and the cylindrical camera yield the same

3.62	6.27	2.37	100	34	20

d A	I/I₀	h k l
1.9834	5	420
1.8911	3	332
1.8106	3	422
1.7395	3	431
1.6194	3	521
1.5680	10	440
1.5212	3	530
1.5212	2	433
1.4783	3	600
1.4783	6	442
1.4389	9	532
1.3687	4	541
1.3372	3	622
1.3078	4	631
1.2803	3	444
1.2070	4	721
1.2070	1	552
1.2070	3	633
1.1647	2	730
1.1265	3	651
1.1265	1	732
1.0918	1	741
1.0918	1	811
1.0756	2	820
1.0175	2	662

d A	I/I₀	h k l
1.0043	1	752
0.9917	4	840
.9795	1	833
.9565	1	921
.9565	1	655
.9149	1	763
.9053	2	844
.8783	1	772
.8783	1	1011
.8535	1	1022
.8457	3	952
.8457	1	1031
.8457	1	765
.8307	1	774
.8307	1	871
.8236	2	1040
.8236	3	864
.8165	2	961
.7902	1	1121
.7902	2	963
.7902	1	1051

$Na_4Si_3Al_3O_{12}Cl$

Sodium Aluminum Chloride Silicate   Sodalite

Ref. National Bureau of Standards, Mono. 25, Sec. 7
(1969)
(Specimen from Bolivia, n=1.483)

Sys. Cubic   S.G. P4̄3n (218) Dx 2.306 Z 2
a₀ 8.870±0.004   b₀   c₀
α   β   γ
Ref. Lons and Schulz, Acta Cryst., 23 434-36 (1967)

Scale factor (Integrated Intensities) 1.130 x 10⁵
λ 1.5405

d A	I/I₀	h k l	d A	I/I₀	h k l
6.27	34	110	2.561	18	222
4.44	7	200	2.371	20	321
3.97	1	210	2.217	1	400
3.62	100	211	2.091	13	330
2.805	8	310	2.091	18	411

FORM CI
B

548

Fig. 2-33: X-ray data card for sodalite. (By permission Joint Committee on Diffraction Standards.)

data, the diffractometer is generally considered the more advantageous. Exposure time in the cylindrical camera usually is about two hours, which is close to the same time for obtaining a diffractogram from 0° to over 100° of 2θ. But X-ray film must be developed, fixed, washed, and dried before it can be measured, and these processes require several hours. A diffractogram can be measured immediately upon completion of the scanning run. In addition, peak positions and intensities are more easily and more quickly obtained than lines in the film method. Figure 2-34 shows a diffractogram and a powder photograph for comparison purposes.

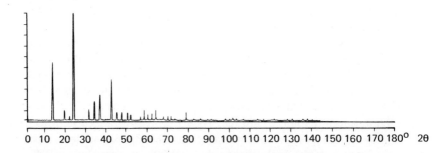

0 10 20 30 40 50 60 70 80 90 100 110 120 130 140 150 160 170 180°   2θ

Fig. 2-34: Diffractogram and powder photograph of sodalite for comparison purposes.

Chapter Three

Nature of Minerals

In Chapter 1 we provided a brief definition of a mineral. Implication was made that a hard and fast definition was difficult to adhere to because of the variety existing among these earth materials and because numerous substances simply did not fit into the definition. A mineral was defined as a natural inorganic crystalline solid whose composition is expressible by a formula and whose crystalline structure is characteristic of itself. Thus, a mineral is usually characterized by two basic features:

(1) chemical nature, involving its composition and various other chemical aspects and

(2) crystalline nature, involving its atomic structure and all related physical properties.

A few minerals are simple in their chemical makeup, consisting of only one element such as sulfur. Others are quite complex and consist of several different elements. Minerals are those chemical substances found in nature, and they are given mineralogical names such as calcite. But if the same substance is made in the laboratory, it is not a mineral, and it is referred to by its chemical or synthesized name, calcium carbonate. In fact, modern technology has learned to synthesize a number of minerals principally in order to improve upon the natural substances. Some of the better known synthesized materials are diamond, mica, ruby, sapphire, rutile, and quartz.

Chemical Aspects

The science of mineral chemistry is founded on a vast knowledge of the compositions of minerals, as chemical analyses have accumulated for nearly 200 years. The term *mineral chemistry* includes two important aspects. First, it is concerned with the exact composition of the mineral, and second, it is concerned with determining the formula and if impurities are present. To learn the exact composition of a mineral, i.e., which elements are present and their proportions, requires some sort of a chemical analysis. Since a large number of chemists, mineralogists, ceramicists, and others are capable of making an analysis, it stands to reason that some variation in accuracy of analyses exists among these people. Two analyses of a sample of a mineral, one made by a chemist and one made by a mineralogist, would not show exactly the same results as far as the quantities of the constituent elements are concerned. There would be variation in the quantity of each constituent, for example, iron (Fe), as determined by the chemist from that determined by the mineralogist. Both quantities would probably be very close to each other, but a difference of 0.5 to 1.0 percent would not be unusual for major components.

A chemical analysis usually is made to determine the elements present and their proper amounts. Consequently, any analysis should be made as completely and as accurately as possible. Accuracy, of course, depends on the quality of the methods used, on the quality of the analyst's work, and on the type of instruments and equipment used. As a rule, even the best methods used by competent workers may not be completely accurate; there always seems to be some margin of error.

When an analysis is made, the amounts of the constituents are expressed in percentages by weight rather than by volume, because it is much easier to weigh out different portions during the analysis than it is to measure volumes. The percentage total should be 100.0, but any total between 99.5 and 100.5 is acceptable as a good one. This does not mean that the analysis is necessarily accurate, because errors made along the way may be offset by other errors which might compensate to keep the total between 99.5 and 100.5

Determining the Formula

As we saw earlier in the definition of a mineral, the composition must be such that it can be expressed by a formula. So it is of interest here to explain how a mineral's formula is determined. In the formula of cuprite, Cu_2O, there are two coppers (Cu) for each one oxygen (O). But how is this ratio determined? Just how does a mineralogist (or chemist) finally arrive at such a specific formula with the elements (and atomic groups in some cases) in such exact proportions, Cu_2O, instead of Cu_3O, or CuO_2?

When a chemical analysis of a mineral is made, the results are given in weight percentages. Then the weight percentages are converted to *atomic proportions* by dividing the weight percentage of each element by the atomic weight of that element. Atomic weights are obtained from chemical tables of elements. For the mineral cuprite (Cu_2O), the following weight percentages are obtained:

Wt. Percentage	Atomic Wt.	Atomic Proportion	Set Proportions to Ratio
Cu 88.8	63.54	88.80/63.54 = 1.397	1.000
O 11.2	16.00	11.20/16.00 = 0.700	0.500

If the atomic proportion of Cu is set to 1, the atomic proportion of O becomes 0.500. Instead of expressing the formula as $Cu_1O_{0.5}$, it is best expressed in whole numbers as Cu_2O. This means that for each 2 copper atoms, there is 1 oxygen atom.

Conversely, the weight percentage of each element can be computed from the formula. For cuprite the formula is Cu_2O. The atomic weight of copper (Cu) is 63.54, and for oxygen (O) it is 16.00; consequently, the gram-formula weight of cuprite is $(63.54 \times 2) + 16.00 = 143.00$. The weight percent of Cu is calculated as $(127.08/143.00) \times 100 = 88.8$ and the weight percent of O as $16.00/143.00 \times 100 = 11.2$.

Interpretation of Analyses

Analyses of cuprite, regardless of where the samples come from, nearly always show that only Cu and O are present, and that practically never is there any other element present as an impurity. However, very few minerals have such a fixed chemical composition. Most of the minerals have a composition which allows limited substitutions to take place among certain elements in the mineral. For example, in sphalerite, zinc sulfide, ZnS, a certain amount of iron (Fe) can substitute for zinc (Zn). Specimens of sphalerite are known which range from pure ZnS to sphalerite with 36 percent of the Zn being substituted for by Fe. Nevertheless, the Fe-bearing zinc sulfide is still sphalerite because it has all the physical properties and most of the chemical properties of sphalerite. The formula for such a mineral would be written (Zn,Fe)S, to show that Fe substitutes for Zn. So Zn is placed first, followed by that element, iron (Fe), which does the substituting. Occasionally, other elements such as manganese (Mn), as well as the Fe, also may substitute for Zn. The formula then would be written (Zn,Fe,Mn)S to indicate this substitution. This phenomenon of atomic substitution is fairly common in the minerals.

It is not to be assumed by the reader that a chemical analysis of a mineral is made as a daily routine simply to find out its composition. Analyses are too expensive for that. So they are made as needed, e.g., when an unknown mineral must be identified, a new mineral must be named, or when a crystal analysis is necessary for part of a research project which might have important geological connotation.

The chemical composition has been used alone extensively in the past in the classification of minerals. But today's classification considers crystal structure as well as chemical composition. Consideration in classification is also given to the fact that atomic substitution may take place in a given structure.

Atomic Substitution and Related Phenomena

Many minerals are somewhat variable in their chemical composition because an element may be substituted for another. Such atomic or ionic substitution comes about because minerals usually crystallize from natural solutions which contain other atoms or ions in addition to those necessary to form a particular mineral. Consequently, some foreign atoms or ions often become incorporated into the structure, taking the place of some of the essential atoms or ions. In fact, this phenomenon is extremely common among the minerals. Since most of the minerals are ionic structures without molecules, any ion in the structure may be replaced (within certain limits) by another ion of similar size, which may be present, without seriously distorting the structure.

Solid Solution

Substitution between two ions of similar size and similar charge may take place in two ways, both of which are called *solid solution*. In the first way, a certain ion may substitute for another only in limited amounts (usually quite small). In the second way, substitution of one for another may be in unlimited quantities. Regardless of which type of substitution takes place, there is no change in the atomic structure of the mineral.

An example of the first case is noted in the mineral sphalerite, ZnS. In this mineral, iron (Fe) readily substitutes for some of the zinc (Zn). (See the discussion in this chapter under *Interpretation of Analyses*). In fact, up to 36 percent of the zinc (Zn) ions can be replaced by iron (Fe) ions without disrupting the crystals and without forming a new mineral; the mineral is still sphalerite. This is a type of *solid solution*. But 36 percent is the upper limit of substitution of Fe for Zn; no further substitution can take place even at high temperature —the isometric structure of sphalerite will not permit it. If anymore Fe is forced into the sphalerite, the crystals become badly distorted and lose their good isometric arrangement, as shown by careful laboratory experimentation. In the case of sphalerite allowing 36 percent Fe to substitute for Zn, this is an exceptionally large amount of substitution. Most substitution usually takes place in amounts less than 10 percent.

An example of the second type is seen in the mineral group called olivines. In this group, iron (Fe) and magnesium (Mg) may substitute for each other in unlimited amounts. The main components of this group are forsterite, Mg_2SiO_4 and fayalite, Fe_2SiO_4. These two minerals act as end members of the group. Common olivine, the one found most abundantly in the rocks, is called a *solid solution* of the two end members because 15 percent of the Mg is substituted for by Fe. Therefore the formula for common olivine is written $(Mg,Fe)_2SiO_4$. To be more correct, the formula should be written $(Mg_{85}Fe_{15})_2SiO_4$, where the subscripts 85 and 15 really mean percentages or proportions. There are other minerals in this group which have varying amounts of Fe substituting for Mg. No change in structure occurs regardless of how much substitution takes place; all three minerals, forsterite, olivine, and fayalite, and others in the olivine group, have the same structure.

Isomorphism

Minerals which have similar formulas and which also have similarly sized cations and anions often crystallize in similar kinds of structures. This phenomenon is called *isomorphism*, and minerals exhibiting this phenomenon among them are said to be *isomorphous*. In fact, such a similarity among minerals is used as one of the features in the classification. The term isomorphism actually embraces two different concepts. The first involves a similarity of formulas, ionic size, and structure, and also substitution of one ion for another. This was shown in the preceding section as solid solution using olivine as an example. The second involves a similarity of formula, similarity of ionic sizes, and similarity of structures, but without ionic substitution being necessary. Consequently, no isomorphous group exists.

A good example of the second type of isomorphism is between the two carbonate minerals strontianite ($SrCO_3$) and cerussite ($PbCO_3$). Note that the formulas are quite similar—$SrCO_3$ and $PbCO_3$. Both of these minerals crystallize in the orthorhombic system in very similar structures. The reason for this is that Sr^{2+} and Pb^{2+} have similar ionic radii, 1.12 Å and 1.20 Å, respectively. But substitution of Pb by Sr (or vice versa) is not required. Yet strontianite and cerussite are isomorphous, and no isomorphous series exists between these two minerals.

There are two different isomorphous groups among the carbonate minerals, shown in Table 3-1, each revealing complete or incomplete solid solution series, which will serve to exemplify the phenomenon even further.

A variation of this second type of isomorphism, in which similar formulas, similar ionic sizes, and similar structures exist but substitution of ions is not required, is exhibited even among minerals having different compositions. For example, soda niter (sodium nitrate, $NaNO_3$) and calcite (calcium carbonate, $CaCO_3$) are isomorphous and display isomorphism because both crystallize in the hexagonal system. Also, niter (potassium nitrate, KNO_3) is isomorphous with aragonite (calcium carbonate, $CaCO_3$), both crystallizing in the orthorhombic system.

Of utmost importance in the phenomenon of isomorphism as shown in an isomorphous group is the similarity in the sizes of the different ions which occupy similar positions in the isomorphous structures. Many isomorphous

Table 3-1: Isomorphous Carbonate Groups

Aragonite Group (Orthorhombic System)		
Composition	Name	Radius of Cation (Å)
$BaCO_3$	witherite	1.34
$PbCO_3$	cerussite	1.20
$SrCO_3$	strontianite	1.12
$CaCO_3$	aragonite	0.99
Calcite Group (Hexagonal System)		
Composition	Name	Radius of Cation (Å)
$CaCO_3$	calcite	0.99
$MnCO_3$	rhodochrosite	0.80
$FeCO_3$	siderite	0.74
$ZnCO_3$	smithsonite	0.74
$MgCO_3$	magnesite	0.66

groups exist among the minerals, in addition to the olivines and the carbonates. The plagioclase feldspars, the pyroxenes, the amphiboles, and still others also display isomorphic type of solid solution.

Polymorphism

In the previous section on isomorphism, it was pointed out that the compound $CaCO_3$ (calcium carbonate) sometimes crystallized in the orthorhombic system and sometimes in the hexagonal system, depending on the temperature, pressure, and chemical environment during formation. This characteristic of a compound (minerals or native elements too) to exist in more than one atomic arrangement is called *polymorphism*, a term which means "many forms." Each form of the substance has its own set of physical properties and its own distinct crystal structure. For example, calcite has different physical properties and a different crystal structure from aragonite. Yet both are the same compound ($CaCO_3$). A polymorphous substance may be described as dimorphic, trimorphic, quadrimorphic, pentamorphic, etc., according to the number of distinct crystalline forms it may develop. The phenomenon of polymorphism indicates that the chemical composition does not control the crystal structure. It further indicates that there is more than one geometrical arrangement into which the same atoms or ions may be placed in proper proportions, depending on the environment of formation.

Other common examples of polymorphism may be mentioned: carbon (C) forms diamond (isometric) and graphite (hexagonal); iron disulfide (FeS_2) forms pyrite (isometric) and marcasite (orthorhombic); silicon dioxide (SiO_2) forms quartz (hexagonal) and tridymite (also hexagonal, but different structure) and also cristobalite (isometric); titanium dioxide forms rutile (tetragonal), brookite (orthorhombic), and anatase (tetragonal).

After many laboratory experiments on polymorphism have been completed over the past thirty or forty years, there is a general belief that most substances may show polymorphic relations if the conditions of temperature and pressure are systematically varied. The reason for studying polymorphism in the laboratory is to obtain data on the actual temperatures and pressures at which the various mineral polymorphs form, remain stable, and become unstable. This type of information is then applied to minerals and rocks in an attempt to determine the actual temperatures and pressures of formation in the Earth's crust. Even though it is not always possible to determine the exact temperatures and pressures of mineral formation, the experimental data assist us in making estimates.

Noncrystalline Minerals

One of the specifications for a substance to be technically called a mineral is that is must be crystalline. The property of crystallinity is a prerequisite, as given in our original definition of a mineral. However, there are a few naturally occurring solids which are not crystalline, but which are generally considered as minerals because they are fairly common and because it is not always practical to adhere strictly to a hard-and-fast definition. Actually, there are two types of these noncrystalline minerals: *metamict minerals*, in which the original crystallinity has been destroyed, and *amorphous minerals*, which were hardened slowly from gelatinous matter or were supercooled from a hot molten mass and never obtained crystallinity.

Metamict minerals are usually glassy or pitchy in appearance and show no cleavage. The breakdown of original crystallinity to produce a metamict results from bombardment by alpha particles from the radioactive elements uranium or thorium which are usually present in small quantities.

Amorphous minerals include mostly gels. A gel is formed when colloidal solutions solidify. In chemical terms, a colloidal solution is intermediate between true solutions and true suspensions, and the particles in a colloidal solution range from about 1/100 mm to 1/100,000 mm. In some cases, these particles in the solutions have a somewhat crystalline character and in some cases they do not, and they are truly amorphous. Such colloidal solutions solidify into a gel either after they are cooled or after they lose water.

One of the most common gel minerals is opal, formed by the solidification of colloidal solutions of silica (SiO_2). Opal contains a variable amount of water along with the silica, the water varying from 3 to 10 percent by weight, and its formula is thus written $SiO_2 \cdot nH_2O$.

Pseudomorphism

Another phenomenon among minerals is called *pseudomorphism*. The term means "false form." It is a well-known fact that a mineral can be replaced by another mineral by various processes of weathering or oxidation.

If the new replacing mineral maintains the outward crystal form of the earlier replaced one, the phenomenon is known as *pseudomorphism*, even though the replacing mineral actually crystallizes in a different system. The replacing mineral is thus called a *pseudomorph*. Even though the outward form of the first mineral is maintained by the second, the second has internal structure which is characteristic of itself and not of the first. In pseudomorphism, it is only the external form which is maintained.

Pseudomorphism is of two types. The first is one in which there is no change of substance, such as hexagonal calcite ($CaCO_3$) pseudomorphous after orthorhombic aragonite ($CaCO_3$); the new calcite has the outward appearance of aragonite. The other type is one in which there is addition or removal of elements (partial or complete), such as monoclinic malachite, $Cu_2(CO_3)(OH)_2$, after isometric cuprite, Cu_2O; or orthorhombic goethite, $HFeO_2$, after isometric pyrite, FeS_2. The development of a pseudomorph suggests that the original mineral was unstable under a new environment of temperature, pressure, or chemical conditions, and it reverted to a more stable structure, as in the case of the first type, or was chemically altered, as in the case of the second type. The study of pseudomorphism sometimes yields valuable information regarding the geological conditions at the time the pseudomorphs were formed.

Physical Aspects

The physical properties of minerals are used chiefly to distinguish minerals one from another. Since each mineral has its own set of physical properties, identification can be quite easily achieved. But the physical properties also play an important role in whether or not the substance has industrial application.

Although quite rare, minerals with well-developed (euhedral) crystals are easier to identify than those with poorly developed crystals. Euhedral crystals are rare because special and limited conditions must exist in nature for good development of crystals to be achieved. The percentage of euhedral crystals in the rocks is quite small; most mineral specimens are irregular masses with poorly developed faces. Little wonder than that collectors cherish euhedral crystals so much.

The physical properties often determine if a mineral has technical applications in industry. For example, the relatively high hardness of quartz makes it readily usable as an abrasive (as in sandpaper), and the white color of kaolin clay makes it usable as a paper whitener. Physical properties, as a rule, are quite easily determined, and they serve to identify each mineral. It may be useful to list and discuss in some detail the main physical properties of minerals into six main groups as follows (modified after Simpson, pp. 42–43):

Group (1) Properties dependent on light
        a. color   d. diaphaneity
        b. streak   e. luminescence
        c. luster

Group (2) Properties dependent on crystalline aggregation
        a. habit   c. tenacity
        b. cleavage and fracture   d. hardness

Group (3) Properties dependent on taste, smell, touch
        a. flavors
        b. odors
        c. feel

Group (4) Properties due to magnetism, electricity, radioactivity
        a. magnetic properties
        b. electrical properties
        c. radioactivity

Group (5) Thermal properties

Group (6) Specific gravity and density

Properties Dependent on Light

Color

The most obvious physical feature of a mineral is its color. Among the minerals, nearly every color is exemplified, from colorless, white, red, green, blue, brown, yellow, orange, to black. In some cases, color is related to the chemical composition of the mineral, i.e., the nature of the elements making it up. In other cases, color is due to impurities. The color of a mineral is actually produced as a result of the absorption of certain wavelengths of light energy which comprise white light, and the actual color seen is the sum effect of the remaining unabsorbed wavelengths. Since white light consists of energy of a great number of wavelengths, nearly complete absorption of white light would cause the mineral to be a dark color. Conversely, a mineral which absorbs only few or no wavelengths would have a very light color or would be white. Color is a useful and important property in mineral identification, but a certain amount of caution must be employed in using it because colors of many minerals may vary somewhat. In fact, the causes of mineral coloration are rather varied and complex.

Regardless of the cause of color, minerals often show variation. For example, fluorite may be colorless, purple, blue, yellow, green, or pink. Any mineral which shows such color variation is said to be *allochromatic*. On the other hand, there are some minerals which have a characteristic color and show no variation. For example, azurite is always blue. Such minerals with constant color are said to be *idiochromatic*.

Sometimes color can be related to the composition of a mineral. This is usually due to the presence of one of a series of eight elements which are: titanium (Ti), vanadium (V), chromium (Cr), manganese (Mn), iron (Fe), nickel (Ni), cobalt (Co), and copper (Cu). This group of elements, shown in Table 3-2, is called *transition elements* and is responsible for much color in both allochromatic and idiochromatic minerals.

Table 3-2: COLORS ASSOCIATED WITH TRANSITION ELEMENTS

Element	Color Imparted	Examples
Titanium	Blue	Corundum, rutile
Vanadium	Several	Several
Chromium	Green, red	Emerald, ruby
Manganese	Pink, red	Rhodochrosite, rhodonite
Iron	Red, brown	Hematite, limonite
Cobalt	Pink	Erythrite
Nickel	Green	Garnierite
Copper	Green, blue	Malachite, azurite

Ions or groups of ions which produce color are called chromophores or color centers. These strong color centers are characterized by an unusual arrangement of electrons in electron shells. In the transition elements, the third electron shell contains less than eighteen electrons. In other elements there are eighteen. Any extra, one or two, is located in a fourth shell seemingly lost from the third shell. Coloring properties arise from the ease with which these electrons in the fourth shell vibrate under impulses of light and thus absorb much of the energy of the light. Fairly intense coloration can often be imparted to a mineral by only a very small or sometimes negligible percentage of any of these transition elements. For example, chromium is present in very small quantities in the mineral called emerald (actually beryl, $Be_2Al_2Si_6O_{18}$), but the chromium still imparts a very intense green color to an otherwise white-to-very-pale mineral. Also, when pure, quartz (SiO_2) is colorless; but a very small amount of manganese (Mn) or titanium (Ti) produces the delicate pink of rose quartz. Sometimes a mineral appears to have a different color from what it is supposed to have due to the presence of minute crystals of a second mineral which carries the color. Colors commonly seen in minerals are presented in Table 3-3.

Streak

Streak is the color shown by the mineral in a finely powdered form. The streak can be seen when a mineral is rubbed on a piece of unglazed porcelain (streak plate), which causes the mineral to be finely powdered by abrasion. A porcelain streak plate has a hardness of 7.0, so it can be used only on minerals of hardness less than 7.0. The streak can also be obtained by crush-

Table 3-3: SOME COMMON COLORS

Color	Examples
White	Gypsum, halite, calcite
Green	Chlorite, serpentine, malachite
Yellow	Sulfur, carnotite, orpiment
Blue	Covellite, azurite, kyanite
Brown	Siderite, goethite, zircon
Violet	Fluorite, amethyst
Pink-rose	Rhodochrosite, orthoclase
Orange-red	Realgar, crocoite
Red	Cinnabar, hematite, zincite
Gold-brass	Gold, chalcopyrite, pyrite
Silver	Silver, platinum, arsenopyrite
Lead gray-black	Graphite, galena, magnetite

ing, filing, or scratching the mineral. Each mineral has its own characteristic *streak*, and for most minerals the streak is nearly constant. Some examples are given in Table 3-4. Among minerals, streaks vary from white to various tones of red, blue, green, yellow, brown, orange, gray, and to black. In many cases, though, the streak of a mineral is somewhat different from the

Table 3-4: SOME COMMON STREAKS

Streak	Examples
White	Calcite, halite, fluorite
Green	Malachite
Yellow	Gold, autunite, carnotite
Orange	Realgar, thorite
Brown	Sphalerite, goethite, chromite
Red	Cinnabar, hematite, cuprite
Silver white	Silver
Greenish black	Chalcopyrite, pyrite
Lead gray	Galena, stibnite, molybdenite
Black	Chalcocite, arsenopyrite, magnetite

color of the mineral itself. For example, pyrite has a brassy yellow color, but its streak is gray-brown. On the other hand, turquoise has a sky-blue to bluish-green to apple-green color, all of which are quite pale and light colors. Is it any wonder that its streak is white to pale green? To further exemplify this physical property, crocoite, a red to orange-red to orange-colored mineral, has an orange-yellow streak.

The similarity of color and streak in minerals is particularly noticeable in white and slightly tinted ones. Those with hues which are very pale usually have white or very pale streaks. However, this similarity of color and streak is also quite noticeable among the dark minerals.

Luster

The luster of a mineral is the surface appearance as light is reflected from it. The intensity of the luster is related to the quantity of reflected light but is independent of the color of the mineral. In addition, the intensity of the luster depends on the transparency and surface reflectivity. Universal agreement is not obtained for all lusters; different lusters may appear different to different people. Luster is generally classified into two main groups: metallic and nonmetallic. The various types of lusters are given in Table 3-5.

Table 3-5: SOME COMMON LUSTERS

Luster	Examples
Metallic	Gold, pyrite, franklinite
Submetallic	Rutile, enargite, pyrargyrite
Vitreous	Fluorite, quartz, garnet
Adamantine	Wulfenite, diamond, cerussite
Resinous	Realgar, sulfur, sphalerite
Greasy	Nepheline
Silky	Satin spar gypsum, ulexite
Pearly	Talc, copiapite
Dull (earthy)	Limonite, magnesite, kaolinite
Waxy	Cerargyrite, sard, turquoise

Fig. 3-1: Crystal of pyrite from the Isle of Elba, showing high metallic luster, opacity, and good pyritohedral habit. (Courtesy Ward's Natural Science Establishment.)

Fig. 3-2: Dark reddish brown crystals of rutile from near Alexander City, North Carolina, showing submetallic luster.

Metallic luster is seen in minerals which strongly absorb light. When light is strongly absorbed into the minerals, very little or no light passes through them; hence, the minerals are opaque or nearly opaque. Even though the minerals are opaque to light, the metallic minerals have very high brilliant surface reflectivity (Fig. 3-1). As an example, an aluminum pan looks like, and is, metal, and hence has a metallic luster. Some minerals such as the native metals gold and silver, as well as many of the sulfides like galena and pyrite, also look like metal. There is no sharp division between metallic and nonmetallic lusters, and some minerals actually have an intermediate luster because they are only moderately absorptive of light. These are said to be *submetallic* (Fig. 3-2).

Nonmetallic luster is developed in the majority of the minerals because much of the light is reflected from the surface in varying intensities and manners. Very little light is absorbed and much of the light passes through. Since there are so many intensities and manners of reflection, as controlled by the transparency, reflectivity, and surface structure, the nonmetallic lusters are further subdivided as follows:

Adamantine luster: the brilliant luster of diamond, with appearance of being very hard (Fig. 3-3). Minerals with yellow or brown color in association with an adamantine luster develop a *resinous* luster, similar to amorphous secretions (resins) from plants.

Fig. 3-3: Yellow to orange crystals of wulfenite showing adamantine (brilliant) luster, translucency, and tabular crystal aggregates.

Greasy luster: the appearance developed as if grease had been applied and then wiped off with a cloth, but as though a thin coating still remained (Fig. 3-4).

Fig. 3-4: Nepheline in massive form showing a greasy luster, as if a very thin film of grease covered the surface.

Waxy luster: surface appears coated with a thin film of wax (Fig. 3-5).

Fig. 3-5: Brown variety of cryptocrystalline quartz, called sard, revealing a waxy luster.

Silky luster: appearance of shiny fibrous structures, particularly of parallel-fibrous aggregates, and resembles shiny silk cloth (Fig. 3-6).

Fig. 3-6: Silky luster displayed by satin spar gypsum in fibrous aggregates.

Pearly luster: surface shows irridescence as the result of good cleavage or foliation (Fig. 3-7).

Fig. 3-7: Talc, in foliated masses, from Van Horn, Texas, showing a pearly luster.

Vitreous luster: the luster of common glass, either smooth or broken. Nearly 70 percent of the minerals have varying degrees of vitreous luster (Fig. 3-8).

Fig. 3-8: Pale yellow crystals of fluorite showing excellent octahedral habit, high vitreous luster, and translucency.

Dull (Earthy) luster: surface shows no reflection at all, giving an earthy appearance (Fig. 3-9).

In many cases, the true luster of certain minerals is masked by scums or tarnishes on the surfaces. These scums or tarnishes result readily from chemical alteration and often produce very thin irridescent films on mineral surfaces. Such tarnishes are often seen in some sulfides.

Luster of certain minerals may contribute to the economic value, particularly in gemstones. The value of a gemstone is partly due to color and

Fig. 3-9: Trapezohedral habit and dull luster shown by leucite crystals from near Rome, Italy.

transparency, as well as luster. In many cases, it is the luster which is responsible for the brilliance of gemstones.

Diaphaneity

Diaphaneity is the capacity of a mineral to absorb or transmit light. This property is related to the atomic structure, to the atomic packing, and to the density of minerals. Since there is a wide range of structural arrangements and densities among minerals, there is a range of diaphaneity. However, only three general terms are used:

(a) *transparency:* when a crystal is clear and an object may be clearly seen through it;

(b) *translucency:* when a crystal is somewhat clear, and an object may be seen through it but its outlines are fuzzy and indistinct;

(c) *opacity:* when a crystal is not clear at all, and it absorbs most of the light as a result of atomic packing and density (Fig. 3-10).

Fig. 3-10: Crystals of galena showing good cubic habit, opacity, and metallic luster. (Photo courtesy Field Museum of Natural History.)

The property of diaphaneity varies among the minerals, and there is also quite a lot of variation even among different specimens and varieties of the same mineral. For example, rock crystal quartz is transparent, but rose quartz is usually translucent, and milky quartz is nearly opaque.

Luminescence

Luminescence is the property of minerals to emit light from various points in the ionic network. The term luminescence actually applies to the emission of the light, but this is related to some property of the mineral. Luminescence is produced when a mineral is flooded by ultraviolet light which causes some ions to be excited; luminescence begins as ions reach high enough excitation levels to cause emission of light.

Only a small number of minerals are luminescent. And not all specimens of a particular mineral are luminescent. In addition, luminescence may act differently in different minerals. If the mineral emits light only while it is being irradiated by ultraviolet, the mineral shows *fluorescence*. But if the mineral continues to emit light after the irradiation has ceased, the mineral shows *phosphorescence*.

The mineral fluorite, from which the term fluorescence is derived, generally fluoresces in blue as a result of the presence of a small quantity of rare-earth elements. Calcite fluoresces red, pink, or yellow, due to a small quantity of manganese (Mn). Scheelite fluoresces white or bluish-white but becomes yellow as the quantity of molybdenum (Mo) substituting for tungsten (W) increases significantly. The light being emitted during fluorescence or phosphorescence has wavelengths which are longer than the light used to excite it. If visible light were used to produce luminescence, the luminescent glow would not be visible. Consequently, short wavelength ultraviolet light is used to excite the luminescence of minerals so that the emitted glow will be visible.

Properties Dependent on Crystalline Aggregation

Minerals rarely are formed with well-developed faces and shapes. Most minerals commonly crystallize with poor crystal form and imperfect shape because very special conditions must prevail for these characteristics to obtain perfection, and such special conditions are not encountered in nature except very rarely.

Habit

Regardless of whether a mineral displays good crystal form or not, it usually develops a particular "habit" or general shape. Several terms are used to describe the various habits found among minerals, and those given in Table 3-6 are commonly used. Although a mineral may be found in crystals

Table 3-6: SOME COMMON HABITS

Habit	Examples
Cubic	Fluorite, halite, pyrite
Octahedral	Magnetite, diamond, fluorite
Dodecahedral	Garnet
Trapezohedral	Leucite
Rhombohedral	Calcite, siderite, rhodochrosite
Scalenohedral	Calcite
Prismatic	Apatite, beryl, staurolite
Pinacoidal	Biotite
Tetrahedral	Tetrahedrite
Pyritohedral	Pyrite

that exhibit other forms, they most commonly are found with their characteristic habit. Some common habits are illustrated in Fig. 3-1, and Figs. 3-8–3-14.

Fig. 3-11: Green grossularite garnet showing perfect dodecahedral habit so common in this group.

Fig. 3-12: Rhombohedral habit, vitreous luster, and transparency shown by calcite.

Fig. 3-13: Dogtooth spar variety of white transparent calcite with scalenohedral habit in coarse crystal aggregates.

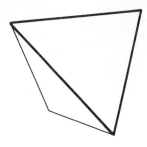

Fig. 3-14: Tetrahedral habit often seen in tetrahedrite and sphalerite.

Crystal Aggregates

Mineral specimens commonly are aggregates of imperfect crystals, and the type of crystal aggregation can oftentimes be useful in identification. A number of terms are used to describe the types of aggregations into which minerals accumulate as listed in Table 3-7. Various crystal aggregations are shown in Figs. 3-15–3-27.

Table 3-7: CRYSTAL AGGREGATIONS

Type	Description	Examples
Radiating	Diverging from a center	Pectolite
Reniform	Rounded masses	Malachite, hematite
Oolitic	Small rounded aggregates	Hematite, calcite
Pisolitic	Pea-sized rounded aggregates	Bauxite
Banded	Alternating layers	Aragonite, anhydrite
Stalactitic	Somewhat like icicles	Limonite, calcite
Bladed	Flattened	Kyanite
Reticulated	Net-like mass	Actinolite
Sheaflike	Bundles	Meta-autunite
Columnar	Long and broad	Hornblende, tourmaline
Tabular	Plate-like groups	Barite
Micaceous	Thin separable flakes	Biotite
Acicular	Needle-like	Pectolite, natrolite
Fibrous	Thin fibers	Asbestos, satin spar
Dendritic	Branching	Pyrolusite, copper
Concentric	Circular layers	Goethite

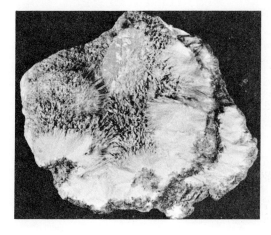

Fig. 3-15: Fine radiating clusters of needlelike (acicular) crystals of white natrolite from Roseburg, Oregon.

Fig. 3-16: Globular (reniform) masses of hematite from England. (Courtesy Ward's Natural Science Establishment.)

Fig. 3-17: Oolitic (small rounded) aggregates so often seen in hematite.

Fig. 3-18: Pisolitic structure (large rounded masses) and dull (earthy) luster shown by bauxite from Arkansas. (Courtesy Ward's Natural Science Establishment.)

Fig. 3-19: Banded aggregates of tiny white anhydrite crystals as mass upon mass accumulated.

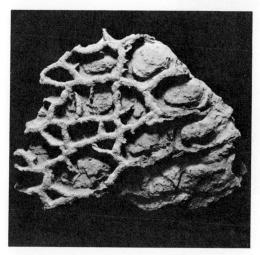

Fig. 3-20: Box-work and stalactitic structure formed by yellowish-brown earthy limonite from Lone Star, Texas.

Fig. 3-21: Long bladed (prismatic) crystal aggregates of kyanite.

Fig. 3-22: Fibrous-like long prismatic crystals of dark green actinolite in somewhat reticulated (criss-cross) aggregates.

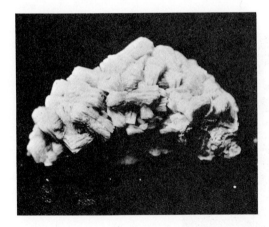

Fig. 3-23: Sheaflike aggregates of long slender autunite (meta-autunite) crystals.

Fig. 3-24: Columnar crystal of hornblende from Colorado.

Fig. 3-25: Micaceous (thin leaf-like) aggregates of phlogopite crystals from Norway, with good basal pinacoidal habit.

Fig. 3-26: Delicate fibrous crystal aggregates of serpentine asbestos from Thetford, Quebec, showing fibrous fracture.

Fig. 3-27: Pyrolusite (black) in dendritic habit on rock surface. (Photo courtesy Field Museum of Natural History.)

Table 3-8: TYPES OF FRACTURES

Type	Characteristic	Examples
Conchoidal	Smooth curved depressions	Quartz
Uneven	Surfaces not flat	Idocrase
Hackly	Surface has sharp points and is rough	Copper
Splintery	Long thin pointed fragments	Spodumene
Fibrous	Threadlike fragments	Asbestos

Cleavage and Fracture

If a crystal of a mineral is broken by a hammer or by some other means, it might break in a very irregular manner or it might break along surfaces which are related to crystal structure. A very important feature of a crystal structure is atomic direction, as can be seen in the positions assumed by faces on a crystal. When a crystal with minimum preferred atomic directional influence is broken, it *fractures*, i.e., it shows *fracture*. But if strong atomic directional influences are present, the crystal *cleaves*, i.e., it displays *cleavage*.

Fracture. Minerals break in a variety of ways which are controlled by the underlying atomic structure. Since there is a fairly wide range of atomic structures among minerals, there are several characteristic types of fractures, as listed in Table 3-8. These various types of fractures are shown in Fig. 3-26, and Figs. 3-28–3-31.

A crystal composed of atoms which are evenly spaced and evenly attracted to each other throughout shows no tendency to fracture in any particular direction because there are no directional influences to control the breakage. A homogeneous mineral of this type will break (or fracture), producing a surface with small smoothly curved shallow depressions called *conchoidal fracture*. A good example of a substance with a conchoidal fracture is a rock known as obsidian, a natural glass. In obsidian, the constituent atoms are

Fig. 3-28: Conchoidal fracture and high vitreous (glassy) luster shown by transparent quartz. Note fracture surface appearing as rounded curving depressions.

Fig. 3-29: Uneven fracture characteristic of idocrase. Broken surface is irregular and slightly rough.

Fig. 3-30: Irregular platy chunk of native copper showing hackly fracture. Broken edges appear sharp and rough.

Fig. 3-31: Splintery fracture shown by prismatic spodumene. Note sharp pointed fragments resembling wood splinters.

arranged at random and are evenly spaced and evenly attracted to each other; there are no preferred directions of breakage. Other natural materials showing conchoidal fracture are opal and amber. Neither obsidian, opal, nor amber ever exhibit exterior crystal faces because they are not crystalline.

Some minerals which are crystalline may also develop good conchoidal fracture. A common example of this is quartz, whose atoms are evenly spaced and relatively (but not completely) evenly attracted to one another. Good conchoidal fracture is displayed in quartz crystals, whether the specimen has developed perfect faces and is flawless or whether it is imperfectly shaped and faces are poorly developed.

In crystals with lesser degrees of homogeneity, fracture surfaces become increasingly guided by atomic directional influences. When fracture is guided wholly by atomic directional influences, the fracture surfaces are known as cleavages. Thus, *cleavage* is a type of fracture controlled by strong atomic directions.

Cleavage. By definition, cleavage is merely a flat fracture guided by atomic structure. In some minerals, the atomic structure provides very strong directional influences, and crystals cannot be fractured except along cleavage planes. In other minerals, only fairly weak directional influences are present, cleavage is not as well developed, and crystals will develop fracture. Usually a cleavage is displayed by plane smooth surfaces, which, in the case of sheet structures, extend for long distances. Cleavage can be related to specific atomic planes of the crystal as well as to exterior faces and directions in the crystal.

A cleavage plane passes between or around ions or ionic groups. It cannot pass through strongly bonded ionic groups. In sheet structures such as mica, the atomic bonding within sheets is very strong, and breaking across sheets is extremely difficult. However, cleavage (and separation) between sheets is very easily accomplished because of very weak forces holding them adjacent to one another.

In minerals which are more uniform in their makeup, directional weaknesses are not so prominent, and it is more difficult to cleave them than in the sheeted minerals. Consequently, cleavages vary from mineral to mineral. When the bonding is weak in certain directions, cleavage is well shown, but when bonding is not weak in any specific direction, cleavage is poorly shown. The mineral will fracture, instead. The presence or absence of cleavage is in no way related to the strength of a crystal. Cleavage simply means that if a mineral breaks, it may break along planes of weak bonding. A mineral may exhibit cleavage in one or more crystallographic directions, and these can usually be distinguished because the cleavage will have somewhat different quality in different directions.

Cleavage is described as *perfect* if flat and smooth surfaces are produced, as in mica. Other terms used to describe the quality of cleavage are *good, fair,* and *poor,* and cleavage actually grades into types of fractures.

A few important points about cleavage may now be made, points which are often difficult for students to understand. The first point is that, when some crystals grow and develop very good shapes and good development of faces, they may not possess good cleavage. For example, quartz crystals often grow and show very good prism faces and rhombohedral faces. But cleavage in quartz is absent or nearly so, and when broken, the fragments display conchoidal fracture. The breakage does not occur parallel to prisms or rhombohedral faces.

A second point is that in those cases when crystals are developed with poor shapes and poor form, they still may reveal a characteristic cleavage, which would be shown if the crystals were broken. For example, a poorly shaped fluorite crystal will reveal its cleavage along octahedral planes if broken. In addition, such octahedral directions in a fluorite crystal may be seen on the exterior surface of the poorly shaped crystal as fine and tiny cracks or traces of cracks. When the crystal is broken with a hammer, breakage (cleavage) will occur along these tiny cracks because these are the cleavage directions. The different types of cleavage are given in Table 3-9, and shown in Fig. 3-32.

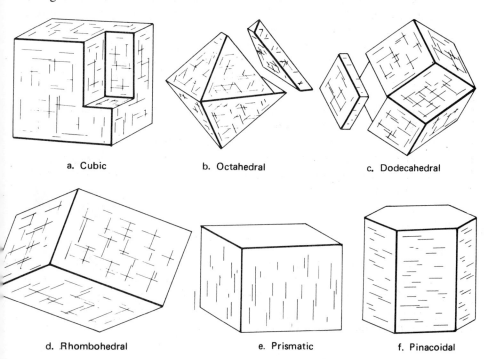

a. Cubic   b. Octahedral   c. Dodecahedral

d. Rhombohedral   e. Prismatic   f. Pinacoidal

Fig. 3-32: Types of cleavage.

Table 3-9: TYPES OF CLEAVAGE

Type	Number of Directions	Examples
Cubic	3	Galena, halite
Octahedral	4	Diamond, fluorite
Dodecahedral	6	Sphalerite
Prismatic	2	Spodumene, hornblende
Basal	1	Muscovite, topaz
Rhombohedral	3	Calcite, magnesite

Cleavage in the crystal systems. Since cleavage is guided by the atomic structure, then cleavage along any one plane (or face) will also be developed along all similar faces on the crystal. For example, galena, which crystallizes in the isometric system, develops cleavage along the cube face. Since there are six faces to a cube, the cleavage is exhibited by each of these six faces arranged in three pairs (front and back, right and left, top and bottom). This type is called *cubic* cleavage. Other common cleavages in the isometric system are called *octahedral* (eight faces) and *dodecahedral* (twelve faces).

Many tetragonal crystals display good *prismatic* cleavage developed along vertical faces, with 90° angles between cleavage surfaces in accordance with the crystallographic directions in the system. Cleavage along *basal pinacoids* (flat tops and bottoms) of tetragonal crystals also are quite numerous.

Crystals in the hexagonal system often display perfect *rhombohedral* cleavage, particularly among the carbonate minerals. *Basal* cleavage and *prismatic* cleavage (along vertical faces) are also easily developed in the hexagonal system.

Basal pinacoid cleavage and *prismatic* cleavage are both developed quite easily in the orthorhombic and monoclinic systems.

Because of the poor symmetry of crystals in the triclinic system, there is not much chance for repetition of faces; consequently, each cleavage direction is parallel to a different set of faces. But *basal pinacoid* and *prismatic* cleavages are developed.

Tenacity

Tenacity is a term which refers to the resistance that any substance offers to breaking, bending, cutting, or crushing. Most minerals are readily crushed by hand in a grinding mortar because they are *brittle*. Such common minerals as calcite, gypsum, halite, quartz, feldspar, and fluorite are quite brittle. On the other hand, certain soft minerals like gypsum, talc, and serpentine can be easily cut with a knife into thin shavings and thus are *sectile*. If a mineral can be bent easily, such as selenite gypsum plates, the property is called *flexibility*, and minerals showing this property are said to be *flexible*. Note in this discussion that gypsum is brittle, sectile, and flexible. A few minerals (mainly native metals such as copper and gold) are *malleable*, which means that their shapes may be deformed, by hammering or grinding, into thin plates or sheets, but the material is not broken.

Hardness

The *hardness* of a mineral is usually defined as its resistance to being scratched and thus is a measure of its cohesiveness, i.e., the strength of atomic bonding. The making of even the smallest groove by scratching produces a separation of atoms as a result of the disruption of the bonding. Needless to say, the great variety of constituent atoms, joined in many combinations in minerals, causes a very wide range of hardnesses among the different mineral species, as shown in Table 3-10.

Table 3-10: SOME COMMON HARDNESSES

Hardness	Examples
1.0	Talc
1.0–1.5	Molybdenite
2.0	Stibnite
2.0–2.5	Cinnabar
2.5	Galena
3.0	Calcite
3.5–4.0	Sphalerite
4.0	Siderite
4.5–5.0	Scheelite
5.0	Apatite
5.5–6.5	Magnetite
6.0	Orthoclase
6.0–6.5	Pyrite
6.5	Olivine
7.0	Quartz
7.5	Zircon
8.0	Topaz
9.0	Corundum
10.0	Diamond

In 1822, the German (Austrian) mineralogist Mohs proposed a scale of hardness by which minerals of various hardnesses could be compared. Although it is an old scale, it is still very much in use because it has not been improved upon. Mohs' scale of relative hardness, consisting of fairly common and easily obtained minerals, is as follows:

(1) talc (softest)

(2) gypsum

(3) calcite

(4) fluorite

(5) apatite

(6) orthoclase (feldspar)

(7) quartz

(8) topaz

(9) corundum

(10) diamond (hardest)

In this scheme, each mineral will scratch any one lower in the scale (lower in number) and can be scratched by any higher in the scale.

The hardnesses of all other minerals are compared to these ten in Mohs' scale. For example, a mineral which scratches apatite, No. 5, but is scratched by orthoclase, No. 6, has a hardness somewhere between 5 and 6, and thus may be designated as $5\frac{1}{2}$. In the Mohs' hardness scale, all hardnesses are relative to each other and do not represent exact uniform steps between them. For practical purposes, though, they are approximately equal. To determine the hardness of a mineral, an estimate may be made according to the following:

(1) Those scratched by the fingernail have a hardness of 2.5 or less.

(2) Those scratched by a penny have a hardness of 3.0 or less.

(3) Those scratched by a knife blade or window glass have a hardness of 5.5 or less.

(4) Those scratched by a good file have a hardness of 6.5 or less.

(5) Those scratched by unglazed porcelain (streak plate) have a hardness of 7.0 or less.

In making the scratch test to determine the hardness, one should not select a granular surface, since scratching may only dislodge granules and give a false impression. A fresh unweathered surface is more reliable and thus should be selected. In testing soft minerals, care should be taken not to mistake smearing for scratching.

The hardness of a mineral is controlled by four factors. Hardnesses among the minerals increase:

(1) as the sizes of the ions decrease, because the smaller ions will fit into various arrangements better and more tightly than large ions;

(2) as the density of packing of ions increases, because closely packed atoms thus are closer together, and more tightly held;

(3) as strength of bonding increases, because it is thus harder to disrupt the atoms by scratching;

(4) as valency (charge) of ions increases, because an ion with a 1+ charge cannot be bonded as strongly as an ion with a 2+ charge.

Properties Dependent on Taste, Smell, Touch

Flavors

In order that a mineral may be tasted, it must be soluble in water or saliva. Very few minerals are so soluble, but various flavors have been determined, and these are presented in Table 3-11.

Table 3-11: EXAMPLES OF FLAVORS

Type	Description	Examples
Sweet metallic	Flavor of metal	Chalcanthite
Alkaline	Flavor of soda	Soda niter
Astringent	Flavor of alum	Borax
Bitter	Flavor of Epsom salts	Epsomite
Bitter salty	Bitter saline	Sylvite
Cooling	Flavor of sodium nitrate	Soda niter
Metallic	Flavor of metal	Pyrite
Pungent	Sharp and biting	Ammonium chloride
Saline	Salty	Halite

Odors

Some minerals emit characteristic odors only when breathed upon, rubbed, scratched, pounded, or heated. Such odors are given in Table 3-12.

Table 3-12 EXAMPLES OF ODORS

Type	Description	Examples
Argillaceous	Odor of clay	Kaolinite
Bituminous	Organic odor of asphalt	Asphalt
Fetid	Odor of rotten eggs	Limestone
Garlic	Odor of garlic	Arsenopyrite
Sulfurous	Odor of sulfur dioxide	Pyrite

Feel

When minerals are touched, they cause a certain type of sensation to the hand or fingers. These are described in Table 3-13.

Table 3-13: EXAMPLES OF FEEL

Type	Description	Examples
Cold	Cool to touch	Silver
Greasy	Slippery to touch	Talc
Harsh	Rough to touch	Alunite
Smooth	No roughness to touch	Sepiolite

Properties Due to Magnetism, Electricity, and Radioactivity

Magnetism

All minerals are in some way affected by a magnetic field. In many cases, special equipment is required to show just what the effect is. Many minerals are not magnetic at all; in fact, they are repelled by a magnet. These are called *diamagnetic*. Minerals which are slightly attracted to a magnet are said to be *paramagnetic*. All iron-bearing minerals are paramagnetic, but some iron-free minerals are also paramagnetic, such as beryl. Minerals which are strongly attracted by a magnet are called *ferromagnetic*.

Only a very few minerals are ferromagnetic, such as magnetite (Fe_3O_4), pyrrhotite ($Fe_{1-n}S$), maghemite (Fe_2O_3), and some specimens of chromite ($FeCr_2O_4$). Sometimes magnetite and maghemite are themselves natural magnets and will attract iron filings. If allowed to swing free on a string, these minerals will align themselves in the Earth's magnetic N-S direction. Such highly magnetic specimens are called *lodestones*.

The magnetic properties are useful in separating minerals from one another, such as when ore is crushed and iron minerals must be separated from others, or when different iron minerals must be separated from one another. The degrees of magnetism of each group thus come into play.

Electrical Properties

Minerals may be divided into two groups according to their ability to conduct electricity: nonconductors and conductors. Most minerals are nonconductors, and the number of conductors is small, being only those characterized by metallic bonding. The conductors include native metals, most of the sulfides, graphite, and a few oxides. All degrees of conductivity exist among conductive minerals. Electrical conductivity varies with crystallographic direction and is thus a vectorial property.

Radioactivity

Comparatively few minerals are radioactive. The property of radioactivity is associated with the presence of uranium and thorium. However, potassium and rubidium also show extremely weak radioactivity, detectable only with very sensitive apparatus. The two strongly radioactive elements, uranium and thorium, disintegrate spontaneously, with the emission of alpha, beta, and gamma rays. Alpha rays are positively charged helium atoms; beta rays are negatively charged electrons; and gamma rays are highly penetrative rays of short wavelength. Radioactivity can be detected with a Geiger counter or by a scintillometer, as well as by its effect on photographic film. The more common radioactive minerals which contain uranium or thorium are uraninite, $2UO_3 \cdot UO_2$; autunite, $Ca(UO_2)_2P_2O \cdot 8H_2O$; carnotite, $K_2O \cdot 2U_2O_3 \cdot U_2O_5 \cdot 2H_2O$; monazite, $(Ce,La,Y,Th)PO_4$; thorianite, ThO_2; and thorite, $ThSiO_4$.

Thermal Properties

Thermal Conductivity

Like most other substances, minerals have some ability to transmit heat. This property is known as *thermal conductivity* and is measured as the amount of heat (in calories) which passes through one square centimeter of surface of a slab of mineral one centimeter thick. The ability to transmit heat is directly related to the structure of the mineral.

Thermal Expansion Coefficient

When a mineral is heated, its volume increases. This property is called the *thermal expansion coefficient* and is measured in percentage volume increase when the temperature is raised from 0°C to 1°C. Thermal expansion coefficients have a fairly wide range among the minerals.

Since a temperature rise is required to detect the thermal expansion coefficient, the amount of heat (in calories) required to raise one gram of mineral through 1°C of temperature is called its *specific heat*.

Specific Gravity and Density

The terms specific gravity and density often are used interchangeably, but there is a distinction between them. The *specific gravity* refers to the weight of a mineral compared to an equal volume of water. To exemplify this, if a quart of molten gold is poured into a mold, cooled, and then the cast piece of gold placed on a scale, 19.3 quarts of water would be required to balance it. The specific gravity of gold is thus 19.3, meaning that, volume for volume, it is 19.3 times heavier than water. Water is used as a comparison standard in all specific gravity determinations, and this comparison is thus built into the term.

The term *density* means nearly the same as specific gravity, but weights and volumes are used, such as grams per cubic centimeter. In the case of gold, the density of gold would be 19.3 gm/cu cm. This number actually reveals the weight (in grams) of each cubic centimeter. But as a rule, densities of minerals are rarely used because the specific gravities are usually sufficient in weight considerations.

There is a wide range of specific gravities among minerals because of the great variety of elements that comprise them as presented in Table 3-14. The

Table 3-14: SOME SPECIFIC GRAVITIES

Mineral	S.G.
Borax	1.71
Sylvite	2.00
Sulfur	2.07
Gypsum	2.32
Orthoclase	2.56
Quartz	2.65
Fluorite	3.18
Diamond	3.50
Siderite	3.96
Rutile	4.25
Ilmenite	4.72
Pyrite	5.01
Hematite	5.26
Scheelite	6.10
Cassiterite	7.00
Galena	7.57
Silver	10.50
Gold	19.30

common elements vary both in atomic size and in the number of electrons and neutrons they contain. And, as we have seen before, there are many differences in atomic sizes according to number of electron shells and number of electrons. The heavy elements give rise to high specific gravities, such as the metals copper, iron, nickel, lead, and tungsten. But minerals like silicates or carbonates usually have low specific gravities because most of the space is occupied by the large and less dense oxygen atoms, even though fairly heavy, small-sized cations are also present.

Fig. 3-33: Jolly balance.

Several methods are used to determine the specific gravity of a mineral, but the best known method is by the Jolly balance (Fig. 3-33). The Jolly balance consists of an upright stand with a graduated scale mirror on the face of the upright. The upright can be extended by a retractable and extendable rod. From a horizontal bar at the top end of the extendable rod is suspended a spring carrying two small pans at the lower end, one above the other. The lower pan is suspended in water. The spring hangs so that a small attached bead above the upper pan casts a reflection in the mirror. A reading is taken by getting the bead and its reflection into coincidence and then reading on the graduated scale the point where the top of the bead is positioned.

Three readings are made, and in each case the lower pan is immersed in water:

(1) with the spring carrying both pans (n);
(2) after placing a small mineral fragment on the upper pan in air (n_1);
(3) after placing the same mineral fragment on the lower pan in water (n_2).

The specific gravity is given by the formula

$$\text{S.G.} = n_1 - \frac{n}{n_1} - n_2$$

A second method of determining specific gravity is with the use of "heavy liquids" with high specific gravities. The standard heavy liquids are miscible with one another, and so intermediate densities can be obtained by diluting one with another. The heavy liquids often used in standard practice are as follows:

(1) Bromoform, $CHBr_3$; S.G. = 2.9.

(2) Tetrabromoethane, $C_2H_2Br_4$; S.G. = 2.96.

(3) Methylene iodide, CH_2I_2; S.G. = 3.3.

(4) Clerici solution; S.G. = 4.2.

A mineral fragment is immersed in one of these liquids. If it floats, the liquid is diluted with one of lower density until the fragment neither sinks nor floats. If it sinks the liquid is made more dense by the addition of a more dense liquid until the mineral neither sinks nor floats. In each case, then, the density of the blended liquid has been previously determined.

Any mineral may vary somewhat in its specific gravity, mainly because ions of other elements may substitute for others in the structure. There is a marked increase in the specific gravities as more heavy ions make up the minerals. However, a pure mineral, which has a constant formula, has a constant specific gravity. For example, quartz has a constant specific gravity of 2.65 because its composition is fixed and no substitution takes place. But in two different minerals made up of the same elements, such as diamond and graphite, the specific gravity is related to the packing of the carbon atoms. Diamond's specific gravity is 3.52, while graphite's is 2.3.

Classification of Minerals

Introduction

Because of the great variety of minerals and their elemental compositions, it is hardly necessary to state that there is also a wide variety of conditions which play a part in mineral formation. However, in very broad aspects, only three factors are important: available material, pressure, and temperature. These three factors vary considerably when the formation of all the different types is taken into consideration.

In any one case where a mass of elements is forming from a solution into minerals, the minerals which form will be determined by the elements present and by the varying temperatures and pressures which prevail during the time the crystals are developing. For example, if metals such as Cu, Zn, Pb, and Fe are present along with sulfur in the solution, then sulfides of these elements will form, such as pyrite, FeS_2, chalcopyrite, $CuFeS_2$, galena, PbS, and sphalerite, ZnS. On the other hand, if metals such as Cu, Zn, Pb, and Fe are present along with carbon and oxygen, then carbonates are likely to form, such as siderite, $FeCO_3$, cerussite, $PbCO_3$, smithsonite, $ZnCO_3$, and malachite, $Cu_2(OH)_2CO_3$. Temperature and pressure come into play as far as controlling the nature of the minerals to form by exerting influence on the solubility and time of precipitation of the different species.

Common Elements of Minerals

Although there are about 80 elements which develop into minerals, there are certain restrictions which prevent unlimited numbers of species to develop. In fact, there are only about 2000 mineral species grouped into only a very few types of minerals. A great many of the mineral species are very rare indeed in the Earth's crust. The number of different minerals that can be considered as common is only about 120 or so. The remaining 1880 are classed as rare or very rare. Actually, over 98 percent of the crust of the Earth is composed of just eight elements making up common minerals, as can be seen in Table 3-15. A close examination of the table will reveal that elements

Table 3-15: ELEMENTS THAT COMPRISE MOST COMMON MINERALS

Element	Symbol	Percent
Oxygen	O	46.40
Silicon	Si	28.15
Aluminum	Al	8.23
Iron	Fe	5.63
Calcium	Ca	4.15
Sodium	Na	2.35
Potassium	K	2.33
Magnesium	Mg	2.09

such as copper, lead, zinc, tin, and nickel are not listed. This is because those metal elements are rare in the Earth's crust, although they are sometimes concentrated in ore deposits which can be mined. Because of the great abundance of oxygen and silicon, the bulk of the minerals in the crust are silicates. The silicates have silicon (Si) and oxygen (O) ions in some type of linked networks, in which one silicon ion is centered between four oxygen ions, making a $(SiO_4)^{4-}$ anion group.

Aluminum, too, is fairly abundant, and in many of the common minerals, it is present along with oxygen and silicon. In a general sense, the numerous silicate minerals are considered as networks of silicate ions containing Al, e.g., $(Si[Al]O_4)$, linked together by cations of the other common elements, Na, K, Ca, Fe, Mg, in various arrangements and proportions to produce the more common species.

Since every silicate mineral consists of a $(SiO_4)^{4-}$ tetrahedral unit as the main part of the framework, the remainder of the structure is made up of cations, oxygens, water molecules, or hydroxyl ions. When any of these combine with the silicate anion group or groups, a stable and electrically neutral structure is produced.

In many of the silicate structures, the aluminum cation has an unusual ability to perform two different roles. First, aluminum, particularly in the tektosilicates but also in other subclasses, may replace some of the silicon in the $(SiO_4)^{4-}$ anion; hence the main anionic unit will be $[(Si,Al)O_2]^{-1}$, $(Al,SiO_4)^{-1}$, or $(Al,Si_3O_8)^{-1}$. Second, aluminum may replace other common cations such as Mg^{2+}, Fe^{3+}, etc. In some subclasses, aluminum may play one or the other role, as in sodalite, $Na_8(Al,SiO_4)_6Cl_2$, or in glauconite, $K(Fe,Mg,Al)_2(Si_4O_{10})(OH)_2$. Aluminum may play both roles in the same mineral, as in chlorite, $(MgFeAl)_6(Al,Si)_4O_{10}(OH)_8$; in hornblende, $NaCa_2(Mg,Fe,Al)_5(Si,Al)_8O_{22}(OH)_2$; and in augite, $Ca(Mg,Fe,Al)(Al,Si)_2O_6$.

Some of the rarer elements have chemical behaviors similar to the more common elements and hence do not form minerals in which they themselves are important framework units in the crystals. Rather, they are found accompanying the common elements. For example, hafnium behaves similarly to zirconium; hence it is found in zircon where it substitutes for zirconium. Also, rubidium, when found, is associated with potassium, an element with similar chemical behavior.

Minerals are usually classified according to the fundamental ion or ion group of which they are composed. As is well known, an ion is a chemical element which has lost or gained electrons. There are ninety-two natural chemical elements, but only about sixteen can be considered as common in the Earth's crust. The rest are considered rare, and although they are found in some minerals, they do not constitute fundamental ions as far as the classification is concerned. In fact, of the ninety-two elements, only about sixteen are used in making up over 99 percent of the Earth's crust.

These sixteen elements, with their chemical symbols, are shown in Table 3-16 in the order of their abundance in the crust. Three other elements, carbon, chlorine, and nitrogen, are added at the bottom because they play particularly important roles in the classification, even though they are not too abundant. The more common ions of these nineteen elements as they are present in minerals are listed.

Since these sixteen elements are the most abundant ones in the Earth's crust, it is to be expected that some of them will serve as fundamental ions or ion groups for the classification. In our earlier discussion of ions, it was pointed out that some ions, the cations, have positive $(+)$ charges, while others, the anions, have negative $(-)$ charges. When two simple ions are in combination to make a new, more complex ion, the new complex ion also has either a positive $(+)$ charge or a negative $(-)$ charge. Because anions are formed by gaining electrons, or form other complex ions, they often become the largest units in the crystal structure; consequently, they control the kind

Table 3-16: COMMON ELEMENTS IN EARTH'S CRUST

Element	Symbol	Abundance or Parts Per Million	Common Ions
Oxygen	O	464,000	O^{2-}
Silicon	Si	281,500	SiO_4^{4-}
Aluminum	Al	82,300	Al^{3+}
Iron	Fe	56,300	Fe^{2+}, Fe^{3+}
Calcium	Ca	41,500	Ca^{2+}
Sodium	Na	23,600	Na^{1+}
Potassium	K	23,300	K^{1+}
Magnesium	Mg	20,900	Mg^{2+}
Titanium	Ti	5,700	Ti^{4+}, TiO_4^{4-}
Hydrogen	H	1,400	H^{1+}
Phosphorus	P	1,050	P^{5+}, PO_4^{3-}
Manganese	Mn	950	Mn^{2+}, Mn^{4+}
Fluorine	F	625	F^{1-}
Barium	Ba	425	Ba^{2+}
Strontium	Sr	375	Sr^{2+}
Sulfur	S	260	S^{6+}, SO_4^{2-}, S^{1-}
Carbon	C	200	C^{4+}, CO_3^{2-}
Chlorine	Cl	130	Cl^{1-}
Nitrogen	N	20	N^{5+}, NO_3^{3-}

Modified from Mason and Berry, 1968

of structure the crystal will develop during crystal growth. Crystals of minerals are usually regarded as structures with large-sized fundamental anions, and the smaller, positively (+) charged metal ions fit in between in a systematic way. The classification, known as the Berzelian system, places minerals into broad classes according to the large-sized negative anions which serve as the fundamental framework unit.

In the formulas, the anions are given to the right and the cations to the left, as shown by the following example for calcite:

cation valence $\quad$ anion valence

$Ca^{2+} \quad CO_3^{2-}$

calcium $\quad$ carbonate

$CaCO_3$
Calcite

Ca = symbol for calcium (one atom)

CO_3 = symbol for carbonate anion

C = carbon (one atom)
O = oxygen (three atoms)

As we noted in a previous section, the formula shows the proportions of the different atoms making up the mineral as indicated by the small subscripts at the lower right of the symbols. A formula is written in the simplest rational manner which shows the proper proportions of the constituent atoms. For example, in the formula for hematite, Fe_2O_3, there are three oxygen atoms for each two iron atoms. This cannot be written in any simpler rational numbers which show the proper proportions. It could be written $Fe_1O_{1.5}$, but to avoid fractions and decimals, the lowest series of whole numbers is used. In addition to showing proportions of atoms, the formula often reveals a little information about the crystal structure. For instance, $CaCO_3$ shows C and O in a complex anion with three oxygens surrounding a central carbon, the well-known carbonate anion. In some cases, the formula reveals nothing about the crystal structure, such as in rutile, TiO_2.

The Berzelian Classification

The system of classification called the Berzelian system contains the native elements first in order to show the more simple minerals. Then follow more complex compounds which contain two or more elements. These are followed by those with radicals (ion groups) containing oxygen, and the classification ends with the silicates, which is the largest and most important group of all. The Berzelian classification is presented in Table 3-17.

Table 3-17: BERZELIAN CLASSIFICATION SCHEME

Class	Main Anions
I. Native elements	Metals, semimetals, nonmetals
II. Sulfides	Sulfur (S)
Sulfosalts	Sulfur and a semimetal (As, Bi, Sb)
III. Oxides	Oxygen (O)
Hydroxides	Hydroxyl radical (OH)
IV. Halides	Halogens (F, Cl, Br)
V. Carbonates	Carbonate radical (CO_3)
Nitrates	Nitrate radical (NO_3)
Borates	Borate radical (BO_3)
VI. Sulfates	Sulfate radical (SO_4)
Chromates	Chromate radical (CrO_4)
Molybdates	Molybdate radical (MoO_4)
Tungstates	Tungstate radical (WO_4)
VII. Phosphates	Phosphate radical (PO_4)
Vanadates	Vanadate radical (VO_4)
VIII. Silicates	Silicate radical (SiO_4)

As mentioned earlier, the various classes of minerals are derived and named according to the large anions (simple or complex) which make up the main framework unit of the structures. There is one exception to this scheme: the class known as *native elements* consists of minerals which are neither cations nor anions, but rather are electrically neutral (no charge) atoms. The remaining classes of minerals are named from the large anion which appears to the right in the formulas. For example, the sulfides show sulfur (S) to the right, but if sulfur is combined with oxygen (O) to make a more complex negative ion $(SO_4)^{2-}$, then a sulfate is indicated. In fact, formulas of various minerals in the same class often are very similar, e.g., calcite, $CaCO_3$; siderite, $FeCO_3$; magnesite, $MgCO_3$. None of the following groups are complete. Only the more common minerals in each group are listed and described.

Descriptions of Common Minerals

Each of the following descriptions of the 114 common minerals contains a summary of that mineral's physical properties. A brief comment is made to aid in distinguishing the mineral from others which appear visually similar. Accompanying each description is a photograph of the mineral, either alone or in association with other minerals and rock material. The photograph is provided to reveal the physical nature of the mineral, and in many cases, specimens have been propped up for best photographic effects. Figure captions also assist in pointing out important physical properties of the minerals. No attempt has been made to provide a full mineralogical discussion; rather, brief but meaningful information is presented.

Native Elements

A number of elements are found in the native state as minerals, but for the most part they occur as small-sized accumulations. However, elements such as native gold, native silver, native copper, native carbon (diamond and graphite), and native sulfur occasionally are found in sufficiently large accumulations to permit mining.

The native metals gold, silver, and copper have similar physical properties. For example, they are heavy and generally malleable, i.e., they can be hammered into thin sheets. These metals also have similar habits, occurring frequently as wires or branching shapes. These similarities are not obtained among the other three native nonmetals. The native elements are each composed of only one type of atom, while all other minerals are composed of atoms of two or more different elements.

The common native elements are:

gold, Au	sulfur, S
silver, Ag	diamond, C
copper, Cu	graphite, C

Sulfides (Also Sulfosalts)

Sulfides comprise a fairly large number of minerals which are compounds of metals and semimetals with sulfur. There may be one or more metals in certain minerals, such as iron (Fe) in pyrite, FeS_2; or both cobalt (Co) and arsenic (As) in cobaltite, CoAsS. In some circumstances, the formula for a mineral may consist of two metals placed in parentheses, e.g., for the mineral sphalerite, (Zn,Fe)S. This means that zinc (Zn) is the abundant metal and that iron (Fe) often substitutes for some of the zinc in the crystal structure. However, the iron is not essential to the ideal formula for sphalerite, ZnS. The absence of parentheses in the formula for cobaltite, CoAsS, means that cobalt (Co) and arsenic (As) both are primary elements in the structures and are not substituting for one another. The bonding in sulfides displays wide variation, ranging from ionic in some, to homopolar in others, and to metallic in still others.

Also included with the sulfides is a smaller group of somewhat similar minerals known as the *sulfosalts*. In the sulfosalts, there are semimetals such as arsenic (As), antimony (Sb), or bismuth (Bi) acting as negatively charged ions (anions) along with sulfur. Consequently, the formulas contain two anions to the right-hand side. This is seen in the formula for pyrargyrite, Ag_3SbS_3. In this silver antimony sulfide, silver is the cation $(+)$ and both antimony and sulfur are anions $(-)$. A few of these sulfosalts, such as pyrargyrite and proustite, are commercially valuable.

The crystal structural arrangements of the sulfosalts are less well known than those of sulfides, but apparently structural similarities exist between these two groups. The common minerals in these various groups are:

Sulfides

argentite	Ag_2S	cinnabar	HgS
chalcocite	Cu_2S	realgar	AsS
bornite	Cu_5FeS_4	stibnite	Sb_2S_3
galena	PbS	pyrite	FeS_2
sphalerite	$(Zn,Fe)S$	cobaltite	$CoAsS$
chalcopyrite	$CuFeS_2$	marcasite	FeS_2
pyrrhotite	FeS	arsenopyrite	$FeAsS$
pentlandite	$(Fe,Ni)_9S_8$	molybdenite	MoS_2
covellite	CuS		

Sulfosalts

pyrargyrite	Ag_3SbS_3
tetrahedrite	$Cu_{12}Sb_4S_{13}$
enargite	Cu_3AsS_4

Oxides (*Also Hydroxides*)

The oxides comprise a much larger part of this class than do the hydroxides. In the oxides, oxygen $(O)^{2-}$ plays the part of the important anion, while in the hydroxides, the hydroxyl $(OH)^{1-}$ ion is important in this role. In a few instances, both oxygen $(O)^{2-}$ and hydroxyl $(OH)^{1-}$ are constituents of the mineral.

The simple oxides contain a metal and oxygen, while multiple oxides contain two metal cations along with oxygen, and in some cases they contain water, H_2O also. A simple oxide, pyrolusite, has a simple formula, MnO_2, showing only one metal cation. Spinel, $MgAl_2O_4$, serves as an example of a multiple oxide with two different metal cations, magnesium (Mg) and aluminum (Al), with oxygen. The multiple oxides (as well as the hydroxides) are usually bonded by ionic bonds and do not have discrete complex anions in the structures such as are present in carbonates and nitrates. In the multiple oxides, the bonds between oxygen and the two metal cations are of the same strength, and there is no preferential linkage of one of the cations to oxygen.

Oxygen is both a very abundant and a very reactive element. Consequently, other elements and minerals which might undergo decomposition will combine with oxygen to form more stable oxides and hydroxides.

The common oxides and hydroxides are:

Oxides

cuprite	Cu_2O	ilmenite	$FeTiO_3$
spinel	$MgAl_2O_4$	pyrolusite	MnO_2
magnetite	Fe_3O_4	psilomelane	$(Ba, H_2O)_2Mn_5O_{10}$
chromite	$FeCr_2O_4$	rutile	TiO_2
corundum	Al_2O_3	cassiterite	SnO_2
hematite	Fe_2O_3	uraninite	UO_2

Hydroxides

brucite	$Mg(OH)_2$	
boehmite	$AlO(OH)$	
gibbsite	$Al(OH)_3$	bauxite minerals
diaspore	$HAlO_2$	
manganite	$MnO(OH)$	
goethite	$HFeO_2$	
(limonite)		

Halides

A group of four elements, referred to as the *halogens*, are responsible for the name *halides*. The halogen elements are fluorine (F), chlorine (Cl), bromine (Br), and iodine (I). In the halide minerals, these elements act as the framework anions. Most of the halide minerals contain chlorine as the anion, and so the largest subgroup contains chlorides. Fluorides, with fluorine (F) anions, are much less common than the chlorides, whereas bromides, with bromine (Br) and iodides, with iodine (I), are extremely rare.

Halite, a normal halide, has a formula consisting of a metal cation (Na) and a halogen anion (Cl), thus NaCl. The halides, a relatively small group, are commonly bonded by ionic bonds. The common halides are:

Halides

halite	NaCl
sylite	KCl
fluorite	CaF_2

Carbonates (Also Nitrates and Borates)

In the carbonates, the carbonate anion $(CO_3)^{2-}$ is the important funda-mental anionic unit. The $(CO_3)^{2-}$ anionic unit is made up of one carbon ion $(C)^{4+}$ at the center of an equilateral triangle of three oxygen ions $(O)^{2-}$ (Fig. 3-34). The ionic type of bonding is well displayed in the carbonates. Most of the carbonate minerals are fairly scarce, but a few are extremely common. The most common carbonate is calcite, which is the main ingredient of limestones, a very common rock. But others are also common. The more common ones can be grouped into the calcite group, the dolomite group, and the aragonite group. These groups contain the important carbonate anion attached to small single cations, to two small cations alternating in position, or to single large cations, respectively. As an example of the first case, the formula of magnesite is $MgCO_3$, where the Mg^{2+} cation is small. In the second case, the dolomite formula is $CaMg(CO_3)_2$, in which Ca and Mg alternate in atomic position in the crystal structure. As an example of the third case, in the formula for strontianite, $SrCO_3$, the Sr^{2+} cation is quite large. Ionic bonding prevails in the carbonates. The common carbonates are:

Calcite Group		Dolomite Group		Aragonite Group	
calcite	$CaCO_3$	dolomite	$CaMg(CO_3)_2$	aragonite	$CaCO_3$
magnesite	$MgCO_3$			strontianite	$SrCO_3$
siderite	$FeCO_3$			cerussite	$PbCO_3$
rhodochrosite	$MnCO_3$	*Other Carbonates*			
smithsonite	$ZnCO_3$	malachite	$Cu_2CO_3(OH)_2$		

The nitrates are grouped along with the carbonates because the main anion, nitrate $(NO_3)^{2-}$ is complex and contains one nitrogen ion $(N)^{4+}$ at the center of an equilateral triangle of three oxygen ions $(O)^{2-}$. This is a structure similar to that of the carbonate $(CO_3)^{2-}$ anion of the carbonates. Ionic bond-ing prevails here also. There are very few nitrate minerals, and these few are

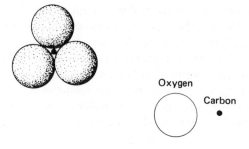

Oxygen

Carbon
•

Fig. 3-34: The carbonate anion.

found only in very arid regions, deposited from surface lakes by extensive evaporation. The most common nitrates are soda niter, $NaNO_3$, and niter, KNO_3, but the former is much more abundant than the latter.

The borates are grouped with the carbonates, too, because the main anionic unit consists of one boron ion $(B)^{3+}$ at the center of an equilateral triangle of oxygen as in the carbonate anion. This simple borate anion $(BO_3)^{3-}$ occurs in some borate minerals, but a somewhat more complex borax anion exists in others. In the more complex borate anion, one boron ion $(B)^{3+}$ is located at the center of a tetrahedron (pyramid-like) with four oxygens at the corners. These tetrahedra usually are linked to one another by sharing an oxygen. Occasionally some of the oxygen positions are occupied by (OH), but this does not alter the classification. The borates are ionic compounds.

Although the most widespread borate is borax, three other borates are also listed:

Borates

kernite	$Na_2B_4O_7 \cdot 4H_2O$
borax	$Na_2B_4O_7 \cdot 10H_2O$
ulexite	$NaCaB_5O_9 \cdot 8H_2O$
colemanite	$Ca_2B_5O_{11} \cdot 5H_2O$

Sulfates (Also Chromates, Molybdates, Tungstates)

The sulfates is a very large group of minerals, and in each one the sulfate radical $(SO_4)^{2-}$ is the important anionic unit of the framework. In this group, there is a sulfur ion $(S)^{6+}$ at the center of a tetrahedron of four oxygens $(O)^{2-}$, making a complex anionic unit, the sulfate anion $(SO_4)^{2-}$ (Fig. 3-35). The

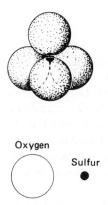

Oxygen

Sulfur

Fig. 3-35: The sulfate anion.

sulfates are commonly divided into three main sections: the anhydrous sulfates without water, hydrated sulfates with water, and anhydrous sulfates with hydroxyl.

The sulfates are bonded by ionic type of bonding, but in most cases the bonding is rather weak. The common sulfates are as follows:

Sulfates

barite	$BaSO_4$
celestite	$SrSO_4$
anglesite	$PbSO_4$
anhydrite	$CaSO_4$
gypsum	$CaSO_4 \cdot 2H_2O$
alunite	$KAl_3(SO_4)_2(OH)_6$

Chromates, tungstates, and molybdates are grouped along with sulfates because of the similarity in the structure of the anionic unit. In the chromates, the chromium ion $(Cr)^{6+}$ is located at the center of a tetrahedron of four oxygens $(O)^{2-}$ to make the chromate anion $(CrO_4)^{2-}$. In the chromates, ionic bonding is prevalent in holding the ions in position. As a rule, chromates are very rare, and only one fairly well-known species occurs. This is crocoite, $PbCrO_4$.

In the tungstates and molybdates, either the tungsten ion $(W)^{6+}$ or the molybdenum ion $(Mo)^{6+}$ is centrally located between four oxygens $(O)^{2-}$ to make a tetrahedral unit. However, in these two ionic groups, the tetrahedron is slightly distorted and is not symmetrical as in the sulfates and chromates

The common tungstates and molybdates are as follows:

Tungstates		*Molybdates*	
wolframite	$(Fe, Mn)WO_4$	wulfenite	$PbMoO_4$
scheelite	$CaWO_4$		

Phosphates (Also Vanadates)

The phosphate group is characterized by having an anionic unit composed of a phosphorus ion $(P)^{5+}$ with four oxygen ions $(O)^{2-}$ to form a $(PO_4)^{3-}$ ion as the fundamental anion (Fig. 3-36). In fact, the vanadates are similar in the structure of their anionic units, with the vanadium ion $(V)^{5+}$ in analogous positions to the $(P)^{5+}$ anionic units. Among the phosphates and vanadates, the phosphates far outnumber the vanadates as far as the common ones are concerned. Collectively, the groups are classified into anhydrous, anhydrous with hydroxyl (OH), hydrated, hydrated with hydroxyl (OH), and additional groups.

These minerals are bonded with ionic type of bonding, and the common ones are as follows:

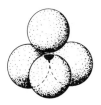

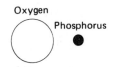

Oxygen

Phosphorus

Fig. 3-36: The phosphate anion.

Phosphates

apatite	$Ca_5(PO_4)_3(F, Cl, OH)$
pyromorphite	$Pb_5(PO_4)_3Cl$
autunite	$Ca(UO_2)_2(PO_4)_2 \cdot 10\text{--}12H_2O$
turquoise	$CuAl_6(PO_4)_4(OH)_8 \cdot 4H_2O$

Vanadates

carnotite	$K_2(UO_2)_2(VO_4)_2 \cdot 3H_2O$

Silicates

The silicates comprise the largest group of minerals. In fact, about 95 percent of the Earth's crust is made up of silicates, predominantly feldspars and quartz. Consequently, the silicates are some of the most important rock-making minerals, as will be seen in a later section when the composition, origin, and classification of rocks are discussed.

In all the silicates, the main part of the framework consists of a small silicon ion $(Si)^{4+}$, centrally located between four large oxygen ions $(O)^{2-}$, arranged in the form of a tetrahedron (four-sided structure with triangular sides) (Fig. 3-37). This complex anionic unit $(SiO_4)^{4-}$ may exist in some silicates as an isolated tetrahedron, i.e., not joined to any other $(SiO_4)^{4-}$ tetrahedron. In others, however, two $(SiO_4)^{4-}$ tetrahedra may be joined through a common oxygen to form anionic units. In still others, the silicate linkage becomes more complicated, forming various types of rings, chains, or continuous networks of tetrahedra, the connected tetrahedra being united through common oxygens. Ionic bonding prevails in the silicates. Silicates are subdivided into six subclasses partly on the basis of the number of joined tetrahedra and partly on the arrangement of the linked $(SiO_4)^{4-}$ tetrahedra, i.e., isolated, paired, ring, chain, sheet, or continuous.

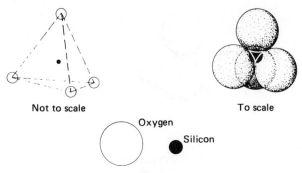

Not to scale   To scale

Oxygen
Silicon

Fig. 3-37: The silicate anion.

The number of natural elements which form silicate minerals are relatively few, but these relatively few form a great multiplicity of minerals because the same elements may combine in different proportions with the $(SiO_4)^{4-}$ tetrahedra. They often have complicated and variable compositions. The variation of composition and silicate tetrahedral linkage were not well understood until X-ray investigations solved these problems and led to the subclassification of the silicates according to $(SiO_4)^{4-}$ tetrahedral linkages and arrangements.

Each $(SiO_4)^{4-}$ tetrahedron in the silicates contains very strong bonds between the one silicon ion and the four oxygen ions, and the tetrahedral units maintain nearly constant dimensions and shape regardless of what the rest of the structure is like. Only the more common ones in each subclass are listed.

Subclass Nesosilicates

This subclass contains tetrahedra which are independent $(SiO_4)^{4-}$ anionic units, not linked to any other $(SiO_4)^{4-}$ unit. Formulas of minerals in this

Common Nesosilicates

olivine	$(Mg, Fe)_2SiO_4$
andalusite	Al_2SiO_5
sillimanite	Al_2SiO_5
kyanite	Al_2SiO_5
staurolite	$Fe_2Al_9O_6(SiO_4)_4(O, OH)_2$
topaz	$Al_2SiO_4(OH, F)_2$
garnet group:	
almandite	$Fe_3Al_2(SiO_4)_3$
pyrope	$Mg_3Al_2(SiO_4)_3$
spessartite	$Mn_3Al_2(SiO_4)_3$
grossularite	$Ca_3Al_2(SiO_4)_3$
andradite	$Ca_3Fe_2(SiO_4)_3$
uvarovite	$Ca_3Cr_2(SiO_4)_3$
zircon	$ZrSiO_4$
sphene	$CaTiSiO_5$

subclass therefore contain SiO_4 to indicate this composition, such as olivine, $(Mg,Fe)_2SiO_4$.

Subclass Sorosilicates

In this subclass, two $(SiO_4)^{4-}$ tetrahedra are joined by sharing one oxygen between them. Since this one oxygen is shared, it really is part of both tetrahedra. Formulas of minerals in this subclass therefore contain Si_2O_7 to indicate this composition, e.g., hemimorphite, $Zn_4Si_2O_7(OH)_2 \cdot H_2O$.

Common Sorosilicates

hemimorphite	$Zn_4Si_2O_7(OH)_2 \cdot H_2O$
idocrase	$Ca_{10}Mg_2Al_4(Si_2O_7)_2(SiO_4)_5(OH)_4$
epidote	$Ca_2(Al, Fe)Al_2O(SiO_4)(Si_2O_7)(OH)$
prehnite	$Ca_2Al(AlSi_3O_{10})(OH)_2$

Subclass Cyclosilicates

Silicates assigned to this subclass contain $(SiO_4)^{4-}$ tetrahedra which are joined through oxygens $(O)^{2-}$ to form three types of ring structures. Ring structures consisting of three linked tetrahedra display formulas containing $(Si_3O_9)^{6-}$ anions. Those containing four linked tetrahedra have formulas with $(Si_4O_{12})^{8-}$ anions. The third type consists of ring structures made up of six linked tetrahedra, shown by formulas with the main anionic unit $(Si_6O_{18})^{12-}$. Examples of each type are: benitoite, $BaTiSi_3O_9$; axinite, $(Ca,Mn,Fe)_3Al_2(BO_3)Si_4O_{12}(OH)$; and beryl, $Be_3Al_2Si_6O_{18}$, respectively.

Common Cyclosilicates

beryl	$Be_3Al_2Si_6O_{18}$
cordierite	$(Mg,Fe)_2Al_3(AlSi_5O_{18})$
tourmaline	$Na(Mg,Fe)_3Al_6(BO_3)_3(Si_6O_{18})(OH)_4$

Subclass Inosilicates

This subclass contains units of linked $(SiO_4)^{4-}$ tetrahedra arranged in indefinitely long chains of two types: single chains and double chains. Because of this property to form continuous chain arrangements, the crystals of silicates in this subclass often are elongated, needle-like, or fibrous. In the single chains, each $(SiO_4)^{4-}$ tetrahedron shares two of its oxygens, one with each of the next linked tetrahedron on each side. Since two oxygens are shared, the formulas display an anionic unit of $(SiO_3)^{2-}$ such as in wollastonite, $CaSiO_3$, or multiples thereof, e.g., (Si_2O_6) as in diopside, $CaMgSi_2O_6$.

Double chains are assembled in a unique manner. Essentially two single chains are cross-linked to each other by sharing some of the remaining oxygens. Actually, alternate tetrahedra in each single chain component share oxygens with each other. Formulas of double chains contain an anionic unit of $(Si_4O_{11})^{6-}$ or multiples thereof, e.g., (Si_8O_{22}). An example of a double chain is hornblende, $NaCa_2(Mg,Fe,Al)_5(Si,Al)_8O_{22}(OH)_2$.

Common Inosilicates

amphibole group:	
tremolite—actinolite	$Ca_2(Mg,Fe)_5Si_8O_{22}(OH)_2$
hornblende	$NaCa_2(Mg,Fe,Al)_5(Si,Al)_8O_{22}(OH)_2$
pyroxene group:	
enstatite—hypersthene	$(Mg,Fe)_2Si_2O_6$
augite	$Ca(Mg,Fe,Al)(Al,Si)_2O_6$
spodumene	$LiAlSi_2O_6$
pyroxenoid group:	
wollastonite	$CaSiO_3$
rhodonite	$MnSiO_3$

Subclass Phyllosilicates

Silicates placed in this subclass contain $(SiO_4)^{4-}$ tetrahedra, each of which shares three of its oxygens with next adjacent tetrahedra to form sheet structures. These sheet structures are actually made up of a series of double-chain structures extending indefinitely in two directions, instead of only one direction as in the inosilicates. Formulas of minerals in this subclass contain an anionic unit of $(Si_2O_5)^{2-}$ or multiples thereof, e.g., (Si_4O_{10}). A typical example of a phyllosilicate is serpentine, $Mg_6Si_4O_{10}(OH)_8$.

Common Phyllosilicates

kaolinite	$Al_4Si_4O_{10}(OH)_8$
serpentine	$Mg_6Si_4O_{10}(OH)_8$
talc	$Mg_3Si_4O_{10}(OH)_2$
montmorillonite	$Al_2Si_4O_{10}(OH)_2 \cdot nH_2O$
vermiculite	$Mg_3Si_4O_{10}(OH)_2 \cdot nH_2O$
phlogopite	$KMg_3(AlSi_3O_{10})(OH)_2$
muscovite	$KAl_2(AlSi_3O_{10})(OH)_2$
biotite	$K(Mg,Fe)_3(AlSi_3O_{10})(OH)_2$
chlorite	$(Mg,Fe,Al)_6(Al,Si)_4O_{10}(OH)_8$

Subclass Tektosilicates

This subclass contains $(SiO_4)^{4-}$ tetrahedra, each of which shares all four of its oxygens with other tetrahedra, thus producing a continuous three-dimensional network. Since all oxygens are shared, the ratio of Si: O is 1: 2, as in quartz, SiO_2, where the positive and negative charges are balanced. However, in this subclass, substitution of Al^{3+} ions for some of the Si^{4+} ions readily takes place, and the main anionic unit actually is $(Si,Al)O_2$ or multiples thereof, e.g., $(Si_7Al_2)O_{18}$. This gives a ratio of 9 : 18, the same as 1 : 2. However, substitution of Al^{3+} for some of the Si^{4+} ions upsets the electrical balance, because a 3^+ ion takes the place of a 4^+ ion. To rebalance the crystal to make it electrically neutral, additional cations are necessary. The necessary cations are usually available in the solution from which the minerals are crystallizing, such as K^+ or Na^+, and thus such minerals as microcline, $KAlSi_3O_8$ [properly written $K(Si_3Al)O_8$] may form. In another situation, perhaps nepheline, $NaAlSiO_4$, may form [properly written $Na(SiAl)O_4$].

Common Tektosilicates

quartz	SiO_2
opal	$SiO_2 \cdot nH_2O$
orthoclase	$KAlSi_3O_8$
microcline	$KAlSi_3O_8$
plagioclase	$(Ca,Na)(Al,Si)AlSi_2O_8$
scapolite	$(Na,Ca)_4[(Al,Si)_4O_8], (Cl,CO_3)$
nepheline	$NaAlSiO_4$
heulandite	$CaAl_2Si_7O_{18} \cdot 6H_2O$
stilbite	$CaAl_2Si_7O_{18} \cdot 7H_2O$
chabazite	$CaAl_2Si_4O_{12} \cdot 6H_2O$
analcite	$NaAlSi_2O_6 \cdot H_2O$

NAME: **Gold**

FORMULA: Au

Color: Yellow-gold
Streak: Yellow-gold
Luster: Metallic   *Cleavage:* None
Hardness: 2.5—3.0 malleable   *S.G.:* 19.3

CRYSTALLOGRAPHY: Isometric. Often in arborescent growths of numerous crystals. Also in scales and nuggets.

DESCRIPTION: Gold occurs chiefly in hydrothermal deposits and also in placer deposits. It is very often associated with quartz, pyrite, arsenopyrite, sphalerite, and galena. The gold color, the malleability, and the high specific gravity are diagnostic characteristics which distinguish gold from "fool's gold" pyrite and chalcopyrite. Pyrite is harder than gold, while chalcopyrite is brittle. Weathered biotite often appears as golden flashes, but it is very brittle. Native gold commonly has a little silver alloyed with it, and if the silver content reaches 20 percent or more, the metal is referred to as *electrum*. The purity (fineness) of native gold is expressed in parts per 1000. Most gold carries a little impurity and thus runs about 900 fine. Gold does not tarnish, and this together with its malleability and sectility permit it to be used in high-quality jewelry.

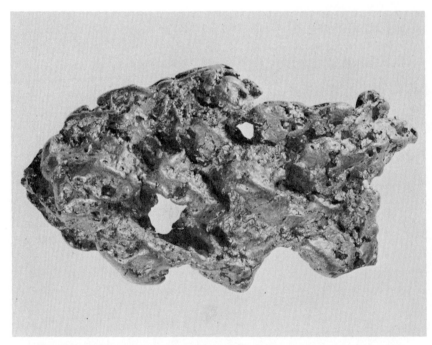

Fig. M-1: Gold nugget showing high metallic luster and irregular shape. This nugget was probably found in a placer deposit. (Photo courtesy of Smithsonian Institution.)

NAME: **Silver**

FORMULA: Ag

Color: Silver-white
Streak: Silver-white
Luster: Metallic
Hardness: 2.5—3.0 malleable and ductile

Cleavage: None
S.G.: 10.5

CRYSTALLOGRAPHY: Isometric. Usually in rude masses or wires. Also in plates and scales.

DESCRIPTION: Silver has a diagnostic silver-white color, and along with its malleability and ductility, it is fairly easily distinguished from other similar metals. It tarnishes readily with the development of silver sulfide. Native silver is found in primary hydrothermal sulfide deposits and in the oxidized zone of some ore deposits, associated with calcite, barite, quartz, and various sulfides of silver and lead. Silver also is associated with uraninite in certain places. Much of the silver obtained in the United States is associated with galena. Some silver is nearly always present with native gold. Platinum resembles silver but is somewhat harder. Native bismuth also looks somewhat like silver, but bismuth has a pale reddish color on fresh surfaces. Silver is used in coins, in silver plating, and in the photography industry.

Fig. M-2: Small grains and irregular masses of native silver in vein rock from Cobalt, Ontario. Note the high metallic luster and very white color of the silver. (Courtesy Ward's Natural Science Establishment.)

NAME: **Copper**
FORMULA: Cu

Color: Copper-red
Streak: Pale red (metallic)
Luster: Metallic   *Cleavage:* None
Hardness: 2.5—3.0   *S.G.:* 8.95

CRYSTALLOGRAPHY: Isometric. Often in arborescent, nodular masses. Plates and twisted forms also.

DESCRIPTION: Native copper is most often associated with basic extrusive igneous rocks and is found as cavity fillings. Large masses of native copper in veins in lava flows and in conglomerates are mined in the Keweenaw Peninsula of northern Michigan. Native copper is also produced in New Mexico and Arizona, where it is found in the oxidized portions of hydrothermal deposits associated with malachite, azurite, bornite, chalcopyrite, and calcite. Native copper alters quite easily to other copper minerals. The copper-red color, the sectility, the metallic luster, and the hackly fracture distinguish copper from other minerals. Niccolite resembles copper in color but is harder than copper, and niccolite has a brownish-black streak. Copper tarnishes rather easily; however, it is used extensively in alloys and in electrical wire. Copper is alloyed with zinc to make brass and with tin to make bronze.

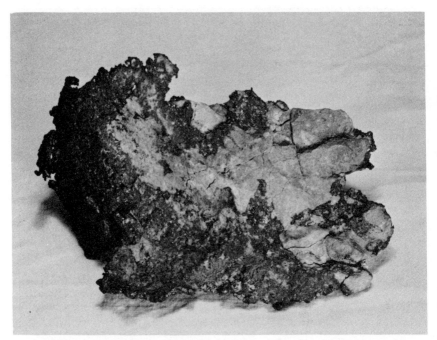

Fig. M-3: Irregular platy bent mass of native copper from Bisbee, Arizona, showing dull tarnish over metallic luster. Note hackly fracture of this copper-red native element.

NAME: **Sulfur**
FORMULA: S

Color: Yellow to yellowish-brown
Streak: White
Luster: Resinous to greasy   *Cleavage:* None
Hardness: 1.5—2.5   *S.G.:* 2.07

CRYSTALLOGRAPHY: Orthorhombic. In fine pyramidal, tabular crystals. Also earthy, massive, and reniform.

DESCRIPTION: Sulfur commonly is associated with volcanic activity, deposited by fumarolic gases. It also is found in sedimentary rocks associated with gypsum and limestone, and also with celestite and aragonite. Sulfur occasionally is found in hydrothermal veins. The transparent, brittle, soft, easily fusible yellow crystals make sulfur easy to identify. The absence of cleavage distinguishes it from orpiment. Sulfur melts at 113°C and burns at 270°C, properties usable for identification. Most of the commercial sulfur used in the United States comes from salt domes in Texas and Louisiana. A hole is drilled into the upper part of the salt domes where the sulfur is located, hot steam is pumped down to melt the sulfur, and then the liquid sulfur brine is returned to the surface. This method is called the Frasch process. Sulfur is used in the manufacture of H_2SO_4, in rubber, in the treatment of wood pulp, and in insecticides.

Fig. M-4: Sulfur from Nevada in imperfectly crystallized bright yellow mass, showing conchoidal to uneven fracture and vitreous luster.

NAME: **Diamond**

FORMULA: C

Color: Colorless or pale tones
Streak: White
Luster: Adamantine to greasy   *Cleavage:* Perfect octahedral
Hardness: 10.0   *S.G.:* 3.50

CRYSTALLOGRAPHY: Isometric. In individual octahedral or cubic crystals which may appear rounded.

DESCRIPTION: Diamond is associated with ultrabasic igneous rock called "kimberlite." It is also found in certain gravels because it is inert, heavy, and hard. In fact, most diamonds are found in gravel placers. Only a few localities are producers. Diamond is the hardest known substance. The great hardness, the brilliant luster, and the octahedral cleavage are diagnostic features. Some quartz resembles diamond, but quartz is much softer and has no cleavage. Both topaz and clear corundum look similar to diamond, but diamond is harder than both. All diamonds are not colorless; many are tinted deep tones and even black. Gem quality diamonds are rare; most are of industrial grade. Cut diamonds have a high refractive index, giving them brilliance or "fire." Uncut diamonds look greasy. Other varieties are bort (granular and flawed) and carbonado (gray or black). Diamonds come from the Congo, India, Brazil, and South Africa. Synthetic diamonds are small and not of gem quality.

Fig. M-5: Diamond from South Africa showing somewhat rounded octahedral crystal, adamantine to greasy luster and transparency.

NAME: **Graphite**
FORMULA: C

Color: Black
Streak: Black
Luster: Metallic to dull   *Cleavage:* Perfect basal pinacoid
Hardness: 1.0—2.0   *S.G.:* 2.09—2.23

CRYSTALLOGRAPHY: Hexagonal. Tabular crystals in foliated masses. Also granular.

DESCRIPTION: Graphite is commonly found in metamorphic rocks, usually in marbles, gneisses, schists, etc. Occasionally graphite forms in igneous rocks, hydrothermal veins, and rarely in pegmatites. It is extremely soft, has a greasy feel, has a black color and streak; these properties easily distinguish graphite. It may resemble molybdenite, but molybdenite has a bluish tint and has a more metallic luster and streak. Graphite is infusible and is unattacked by most acids, properties which permit graphite to be used in refractory crucibles in the metallurgical industry. Most of the graphite which is used in industry in the United States is made artificially by treating anthracite coal or petroleum coke in large electrical furnaces. Major production of graphite is from Ceylon, Madagascar, Russia, Korea, and Mexico. It has many uses.

Fig. M-6: Massive graphite from Kragero, Norway, showing submetallic luster and black color. So soft it soils the fingers.

NAME: **Argentite**

FORMULA: Ag_2S

Color: Black
Streak: Blackish lead-gray
Luster: Metallic   *Cleavage:* Cubic, but poor
Hardness: 2.0—2.5   *S.G.:* 7.3

CRYSTALLOGRAPHY: Isometric. In branching or matted groups. Most frequently in massive form.

DESCRIPTION: Argentite is probably the most important primary silver mineral. It is commonly found in low-temperature hydrothermal deposits with other silver minerals, and often with galena and sphalerite. Argentite may also be found as microscopic inclusions in galena. Argentite is best distinguished from other minerals by its color, sectility, and specific gravity. It somewhat resembles galena, but argentite lacks the perfect cleavage shown by galena. Tarnished silver resembles argentite, but silver is silver-white on a fresh surface. Argentite may resemble chalcocite, but chalcocite is less sectile. Argentite is usually bright on a fresh surface, but it easily becomes dull black as the result of an earthy sulfide coating which develops. Argentite is an important ore in the silver districts of Mexico. Other production comes from Colorado, Montana, and Idaho.

Fig. M-7: Massive argentite from near Port Arthur, Ontario, showing dark lead-gray color. The true metallic luster is masked by an earthy sulfide tarnish.

NAME: **Chalcocite**
FORMULA: Cu_2S

Color: Black to lead-gray
Streak: Black
Luster: Metallic to dull   *Cleavage:* Prismatic, but indistinct
Hardness: 2.5—3.0   *S.G.:* 5.5—5.8

CRYSTALLOGRAPHY: Orthorhombic. Usually in fine grained masses. Often massive. Small tabular crystals are rare.

DESCRIPTION: This important ore mineral of copper is found chiefly in hydrothermal sulfide deposits, as both primary and secondary minerals. Chalcocite occurs very frequently in the secondary enriched zones of copper sulfide deposits, associated with malachite, azurite, cuprite, copper, chalcopyrite, and covellite, such as at Ely, Nevada, at Morenci, Arizona, and at Bingham, Utah. However, at Butte, Montana, it occurs as primary hydrothermal sulfide mineral. Its black color and association with other copper minerals serve to distinguish chalcocite from other minerals. Chalcocite resembles argentite but is more brittle and more sectile. These two minerals are differentiated by crystal form also, argentite being isometric. Chalcocite may resemble enargite, but enargite is brittle. Covellite is distinguished from chalcocite by the indigo-blue color of the former.

Fig. M-8: Massive dark gray to black chalcocite from Butte, Montana, revealing conchoidal fracture and metallic luster. An important copper ore mineral.

NAME: **Bornite**
FORMULA: Cu_5FeS_4

Color: Brownish-bronze, but tarnishes easily to purple and blue
Streak: Grayish-black
Luster: Metallic   *Cleavage:* Octahedral, but poor
Hardness: 3.0   *S.G.:* 5.06—5.08

CRYSTALLOGRAPHY: Isometric. Usually in massive form. Rarely in crystals with cubes, dodecahedrons, or octohedrons.

DESCRIPTION: Bornite is usually found in hydrothermal deposits associated with other copper minerals such as chalcocite, chalcopyrite, covellite, and enargite, along with other sulfides, especially pyrite and arsenopyrite. Occasionally it is found disseminated in basic igneous rocks, in contact metamorphic rocks, and less frequently in pegmatites. Bornite alters easily to chalcocite and covellite. Good crystals of bornite are rather rare; most of the bornite which is mined as copper ore is in massive form. The best identifying feature of bornite is its purplish-blue tarnish over a bronze color ("peacock" ore). Otherwise bornite somewhat resembles chalcocite. Bornite is distinguished from covellite by the indigo-blue color of covellite. Bornite is found at Butte, Montana, and at Superior, Arizona. Other notable localities are Cornwall, England, Mexico, Peru, and Tasmania. The chief use for this mineral is as an ore of copper.

Fig. M-9: Massive bornite, from Butte, Montana, showing metallic luster. This brownish-bronze mineral tarnishes easily to purples and blues.

NAME: **Galena**
FORMULA: PbS

Color: Lead-gray
Streak: Lead-gray
Luster: Metallic   *Cleavage:* Perfect cubic
Hardness: 2.5   *S.G.:* 7.57

CRYSTALLOGRAPHY: Isometric. Coarsely to finely crystalline. Usually in good cubes modified by octahedrons.

DESCRIPTION: Galena is a very common sulfide mineral and is the most important lead mineral. It is chiefly found in hydrothermal sulfide deposits, associated with pyrite, sphalerite, and chalcopyrite, and also with silver minerals, barite, fluorite, and quartz. The perfect cubic cleavage, the softness, the high specific gravity, and the lead-gray streak are its distinguishing features. Galena somewhat resembles argentite, but the perfect cubic cleavage of galena is diagnostic. Galena is sometimes confused with stibnite, but again galena's cubic cleavage and darker color tell the difference. Galena usually carries a little silver, probably either as admixtures of argentite or tetrahedrite, and thus is an important silver ore. Most domestic production of silver is from galena. The deposits of southeast Missouri are very large producers.

Fig. M-10: Good cubic crystals of galena from Galena, Kansas, showing high metallic luster and cubic cleavage.

NAME: **Sphalerite**

FORMULA: ZnS

Color: Brown to yellow, even black

Streak: Brown to yellow

Luster: Resinous   *Cleavage:* Perfect dodecahedral

Hardness: 3.5—4.0   *S.G.:* 3.9

CRYSTALLOGRAPHY: Isometric. Usually fine to coarse crystal masses. Also occurs in compact masses.

DESCRIPTION: Sphalerite is the most important zinc mineral. It is found associated with galena and other sulfides in hydrothermal deposits, usually as replacement masses in limestones. Common associates are pyrite, chalcopyrite, smithsonite, and hemimorphite. Sphalerite has a striking resinous luster and perfect cleavage; these two properties, along with its hardness, serve to distinguish it. The color is too variable to be reliable. Siderite may resemble sphalerite, but siderite's rhombohedral cleavage is diagnostic. Sphalerite may resemble cassiterite, but cassiterite is heavier (7.0) and harder (6–7). Sphalerite is a fairly common mineral and has been mined in Missouri, Kansas, Oklahoma, and other places. Most domestic production of this zinc ore comes from the eastern part of the country, chiefly from Tennessee, New York, and New Jersey.

Fig. M-11: Coarse cleavable mass of sphalerite from New York showing resinous to submetallic luster of this yellowish-brown specimen. Sphalerite is translucent.

NAME: **Chalcopyrite**
FORMULA: $CuFeS_2$

Color: Brassy-yellow, often tarnished
Streak: Greenish-black
Luster: Metallic   *Cleavage:* Prismatic, but not good
Hardness: 3.5—4.0   *S.G.:* 4.1—4.3

CRYSTALLOGRAPHY: Tetragonal. Usually in massive compact form. Crystals are sphenoids but resemble tetrahedrons.

DESCRIPTION: Chalcopyrite (fool's gold) is quite easily distinguished from pyrite by its more yellow-gold color and lower hardness. Its greenish-black streak distinguishes chalcopyrite from gold. Chalcopyrite somewhat resembles pyrrhotite but has brighter yellow color. Chalcopyrite, the most widely distributed copper mineral, is associated in hydrothermal sulfide deposits with other important copper sulfides, such as chalcocite, bornite, and covellite, in veins or in disseminated deposits. Chalcopyrite is one of the most important ore minerals of copper. It is the principal ore mineral in the "porphyry copper" deposits. Chalcopyrite also occurs disseminated in igneous rocks, in pegmatites, in contact metamorphic rocks, and even in other schistose rocks. It may carry a little gold or silver. Chalcopyrite alters readily to malachite, azurite, covellite, chalcocite, and cuprite. It is found at Butte, Montana, and Bingham, Utah.

Fig. M-12: Chalcopyrite (brassy-yellow) in massive form showing good metallic luster.

NAME: **Pyrrhotite**
FORMULA: FeS

Color: Bronze-yellow to brownish-bronze
Streak: Gray-black
Luster: Metallic
Hardness: 3.5—4.0

Cleavage: None
S.G.: 4.60—4.65

CRYSTALLOGRAPHY: Hexagonal. Usually massive or granular. Crystals are often tabular, but good crystals are rare.

DESCRIPTION: Pyrrhotite is a common accessory mineral of igneous rocks, associated with pentlandite and chalcopyrite. It is occasionally found in hydrothermal deposits, in pegmatites, and in contact metamorphic rocks. Its brownish-yellow color, its slight magnetism, and its hardness distinguish pyrrhotite from chalcopyrite (brassy-yellow) and from pyrite (pale brassy-yellow color and hardness of 6.0). Pyrrhotite is very hard to distinguish from pentlandite, but pentlandite is not magnetic and also has a light bronzy streak. At Sudbury, Ontario, large masses of pyrrhotite are intermixed with pentlandite and chalcopyrite, and these are mined for copper, nickel, and platinum. Unfortunately, these three minerals are difficult to distinguish from one another. Pyrrhotite alters easily to limonite, to iron sulfates, and even to siderite. Pyrrhotite is found in Lancaster County, Pennsylvania, and at Ducktown, Tennessee.

Fig. M-13: Coarsely crystalline pyrrhotite from La Paz, Bolivia, showing metallic luster of this brownish-bronze mineral.

NAME: **Pentlandite**
FORMULA: $(Fe,Ni)_9S_8$

Color: Light bronze-yellow
Streak: Light bronze-brown
Luster: Metallic   *Cleavage:* None
Hardness: 3.5—4.0   *S.G.:* 4.6—5.0

CRYSTALLOGRAPHY: Isometric. Usually in granular masses. Often in massive form. Usually mixed with pyrrhotite.

DESCRIPTION: Pentlandite is the principal ore mineral of nickel. It is found chiefly in basic igneous rocks associated with pyrrhotite. Pentlandite resembles pyrrhotite but is nonmagnetic. The degree of magnetism in pyrrhotite is variable, so magnetic properties are not always diagnostic. However, pentlandite is lighter in color than pyrrhotite. Pentlandite may also resemble chalcopyrite, but chalcopyrite is more yellow in color. Pentlandite is found at scattered localities, but it is the chief source of nickel at Sudbury, Ontario, where it is intimately associated with pyrrhotite and chalcopyrite. Another important Canadian deposit is at Lynn Lake near Thompson in northern Manitoba. Nickel alloys of copper, chromium, and iron are strong, have good heat and corrosion resistance, and have good ductility. Nickel metal is also used in coins, in nickel plating, and other aspects of industry.

Fig. M-14: Granular aggregates of pentlandite (large light areas) from Sudbury, Ontario, showing metallic luster of this yellowish-bronze sulfide mineral.

NAME: **Covellite**
FORMULA: CuS

Color: Indigo-blue or darker, often with yellow or red iridescence
Streak: Lead-gray to black
Luster: Metallic   *Cleavage:* Perfect basal pinacoidal
Hardness: 1.5—2.0   *S.G.:* 4.60—4.76

CRYSTALLOGRAPHY: Hexagonal. Usually massive or thin tabular crystals. Often as coatings or crusts.

DESCRIPTION: This copper sulfide mineral is commonly found in hydrothermal deposits associated with other copper sulfides, such as chalcocite, chalcopyrite, bornite, and enargite. Covellite often forms by the alteration of these other sulfides by secondary enrichment processes, but it may be a primary mineral in the deposits as well. Covellite tarnishes rather easily to yellow and red iridescent colors. Bornite may resemble covellite, but bornite has a purplish-blue tarnish and is harder. Chalcocite may look similar to covellite, but chalcocite is black and is much harder. The perfect mica-like cleavage and the blue color are good identifying features. Covellite is not an important mineral, but it is found at Butte, Montana, at Summitville, Colorado, and at La Sal, Utah. Very fine crystals of covellite are reported at Alghero, Sardinia. It is rather rare in the United States. Covellite is used as an ore of copper.

Fig. M-15: Covellite from Butte, Montana, showing submetallic luster and platy masses that tend to split into thin flexible flakes.

NAME: **Cinnabar**
FORMULA: HgS

Color: Vermillion-red
Streak: Scarlet
Luster: Adamantine to earthy   *Cleavage:* Perfect prismatic
Hardness: 2.0—2.5   *S.G.:* 8.09

CRYSTALLOGRAPHY: Hexagonal. Usually fine granular; earthy. Often occurs as dusty incrustations on rock fractures.

DESCRIPTION: Cinnabar, the most important ore mineral of mercury, usually is found in hydrothermal deposits of low temperature volcanic areas. It commonly is associated with other sulfides, such as pyrite and stibnite, and also with quartz and calcite. The bright vermillion-red color and scarlet streak, together with good cleavage and high specific gravity, are its diagnostic features. Sometimes cinnabar is mistaken for realgar, but realgar does not have the same luster or a scarlet streak. Also realgar is more orange in color and is much lighter in weight than cinnabar. Cuprite may resemble cinnabar, but cuprite is darker than cinnabar and is usually associated with oxide-type copper minerals. Hematite may appear similar to cinnabar, but the streaks are diagnostic: dark red for hematite, vermillion-red for cinnabar. United States' production comes from the New Idria and New Almaden areas in California. The world's leading producers are Almaden, Spain, and Idria, Yugoslavia.

Fig. M-16: Bright adamantine rhombohedral crystals of cinnabar, associated with fine grained quartz. (Photo courtesy Smithsonian Institution.)

NAME: **Realgar**
FORMULA: AsS

Color: Aurora-red to orange-yellow
Streak: Orange-red
Luster: Resinous to greasy *Cleavage:* Good pinacoidal
Hardness: 1.5—2.0 *S.G.:* 3.56

CRYSTALLOGRAPHY: Monoclinic. Small or large vertically striated crystals. Often earthy; incrustations.

DESCRIPTION: Realgar is found chiefly in low-temperature hydrothermal deposits associated with other low-temperature sulfides. The orange-red color, low hardness, resinous luster, and association with other arsenic minerals are realgar's distinguishing features. Realgar resembles cinnabar somewhat, but has a resinous luster, orange-red streak, and lower specific gravity. It somewhat resembles crocoite, but its crystals are distinctive, and realgar is much softer. Realgar is nearly always associated with lemon-yellow orpiment, another arsenic sulfide much rarer than realgar. In fact, much orpiment forms as an alteration product of realgar. Realgar is also associated with stibnite and ores of lead, silver, and gold. In the United States, realgar occurs around hot springs in Norris Basin, Yellowstone National Park. It also is found at Mercur, Utah, and at Manhattan, Nevada.

Fig. M-17: Coarsely crystalline translucent orange-red realgar crystals with resinous luster.

NAME: **Stibnite**
FORMULA: Sb_2S_3

Color: Lead-gray
Streak: Lead-gray
Luster: Metallic   *Cleavage:* Perfect prismatic
Hardness: 2.0   *S.G.:* 4.63

CRYSTALLOGRAPHY: Orthorhombic. Slender prismatic crystals in aggregates. Also granular to massive.

DESCRIPTION: This soft, gray, metallic mineral can be distinguished from other similar minerals by its long slender striated prismatic crystals. Stibnite is sometimes mistaken for galena, but the slender prismatic crystals and low specific gravity tell the difference. Stibnite is usually found in low-temperature hydrothermal sulfide deposits, but also in replacement deposits and in hot springs deposits. Stibnite is commonly associated with other minerals which contain antimony and which have been formed by the decomposition of stibnite. It is also found with galena, cinnabar, barite, realgar, orpiment, and sphalerite. Stibnite is the chief ore mineral for the metal antimony. Stibnite is not too abundant in the United States, but it is found in California, Nevada, and Utah. Large crystals are found on the island of Shikoku, Japan. However, the world's leading producing area is Hunan, China.

Fig. M-18: Stibnite showing slender prismatic crystals, some curved, in crude radiating groups. Note high metallic luster of this lead-gray mineral.

NAME: **Pyrite**

FORMULA: FeS_2

Color: Pale brassy-yellow
Streak: Greenish- to brownish-black
Luster: Metallic   *Cleavage:* Fair cubic
Hardness: 6.0—6.5   *S.G.:* 5.01

CRYSTALLOGRAPHY: Isometric. Usually in good cubic crystals, with cube faces striated. Also massive; granular.

DESCRIPTION: The high hardness, brassy-yellow color, and cubic crystal habit serve to identify pyrite from other minerals of similar appearance. such as chalcopyrite, marcasite, and gold. Pyrite is harder and paler then chalcopyrite, and it has different crystal form and darker color than marcasite. Pyrite is also harder than gold. Pyrite somewhat resembles pyrrhotite and pentlandite, but both of these are browner and softer than pyrite. Striations on pyrite crystals are very distinctive. Pyrite is the most common sulfide mineral. It is found chiefly in hydrothermal deposits associated with sphalerite, galena, and chalcopyrite. Pyrite is also found in igneous rocks, in pegmatites, in contact metamorphic rocks, and in volcanic areas. Pyrite is called fool's gold. Large deposits of pyrite are at Rio Tinto, Spain. In the United States, pyrite occurs in Virginia, New York, and Massachusetts. It is used as a source of sulfur.

Fig. M-19: Crystals of pyrite showing striated cube faces of this pale brassy-yellow mineral. Note high metallic luster.

NAME: **Cobaltite**

FORMULA: CoAsS

Color: Silver-white with pink tint

Streak: Gray-black

Luster: Metallic

Hardness: 5.5

Cleavage: Perfect cubic

S.G.: 6.33

CRYSTALLOGRAPHY: Isometric. Usually in good cubic crystal aggregates. Often in granular masses.

DESCRIPTION: Cobaltite is usually found in high-temperature deposits and in hydro-thermal deposits with other cobalt and nickel minerals. It is usually associated with niccolite, skutterudite, and silver, as well as with bismuth, siderite, and calcite. The very perfect cubic cleavage and the silver-white color with a pink tint are enough to distinguish this important ore mineral of cobalt. Cobaltite somewhat resembles pyrite, but cobaltite's silver-white color tells the difference. Cobaltite also may look like skutterudite, but cobaltite is usually in cubes and pyritohedrons, and not octahedrons as skutterudite is. Massive cobaltite resembles a number of other fairly rare iron, cobalt, and nickel sulfides and arsenides. Cobaltite occurs at Tunaberg, Sweden, and at Cobalt, Ontario. It is used as an ore of cobalt, a metal used in high-speed steels and in X-ray machines. Cobalt oxide finds use as a blue pigment.

Fig. M-20: Coarse granular cobaltite from Cobalt, Ontario, showing silver-white color, metallic luster, and peculiar mosaic of crystals.

NAME: **Marcasite**

FORMULA: FeS_2

Color: Pale bronze-yellow to tin-white
Streak: Grayish-black
Luster: Metallic   *Cleavage:* Distinct pinacoidal
Hardness: 6.0—6.5   *S.G.:* 4.89

CRYSTALLOGRAPHY: Orthorhombic. Usually tabular crystals in masses, in radiating forms, and in cockscomb groups.

DESCRIPTION: Marcasite is formed in low-temperature hydrothermal deposits associated with galena and sphalerite, and also in sedimentary rocks such as limestone or clay. The long wedge-shaped crystals in a "cockscomb structure," the pale brassy-yellow color, and the hardness are useful distinguishing features. The paler color and the wedge-shaped crystals are diagnostic from similar-looking pyrite. Marcasite decomposes in air more readily than pyrite does, and specimens tend to fall apart. Millerite somewhat resembles marcasite, but millerite is usually in long fibrous-like crystals. Marcasite may invert to pyrite which is more stable, but marcasite also may alter to limonite. Marcasite is not as common as pyrite. It is found in the lead and zinc deposits near Joplin, Missouri, as well as at Mineral Point, Wisconsin, and at Galena, Illinois. Large amounts are found in Bohemia and in England.

Fig. M-21: Radiating clusters of marcasite from Ottawa County, Oklahoma, showing common cockscomb structure developed by tabular crystals. Note metallic luster.

NAME: **Arsenopyrite**
FORMULA: FeAsS

Color: Silver-white
Streak: Black
Luster: Metallic   *Cleavage:* Distinct prismatic
Hardness: 5.5—6.0   *S.G.:* 6.07

CRYSTALLOGRAPHY: Orthorhombic. Prismatic crystals in masses, often radiating. Also in granular masses.

DESCRIPTION: This important arsenic mineral is usually formed in medium- to high-temperature hydrothermal deposits. The distinguishing characteristics are its silver-white color and its crystal form. Somewhat resembles marcasite, but is whiter. Arsenopyrite may resemble smaltite, but its orthorhombic crystal form is decisive from smaltite's isometric shapes. Arsenopyrite closely resembles both cobaltite and skutterudite, but the latter two are both isometric. It also looks similar to a number of rare silver-colored cobalt, nickel, and iron arsenides and sulfides, and detailed X-ray study and chemical tests are required to tell them apart. Arseno-pyrite alters quite readily to scorodite. Often associated with arsenopyrite are such minerals as wolframite, scheelite, cassiterite, gold, galena, chalcopyrite, and pyrite. In the United States, arsenopyrite is produced at Lead, South Dakota. It is used as the chief ore of arsenic.

Fig. M-22: Good prismatic crystals of arsenopyrite showing metallic luster and silver-white color.

NAME: **Molybdenite**
FORMULA: MoS_2

Color: Lead-gray with slight blue tint
Streak: Gray
Luster: Metallic   *Cleavage:* Perfect basal pinacoidal
Hardness: 1.0—1.5   *S.G.:* 4.62—4.73

CRYSTALLOGRAPHY: Hexagonal. Tabular crystals in masses and scales. Usually in massive form.

DESCRIPTION: Molybdenite is the most common molybdenum mineral and is the chief ore mineral for the metal molybdenum. It is found chiefly in high-temperature hydrothermal deposits, but it is sometimes found in contact metamorphic deposits. Molybdenite is commonly associated with pyrite, chalcopyrite, arsenopyrite, scheelite, wolframite, beryl, quartz, and fluorite. Molybdenite often alters to ferrimolybdite and powellite. Graphite may look like molybdenite, but the bluish tint, higher luster, and higher specific gravity of molybdenite are decisive properties. Also, graphite's black streak and molybdenite's greenish-gray streak are distinctive. Galena in scales may resemble molybdenite in scales, but the blue tint of molybdenite tells the difference. Molybdenite is produced in the United States at Climax, Colorado, at Questa, New Mexico, and at Bingham, Utah. It is also found with tin ores in Bohemia, Norway, England, China, and Mexico.

Fig. M-23: Massive molybdenite showing shiny metallic luster. This soft bluish-gray mineral has a greasy feel; its good micaceous-like cleavage is not apparent.

NAME: **Pyrargyrite-proustite**

FORMULA: Ag_3SbS_3-Ag_3AsS_3

Color: Deep red-scarlet-vermillion
Streak: Red-vermillion
Luster: Adamantine   *Cleavage:* Distinct rhombohedral
Hardness: 2.5   *S.G.:* 5.85—5.58

CRYSTALLOGRAPHY: Hexagonal. Prismatic crystals as masses or grains. Often massive and compact.

DESCRIPTION: These two minerals, called ruby-silvers, are very similar and often are intermixed. They are found in low-temperature silver deposits of hydrothermal origin. The red color of these minerals distinguishes them from other minerals. They are distinguished from cuprite by crystal form, cuprite being isometric with common octahedral crystals, and by cuprite's association with copper minerals. The slight variation in red tone identifies pyrargyrite from proustite, but proustite is also more translucent than pyrargyrite. Pyrargyrite is considerably more abundant than proustite, but both are commonly associated with silver minerals such as native silver, argentite, polybasite, and stephanite, as well as with tetrahedrite, calcite, and quartz. Pyrargyrite and proustite usually alter to silver and argentite. These minerals are found in the United States in Colorado, Nevada, New Mexico, and Idaho.

Fig. M-24: Blackish-red crystalline mass of pyrargyrite from Lake City, Colorado, showing submetallic luster.

NAME: **Tetrahedrite**

FORMULA: $Cu_{12}Sb_4S_{13}$

Color: Flint gray
Streak: Black
Luster: Metallic   *Cleavage:* None
Hardness: 3.0—4.5   *S.G.:* 4.6—5.1

CRYSTALLOGRAPHY: Isometric. Frequently in groups of tetrahedral crystals. Also massive; granular.

DESCRIPTION: The gray color, metallic luster, and tetrahedral crystals offer diagnostic criteria. It is softer than magnetite. Lack of cleavage and lighter color distinguish it from enargite. Tetrahedrite is an important copper mineral, found most commonly in low- to moderate-temperature hydrothermal deposits, along with other sulfides such as pyrite, chalcopyrite, sphalerite, and galena. In fact, tetrahedrite crystals often are coated with tiny yellow chalcopyrite crystals. Tetrahedrite also occurs with bornite, argentite, pyrargyrite, calcite, siderite, barite, and quartz. It sometimes alters by weathering processes to azurite or malachite. Sometimes tetrahedrite carries enough silver to be an ore of silver. In the United States, tetrahedrite is produced in the silver and copper districts of Colorado, Montana, Utah, Arizona, and Nevada. It comes also from Cornwall, England.

Fig. M-25: Tetrahedrite in massive form showing metallic luster. This dark gray mineral is from Silver Plume, Colorado.

NAME: **Enargite**
FORMULA: Cu_3AsS_4

Color: Grayish to iron-black
Streak: Black
Luster: Metallic   *Cleavage:* Perfect prismatic
Hardness: 3.0   *S.G.:* 4.45

CRYSTALLOGRAPHY: Orthorhombic. Long prismatic crystals in bladed masses. Also in massive form.

DESCRIPTION: Enargite is commonly found in moderate- to high-temperature hydrothermal deposits and occasionally in low-temperature veins. It is usually associated with other sulfide minerals such as chalcocite, chalcopyrite, pyrite, bornite, covellite, sphalerite, and galena. The very good cleavage, the color, and the striated crystals are diagnostic features. Enargite may resemble chalcocite, but chalcocite has poor cleavage and is not as brittle as enargite. Enargite's excellent cleavage and darker color serve to distinguish it from tetrahedrite. Stibnite may resemble enargite, but stibnite is lighter in color. Enargite's association with the other sulfides and also with quartz may be helpful in distinguishing this copper arsenic sulfide from others of similar appearance. Enargite is an ore of copper. It is not too common, but it is found at Butte, Montana, and at Bingham, Utah.

Fig. M-26: Enargite from Cerro de Pasco, Peru, showing elongated tabular crystals with good prismatic cleavage. Note grayish-black color and metallic luster.

NAME: **Cuprite**
FORMULA: Cu_2O

Color: Red
Streak: Brownish-red
Luster: Adamantine to submetallic   *Cleavage:* Poor octahedral
Hardness: 3.5—4.0   *S.G.:* 6.10

CRYSTALLOGRAPHY: Isometric. Usually in octahedral crystals. Also granular; massive. Often in hair-like crystals.

DESCRIPTION: Cuprite is commonly found in the oxidized portions of hydrothermal deposits associated with other copper minerals such as native copper, malachite, azurite, and chrysocolla, and also with limonite. Cuprite sometimes forms by the alteration of the copper minerals tetrahedrite and chalcopyrite. On the other hand, cuprite itself alters easily to malachite and native copper. The crystal form, the high luster and the streak are helpful in diagnosis. Cuprite is distinguished from other red minerals by being softer than hematite and harder and heavier than cinnabar. Cuprite somewhat resembles pyrargyrite in streak, color, and luster, but the crystal forms are decisive. Cuprite is isometric and pyrargyrite is hexagonal. Cuprite is a fairly widespread mineral and is an important ore of copper. It is found abundantly at Bisbee, Globe, and Morenci, Arizona.

Fig. M-27: Dark red cuprite in massive form with fairly dull luster.

NAME: **Spinel**

FORMULA: $MgAl_2O_4$

Color: Variable: red, blue, green, brown, colorless, black
Streak: White
Luster: Vitreous   *Cleavage:* None
Hardness: 7.5—8.0   *S.G.:* 3.58

CRYSTALLOGRAPHY: Isometric. Octahedral crystals in coarse aggregates. Also in massive form.

DESCRIPTION: Spinels are high-temperature minerals, often found in basic igneous rocks, in metamorphic rocks, in contact-metamorphic rocks, and also in placers. The vitreous luster, the high hardness, and the crystal shapes are distinguishing properties. Black spinel is distinguished from magnetite by its nonmagnetic character and white streak. Dark red spinel resembles ruby, but ruby is hexagonal. Octahedral crystals of high hardness are decisive for spinel. All the red spinels are referred to as "ruby spinel." Spinel is commonly associated with such minerals as scapolite, calcite, garnet, wollastonite, corundum, zircon, graphite, phlogopite, magnetite, augite, and idocrase. The spinels are very resistant to alteration, but they sometimes alter to talc, mica, or serpentine. Transparent and colored spinels, both natural and synthetic, are used for gemstones.

Fig. M-28: Medium to dark green spinel in massive form from Black Lake, Ontario. Note vitreous luster and translucent appearance.

NAME: **Magnetite**
FORMULA: Fe_3O_4

Color: Black
Streak: Black
Luster: Dull metallic   *Cleavage:* None
Hardness: 5.5—6.5   *S.G.:* 5.20

CRYSTALLOGRAPHY: Isometric. Usually coarse or fine granular massive. Often in octahedral crystals.

DESCRIPTION: Magnetite is a widespread mineral occurring in igneous rocks, in contact-metamorphic rocks, in hydrothermal deposits, and in river gravels. The strong magnetism, the black color, and the hardness identify this important oxide. It may resemble black spinel, but spinel is nonmagnetic and has a white streak. Magnetite may also resemble chromite, but the latter's brown streak is diagnostic. Also, magnetite is harder than tetrahedrite. Ilmenite is differentiated from magnetite by its crystal form and by being less magnetic. Magnetite resembles franklinite, but the brown streak of the latter tells the difference. Magnetite usually alters to hematite and limonite. Magnetite is an important ore of iron. The largest deposit in the world is at Kiruna, Sweden. It is also produced in northern New York, in Pennsylvania, and in South Africa. Magnetite is found in iron meteorites as well.

Fig. M-29: Granular massive magnetite from Cornwall, Pennsylvania, showing metallic luster and black color.

NAME: **Chromite**

FORMULA: $FeCr_2O_4$

Color: Black
Streak: Brown
Luster: Metallic to pitchy   *Cleavage:* None
Hardness: 5.5   *S.G.:* 5.1

CRYSTALLOGRAPHY: Isometric. Usually massive; also granular. Commonly in octahedral crystals.

DESCRIPTION: Chromite is associated with ultrabasic rocks such as peridotites and serpentinites. It is also found in placers. Chromite is often associated with olivine, pyroxene, serpentine, spinel, pyrrhotite, magnetite, and corundum. It is weakly magnetic. The brown streak and the pitchy to metallic luster are usually the identifying features. Magnetite resembles chromite, but it is much more magnetic and has a black streak. Uraninite appears similar to chromite but is highly radioactive. Franklinite looks a lot like chromite, but franklinite is intimately associated with red zincite. Commercial deposits of chromite are not too numerous, the larger ones being in Russia, South Africa, Southern Rhodesia, and Turkey. This chromium ore mineral is rare in the United States, but a fair-sized deposit was mined recently at Stillwater, Montana. The mineral chromite is used in making refractory brick. Chromium metal is corrosion resistant and tarnish resistant, it is very hard and is used for chrome-plating and in making stainless steel.

Fig. M-30: Granular chromite from the Bushveld Complex of South Africa, showing pitchy luster of this black ore mineral of chromium.

NAME: **Corundum**
FORMULA: Al_2O_3

Color: Variable: blue, pink, red, yellow, brown, green, purple, white
Streak: White
Luster: Adamantine   *Cleavage:* None
Hardness: 9.0   *S.G.:* 4.0

CRYSTALLOGRAPHY: Hexagonal. Usually in tabular or barrel-shaped crystals. Sometimes massive.

DESCRIPTION: Corundum is found in metamorphic rocks such as marbles, schists, and gneisses. It is also found in certain igneous rocks. Corundum is commonly associated with such minerals as magnetite, kyanite, spinel, tourmaline, mica, garnet, and feldspars. Corundum alters to other aluminum minerals such as kyanite, sillimanite, margarite (mica), or zoisite. The high hardness, the barrel-shaped crystals, the high luster, and the specific gravity are useful identifying features. Several varieties of corundum exist: emery (mixed with magnetite), ruby (deep red), and sapphire (blue). And there are still others of gem quality. Synthetic rubies and sapphires have been manufactured since 1902 by fusing aluminum oxide in flame, and today both star rubies and star sapphires are synthesized. Very fine natural rubies come from Burma. Both rubies and sapphires are found in Siam.

Fig. M-31: Six-sided brown barrel-shaped crystals of corundum from Zoutpansberg, South Africa. Note common tapering pyramids and flat ends.

NAME: **Hematite**

FORMULA: Fe_2O_3

Color: Reddish-brown to black
Streak: Dark red
Luster: Metallic or dull   *Cleavage:* None
Hardness: 5.5—6.5   *S.G.:* 5.26

CRYSTALLOGRAPHY: Hexagonal. Crystals are tabular or platy. Often compact, earthy, reniform, or in foliated masses.

DESCRIPTION: This most important ore mineral of iron is found most frequently and abundantly as sedimentary rocks in beds of varying thicknesses. Hematite is also found as accessory minerals in igneous rocks. It is commonly associated with magnetite, limonite, goethite, and siderite. The dark red streak and the red color of the earthy type or the black shiny color of the specular variety are conclusive distinguishing properties. Black hematite is distinguished from magnetite by its dark red streak. Hematite may resemble cuprite, but cuprite is heavier and is associated with copper minerals. Cinnabar may appear like hematite, but the scarlet color and streak of cinnabar are distinctive. Hematite is mined extensively around Lake Superior in Minnesota, Wisconsin, and Michigan, where it occurs as compact, earthy, and reniform masses. At Birmingham, Alabama, the sedimentary hematite is oolitic in nature. It is used in steel manufacture.

Fig. M-32: Dark red hematite from the Mesabi Range, Minnesota, showing dull (earthy) luster.

NAME: **Ilmenite**

FORMULA: $FeTiO_3$

Color: Black
Streak: Black
Luster: Metallic
Hardness: 5.0—6.0

Cleavage: None
S.G.: 4.72

CRYSTALLOGRAPHY: Hexagonal. Thick tabular crystals in aggregates. Also massive or as grains.

DESCRIPTION: Ilmenite is usually associated with magnetite in basic igneous rocks and also is found in masses in metamorphic rocks such as gneiss. Ilmenite may also be associated with hematite, zircon, spinel, and rutile. The black streak and frequent flat tabular crystals are useful in distinguishing ilmenite from other similar minerals. Hematite in the form of black tabular crystals may resemble ilmenite, but hematite has a dark red streak. Ilmenite looks somewhat like magnetite, but magnetite is more strongly magnetic than ilmenite, and crystal forms differ, magnetite being isometric in octahedral crystals. Ilmenite resembles columbite, but columbite crystals are thicker and heavier and are usually associated with mica. Ilmenite is an important ore mineral for titanium. It is mined in New York, Virginia, Florida, Georgia, and New Jersey. The largest deposit of ilmenite in the world is at Allard Lake, Quebec.

Fig. M-33: Ilmenite showing metallic luster and tabular crystal habit.

NAME: **Pyrolusite**
FORMULA: MnO_2

Color: Iron-black
Streak: Iron-black
Luster: Metallic   *Clevage:* Perfect prismatic
Hardness: 2.0 (massive), 6.0—6.5 (crystals)   *S.G.:* 4.7—5.0

CRYSTALLOGRAPHY: Tetragonal. Usually in radiating fibrous crystals. Also granular massive; dendritic.

DESCRIPTION: This mineral is the most common manganese ore and is quite widespread. It is found as bog, lake, or shallow marine deposits and also in ore deposits of hydrothermal origin as a secondary mineral, associated with other manganese minerals such as manganite, psilomelane, rhodonite, rhodochrosite, and hausmannite, as well as with limonite, siderite, calcite, and even hematite. Pyrolusite often forms by the alteration of other manganese minerals. Manganite may resemble pyrolusite, but manganite occurs in striated crystals with a hardness of 4 and dark reddish-brown streak. Pyrolusite has a sooty black streak, is soft, and soils the fingers. The most extensive manganese deposits are in Russia, India, South Africa, and Brazil. The United States does not have large manganese deposits to use as sources of manganese metal which is so important in steel manufacture. However, the United States will continue to import its manganese requirements.

Fig. M-34: Small radiating clusters of fibrous crystals of black pyrolusite from Ironwood, Michigan, revealing metallic luster.

NAME: **Psilomelane**

FORMULA: $(Ba,H_2O)_2Mn_5O_{10}$

Color: Black
Streak: Brownish-black
Luster: Submetallic
Hardness: 5.0—6.0

Cleavage: None
S.G.: 4.7

CRYSTALLOGRAPHY: Monoclinic. Usually massive or reinform. It may appear amorphous. Crystals are extremely rare.

DESCRIPTION: Psilomelane is a secondary mineral formed under low temperature and pressure by weathering processes. Psilomelane is commonly associated with other manganese minerals such as pyrolusite, rhodonite, rhodochrosite, and braunite, but it often is associated with goethite and limonite as well. In fact, psilomelane forms by the alteration of these manganese minerals. The botryoidal habit, the hardness, and the brownish-black streak distinguish psilomelane from other similar appearing minerals. Psilomelane resembles both goethite and limonite, but the former has brownish-black streak while goethite and limonite both have yellowish-brown streaks. Psilomelane is used as an ore of manganese, but the United States does not have any large deposits. The metal manganese is important in the manufacture of steel, some 12 to 13 pounds of the metal being needed to make one ton of steel. Its function is to deoxidize and desulfurize the steel.

Fig. M-35: Hard, black reniform to stalactitic masses of psilomelane from Arizona showing dull to submetallic luster. Note a few scattered shiny crystals of manganite.

NAME: **Rutile**

FORMULA: TiO_2

Color: Reddish-brown
Streak: Pale brown
Luster: Adamantine to submetallic   *Cleavage:* Distinct prismatic
Hardness: 6.0—6.5   *S.G.:* 4.25

CRYSTALLOGRAPHY: Tetragonal. Slender crystals in aggregates. Often in elbow twins. Also in granular masses.

DESCRIPTION: Rutile is a rather widespread mineral occurring in igneous, metamorphic, and sedimentary rocks. It is often found in large quantities as black sands associated with magnetite, zircon, and monazite. Long slender rutile crystals are sometimes enclosed in quartz and mica. The pale brown streak, strong luster, reddish-brown color, and slender striated crystals are useful diagnostic features. Rutile is distinguished from cassiterite by having lower specific gravity. Rutile sometimes forms by the alteration of other titanium minerals such as sphene and ilmenite. And rutile itself may alter to leucoxene, a white mixture of titanium oxides. Rutile is now synthesized, but the synthesized rutile looks pale yellow to colorless. Fairly high temperatures are required to synthesize rutile because it is a high-temperature mineral under natural conditions. Good crystals of rutile are found at Graves Mountain, Georgia.

Fig. M-36: Rutile from Goochland County, Virginia, in compact massive form showing submetallic luster. This dark reddish-brown mineral is fairly heavy.

NAME: **Cassiterite**
FORMULA: SnO_2

Color: Reddish-brown to brownish-black
Streak: White or brownish-white
Luster: Adamantine   *Cleavage:* Perfect prismatic
Hardness: 6.0—6.7   *S.G.:* 7.0

CRYSTALLOGRAPHY: Tetragonal. Usually massive granular. Often in radiating fibrous arrangements.

DESCRIPTION: Cassiterite occurs in high-temperature hydrothermal veins and also in alluvial placers. It commonly is associated with wolframite, topaz, tourmaline, fluorite, muscovite, quartz, apatite, and arsenopyrite. Cassiterite's high resistance to weathering and its high specific gravity permit it to accumulate in placers. This most important ore mineral of tin is distinguished best by its color, specific gravity, hardness, adamantine luster, and light streak. Rutile often appears similar to cassiterite, but cassiterite is heavier. Zircon may look like cassiterite, but cassiterite is both heavier and harder and has much better cleavage. Sphene often resembles cassiterite, but sphene has wedge-shaped crystals, is softer, and has a lower specific gravity. Only very small production of cassiterite comes from the United States—in Nevada and New Mexico. Most of the world's tin comes from Malaysia, Bolivia, Russia, and Thailand. Cassiterite has been mined at Cornwall, England, since 600 B.C.

Fig. M-37: Cassiterite, showing good submetallic luster and good development of prisms and dipyramids. (Photo courtesy Smithsonian Institution.)

NAME: **Uraninite**

FORMULA: UO_2

Color: Black
Streak: Brownish-black
Luster: Submetallic to pitchy   *Cleavage:* None
Hardness: 5.0—6.0   *S.G.:* 6.5—8.5

CRYSTALLOGRAPHY: Isometric. Usually in massive form. Sometimes in reniform masses. Good crystals rare.

DESCRIPTION: Uraninite is found in granitic igneous rocks and in pegmatites; it is also found in high temperature hydrothermal deposits. In some places it is found in quartz pebble conglomerates. Uraninite may be associated with such minerals as zircon, tourmaline, and monazite in pegmatites; with cassiterite, chalcopyrite, pyrite, arsenopyrite, copper minerals, sphalerite, and galena in hydrothermal deposits. The pitchy luster, black color, high specific gravity, streak, and radioactivity are its reliable properties. This major ore mineral of uranium often alters to yellow-to-red secondary minerals such as autunite, gummite, curite, or torbernite. Chromite with a pitchy luster may resemble uraninite in color, but uraninite is radioactive and has a higher specific gravity. In addition, chromite is commonly associated with minerals of ultrabasic nature, such as serpentine. Domestic production of uraninite comes from the Colorado Plateau area.

Fig. M-38: Massive uraninite from Great Bear Lake, N.W.T. of Canada, revealing the pitch-like luster of this black radioactive mineral.

NAME: **Brucite**

FORMULA: $Mg(OH)_2$

Color: White to pale green
Streak: White
Luster: Pearly to vitreous   *Cleavage:* Perfect basal pinacoidal
Hardness: 2.5   *S.G.:* 2.4

CRYSTALLOGRAPHY: Hexagonal. Usually foliated or massive. Crystals usually tabular; sometimes fibrous.

DESCRIPTION: Brucite is typically associated with low-temperature hydrothermal deposits. It occurs in highly magnesian metamorphic rocks such as serpentine, dolomite, and magnesite. Its foliated habit, light color, and pearly luster are enough to distinguish brucite. Talc may resemble brucite, but talc has a greasy feel due to its low hardness. Fibrous gypsum may resemble brucite, but the former has silky luster. Brucite may be associated with calcite, dolomite, talc, magnesite, serpentine, periclase, and aragonite. Both serpentine and periclase often alter to brucite, and long fibers of brucite are mixed with some of the serpentine in Quebec. Large deposits of brucite are found at Gabbs, Nevada. Large crystals of brucite have been found at Brewster, New York, and at Lancaster County, Pennsylvania. Brucite is used principally as a refractory material, but it is also used as a source of magnesium metal.

Fig. M-39: Brucite from Gabbs, Nevada, in massive form showing pearly luster and bright white color.

NAME: **Bauxite**
FORMULA: Hydrous aluminum oxides

Color: White, grayish-white, yellowish-brown
Streak: White
Luster: Dull to earthy   *Cleavage:* None
Hardness: 1.0—3.0   *S.G.:* 2.5—3.5

CRYSTALLOGRAPHY: Orthorhombic, Monoclinic. In rounded concretionary grains. Also clay-like, earthy.

DESCRIPTION: Bauxite is a mixture of microscopic-sized crystals of minerals and is composed usually of the following three, any one of which may be dominant: boehmite, $AlO(OH)$, orthorhombic; diaspore, $HAlO_2$, orthorhombic; and gibbsite, $Al(OH)_3$, monoclinic. It often contains amorphous material called cliachite, $Al_2O_3 \cdot nH_2O$. Bauxite actually is a rock formed by tropical to subtropical weathering of aluminum-bearing rocks such as granites and syenites. The weathering leaches away nearly all constituents except the very resistant aluminum oxides and hydroxides. The most useful diagnostic feature is the pisolite (concretionary) structure, but bauxite is not easily confused with any others. Clay resembles massive bauxite, and X-ray tests or heat tests are required to tell the difference. Bauxite is the chief ore of aluminum, and large deposits are located in Arkansas, Georgia, and Alabama. It is used also to make alumina (Al_2O_3) and other abrasives. Bauxite is also used in the refractories industry.

Fig. M-40: Bauxite from Arkansas with good round concretionary structure called pisolites. Pisolites (up to 1/2 inch in size) are red, brown, yellow. Note general earthy appearance.

NAME: **Manganite**

FORMULA: MnO(OH)

Color: Steel-gray to black
Streak: Dark brown
Luster: Submetallic
Hardness: 4.0

Cleavage: Perfect prismatic
S.G.: 4.3

CRYSTALLOGRAPHY: Monoclinic. Usually in bundles or radiating masses. Crystals usually prismatic and striated.

DESCRIPTION: Manganite usually occurs in low-temperature hydrothermal veins, associated with other manganese minerals such as pyrolusite, braunite, and hausmannite, as well as with goethite, siderite, calcite, and barite. Manganite commonly alters to pyrolusite, psilomelane, braunite, and hausmannite. The black color, brown streak, hardness, and prismatic crystals are useful diagnostic features. Pyrolusite somewhat resembles manganite, but pyrolusite has a black streak, is much softer than manganite, and soils the fingers when handled. Enargite in prismatic crystals and fibrous masses may also resemble manganite, but enargite has a black streak. Manganite is used as an ore of manganese, but there are not many deposits in the United States. However, fair amounts of manganite are present at Negaunee, Michigan. Manganite is also found in Virginia, Nova Scotia, and in Germany.

Fig. M-41: Fine fibrous silvery prismatic crystals of manganite from New Mexico associated with dull black pyrolusite.

NAME: **Goethite**
FORMULA: $HFeO_2$

Color: Yellowish-brown to dark brown
Streak: Yellowish-brown
Luster: Adamantine to dull   *Cleavage:* Perfect prismatic
Hardness: 5.0—5.5   *S.G.:* 4.3

CRYSTALLOGRAPHY: Orthorhombic. Usually in long slender radiating crystals. Also massive, reniform, stalactitic.

DESCRIPTION: Goethite is a very common mineral produced under oxidizing weathering conditions. It is also frequently produced in bogs and swamps. Much of the yellow-brown amorphous oxides called "limonite" actually belongs with goethite. Goethite differs from limonite by having a crystalline structure, whereas limonite is amorphous. However, these two minerals are so intimately associated that they are difficult to tell apart. Goethite is often associated with hematite, pyrite, magnetite, siderite, and glauconite. It forms by the alteration of pyrite, magnetite, and siderite. The streak, color, and radial crystal growth serve to identify goethite. Black botryoidal hematite may resemble goethite, but hematite has a dark red streak. Similar appearing psilomelane has a black streak. Goethite (and limonite) is often used as an ore of iron and as a pigment in paint.

Fig. M-42: Reniform (rounded) masses of small radiating crystals of goethite from Ironton, Minnesota.

NAME: **Halite**

FORMULA: NaCl

Color: Colorless, white, gray

Streak: White

Luster: Vitreous

Hardness: 2.5

Cleavage: Perfect cubic

S.G.: 2.16

CRYSTALLOGRAPHY: Isometric. Usually cubic crystals in masses. Also massive, compact.

DESCRIPTION: Halite has a strong salty taste; it is soft and has excellent cubic cleavage. These three properties identify it. Halite is distinguished from sylvite by the latter's bitter taste. Halite occurs chiefly as extensive salt beds formed by evaporation of enclosed bodies of seawater. It often becomes interbedded with other sedimentary rocks such as clays. Halite often becomes associated with anhydrite, gypsum, sylvite, polyhalite, and carnallite. At depth, salt beds often become deformed, which causes the salt to flow upward under pressure along weak zones in the rocks to form salt domes or salt plugs. In the United States, much salt is obtained from salt domes in Texas and Louisiana. Halite is used extensively in the chemical industry to make hydrochloric acid and sodium compounds. Halite is used also in food preparation, in fertilizers, and in the tanning industry.

Fig. M-43: Good colorless cubic crystals of halite from Stassfurt, Germany, showing transparency, cubic cleavage, and vitreous luster.

138

NAME: **Sylvite**

FORMULA: KCl

Color: Colorless or white
Streak: White
Luster: Vitreous   *Cleavage:* Perfect cubic
Hardness: 2.0—2.5   *S.G.:* 2.0

CRYSTALLOGRAPHY: Isometric. Usually coarsely granular cubes. Often in cleavable or fine-grained masses.

DESCRIPTION: Sylvite occurs along with halite in salt beds, but it usually forms after halite during deposition by evaporation. Sylvite is rather hard to tell from halite, but sylvite's bitter taste is most helpful. Sylvite becomes associated with many of the same minerals as does halite, chiefly carnallite, polyhalite, kainite, gypsum, anhydrite, and clay minerals. However, sylvite is rarer than halite. Sylvite often is colored red by hematite inclusions, which also helps distinguish it from halite. This mineral is abundantly associated with the Stassfurt, Germany, halite deposits. Sylvite is also fairly abundant in west Texas and near Carlsbad, New Mexico. The largest deposits of sylvite are in Saskatchewan, Canada, where thick beds are at depths of 3000 feet or more. Sylvite is used as a source of potassium to make potassium compounds, which find multiple uses in industry.

Fig. M-44: Small cubic crystals of red sylvite from Midland, Texas, in granular aggregates, revealing vitreous luster.

NAME: **Fluorite**

FORMULA: CaF_2

Color: Colorless, transparent, yellow, green, blue, violet
Streak: White
Luster: Vitreous   *Cleavage:* Perfect octahedral
Hardness: 4.0   *S.G.:* 3.18

CRYSTALLOGRAPHY: Isometric. Usually in coarse cubic crystals. Also in cleavable masses.

DESCRIPTION: Fluorite usually occurs in high-temperature hydrothermal deposits along with lead and silver minerals, but occurs also in dolomites and limestones. It is commonly associated with calcite, gypsum, celestite, barite, quartz, sphalerite, cassiterite, apatite, topaz, and tourmaline. The excellent octahedral cleavage on cubic crystals, the hardness of 4.0, the vitreous luster, and the fine coloration make fluorite easy to identify. Amethyst quartz resembles purple fluorite but is harder than fluorite and lacks the cleavage of fluorite. Calcite may resemble fluorite, but again cleavage is diagnostic, and calcite effervesces in hydrochloric acid. Fluorite fluoresces violet-blue under ultraviolet light. The term *fluorescent* was derived from this fact about fluorite. The most important use for fluorite is as a fluxstone in steel manufacturing, and it is a much sought after mineral. It also finds use in the manufacture of hydrofluoric acid.

Fig. M-45: Large crystals (1-2 inches) of purple fluorite from Illinois showing vitreous luster, cubic and octahedral cleavages, and penetration twinning.

NAME: **Calcite**

FORMULA: $CaCO_3$

Color: Colorless or white
Streak: White
Luster: Vitreous   *Cleavage:* Perfect rhombohedral
Hardness: 3.0   *S.G.:* 2.71

CRYSTALLOGRAPHY: Hexagonal. Crystal aggregates or coarsely granular. Also compact; stalactitic.

DESCRIPTION: Calcite, one of the most common minerals, occurs as the main ingredient of limestone and marble; it also occurs in sandstones, as cave deposits, and as spring deposits. The hardness of 3.0, the excellent rhombohedral cleavage, the vitreous luster, and the free effervescence in HCl are good diagnostic properties. Calcite is distinguished from aragonite by its rhombohedral cleavage. Dolomite effervesces very slowly in HCl. Over 300 form-combinations have been reported for calcite, but the most common forms are the rhombohedron and the scalenohedron. Several common varieties are: Iceland spar (transparent), dogtooth spar (acute scalenohedrons), Mexican onyx (banded), travertine or tufa (spring deposits), and cave deposits (stalactites and stalagmites). Calcite in various forms is seen around limestone caverns such as Mammoth Caves, Kentucky, Carlsbad Caverns, New Mexico, and Luray Caverns, Virginia.

Fig. M-46: Transparent calcite appearing nearly cubic, but with excellent rhombohedral cleavage.

NAME: **Magnesite**

FORMULA: $MgCO_3$

Color: White to gray
Streak: White
Luster: Vitreous   *Cleavage:* Good rhombohedral
Hardness: 4.0   *S.G.:* 3.0—3.2

CRYSTALLOGRAPHY: Hexagonal. Usually compact, earthy masses. Sometimes in granular masses.

DESCRIPTION: Magnesite usually occurs in veins and masses derived from serpentine. Some magnesite is of metamorphic origin associated with schists and gneisses. Magnesite resembles dolomite, but it is distinguished from it by being heavier than dolomite. Magnesite effervesces in hot HCl, and so can be distinguished from calcite. Microcrystalline magnesite may resemble kaolinite, but it is harder; and kaolinite does not effervesce. Porcelain-like magnesite may appear as chert, but chert is much harder. Magnesite forms mainly by alteration of rocks rich in magnesium, such as serpentinite, dunite (olivine), and peridotite. Magnesite is rarely of hydrothermal origin. Large masses of metamorphic magnesite are commonly associated with talc, chlorite, and mica schists. Magnesite is used chiefly in the manufacture of refractory brick for furnace linings. Large production of magnesite comes from Nevada, California, and Washington.

Fig. M-47: Magnesite from Riverside, California. Notice the white earthy compact appearance which is so common in magnesite.

NAME: **Siderite**

FORMULA: $FeCO_3$

Color: Light to dark brown
Streak: White
Luster: Vitreous   *Cleavage:* Good rhombohedral
Hardness: 4.0   *S.G.:* 3.96

CRYSTALLOGRAPHY: Hexagonal. Usually in coarse crystalline masses. Also reniform (rounded); earthy; compact.

DESCRIPTION: Siderite usually is a sedimentary mineral, in clays, shales, and coal beds, occurring usually in granular massive form. When siderite becomes mixed with clay, it is called clay-ironstone. Siderite may also be associated with sulfides in hydrothermal ores of galena, chalcopyrite, and silver. The brownish color, the specific gravity, and the cleavage usually are sufficient to distinguish siderite from other similar appearing minerals. Siderite is distinguished from dolomite by its brown color. Certain forms of sphalerite may resemble siderite, but siderite's rhombohedral cleavage tells the difference. Siderite is very difficult to distinguish from brown ankerite. Siderite often alters to another iron mineral, particularly to goethite. The chief use for siderite is as an ore of iron. Domestic deposits of siderite are in western Pennsylvania, Connecticut, Colorado, and Arizona.

Fig. M-48: Medium brown siderite from Roxbury, Connecticut, showing rhombohedral cleavage and vitreous luster.

NAME: **Rhodochrosite**
FORMULA: $MnCO_3$

Color: Rose-red
Streak: White
Luster: Vitreous to pearly   *Cleavage:* Good rhombohedral
Hardness: 3.5—4.5   *S.G.:* 3.5

CRYSTALLOGRAPHY: Hexagonal. Usually massive. Not often in good crystals. Sometimes granular.

DESCRIPTION: Rhodochrosite usually is hydrothermal in origin, but it sometimes is sedimentary. It commonly is associated with galena, sphalerite, chalcocite, bornite, and other important hydrothermal sulfides. Occasionally rhodochrosite is associated with garnet, rhodonite, tephroite, and manganese oxides. Rhodochrosite easily alters to black manganese oxides such as pyrolusite and manganite. The rose-red color, the good rhomohedral cleavage, and the effervescence in warm hydrochloric acid are diagnostic. Rhodonite resembles rhodochrosite, but rhodonite has a triclinic crystal form, is harder, and does not effervesce in acid. Very little domestic rhodochrosite exists, but some is obtained at Butte, Montana, as an ore of manganese. A little of this mineral is also found in fair-sized crystals at Leadville, Colorado. Rhodochrosite is also found in the silver deposits in Romania and in Germany.

Fig. M-49: Granular to compact mass of rhodochrosite (pink) from Butte, Montana, showing pearly luster and weak rhombohedral crystals.

NAME: **Smithsonite**

FORMULA: $ZnCO_3$

Color: Tan to dirty brown
Streak: White
Luster: Vitreous   *Cleavage:* Good rhombohedral
Hardness: 4.0   *S.G.:* 3.96

CRYSTALLOGRAPHY: Hexagonal. In reniform (rounded) masses or in crystalline incrustations. Also granular.

DESCRIPTION: Smithsonite is often found in the oxidized portion of lead-zinc deposits or in limestones nearby. It is commonly associated with sphalerite, hemimorphite, cerussite, anglesite, galena, malachite, and calcite. In fact, smithsonite usually forms by the alteration of sphalerite. Smithsonite effervesces freely in hydrochloric acid, has a distinguishing brown color, and a hardness of 4.0. These three properties are usually enough to identify it. Similar appearing prehnite is considerably harder and does not effervesce in hydrochloric acid. Hemimorphite may also resemble smithsonite, but it is lighter in weight and does not effervesce in acid. Good masses of smithsonite have been found at Magdelena, New Mexico. This mineral is used as a zinc ore. James Smithson, founder of the Smithsonian Institute, was honored by having this mineral named after him.

Fig. M-50: Crystalline incrustations of dirty brown smithsonite from Kelly, New Mexico. Note high porous nature.

NAME: **Dolomite**

FORMULA: $CaMg(CO_3)_2$

Color: White, often with pink or flesh tone

Streak: White

Luster: Vitreous   *Cleavage:* Good rhombohedral

Hardness: 3.5—4.0   *S.G.:* 2.85

CRYSTALLOGRAPHY: Hexagonal. Usually in coarse granular cleavable masses. Also fine-grained and compact.

DESCRIPTION: Dolomite effervesces only very slowly in hydrochloric acid. The crystals are usually curved rhombohedrons. These two properties plus the frequent flesh-pink color are useful diagnostic features. Calcite resembles dolomite, but calcite's ease of effervescence in acid gives it away. Dolomite often forms by the alteration of calcite or aragonite by magnesium-bearing solutions. Dolomite is usually associated with sedimentary rocks such as limestone. It also occurs in hydrothermal deposits, associated with galena, sphalerite, barite, fluorite, siderite, calcite, and quartz. Dolomitic marbles are formed by the metamorphism of dolomitic limestones, and scattered crystals of dolomite are also frequently found in serpentine and talc rocks. Beautiful curved crystal groups are found in the Tri-State lead and zinc district near Joplin, Missouri. Dolomite is used as a building stone and as a source of magnesium metal. Magnesium oxide is used in the refractories industry.

Fig. M-51: Pale pink crystals of dolomite showing curved rhombohedrons and pearly luster.

NAME: **Aragonite**
FORMULA: $CaCO_3$

Color: Colorless to white
Streak: White
Luster: Vitreous   *Cleavage:* Good prismatic
Hardness: 3.5—4.0   *S.G.:* 2.93

CRYSTALLOGRAPHY: Orthorhombic. Usually in long crystals in radiating groups. Also reniform and stalactitic.

DESCRIPTION: Aragonite may be deposited from hot springs, in sediments along with gypsum and clay, and in hydrothermal and metamorphic deposits. Thus aragonite may be associated with travertine, calcite, gypsum, clay, limonite, siderite, malachite, smithsonite, or cerussite. Aragonite is a less stable form of $CaCO_3$ than calcite is, and it often converts to calcite. Aragonite is much less common than calcite. Aragonite is distinguished from calcite by being heavier, and it lacks the good rhombohedral cleavage of calcite, having good prismatic cleavage of its own. Aragonite effervesces in cold hydrochloric acid similarly to calcite. Aragonite is generally lighter in weight than both witherite and strontianite, but it may resemble these two minerals otherwise. Aragonite is often the "mother of pearl" on the inside of certain shells. Good crystals are found at Bisbee, Arizona, in the Magdalena district, New Mexico, and at Las Animas, Colorado.

Fig. M-52: Aragonite from near Flagstaff, Arizona, occurring as long slender prismatic gray-white crystals arranged in bands.

NAME: **Strontianite**

FORMULA: $SrCO_3$

Color: Gray to white
Streak: White
Luster: Vitreous   *Cleavage:* Good prismatic
Hardness: 3.5—4.0   *S.G.:* 3.7

CRYSTALLOGRAPHY: Orthorhombic. Usually acicular radiating crystals. Also granular or fibrous.

DESCRIPTION: Strontianite is a low-temperature mineral of hydrothermal deposits, where it is usually associated with celestite, barite, and calcite, and is also associated with limestones and clays. Strontianite effervesces in hydrochloric acid. Its high specific gravity and its effervescence are its most useful identifying properties. It may resemble aragonite, but is heavier. Strontianite is distinguished from celestite by having poorer cleavage and by its effervescence in acid. Much celestite is formed by the alteration of strontianite. Strontianite is used as an ore of strontium, and some commercial deposits are found in California and in Germany. The metal strontium has a variety of uses including medicines, fireworks, rockets, and in sugar refining. However, the demand for strontium is not great. Exploration for strontium minerals will be held at a minimum because they are commonly found with other more important minerals.

Fig. M-53: Acicular crystals of white strontianite from Hamm, Germany, showing vitreous luster.

NAME: **Cerussite**

FORMULA: $PbCO_3$

Color: Colorless to white to gray

Streak: White

Luster: Adamantine to resinous to pearly *Cleavage:* Distinct prismatic

Hardness: 3.0—3.5 *S.G.:* 6.55

CRYSTALLOGRAPHY: Orthorhombic. Usually in aggregates of slender crystals. Also granular, compact, earthy.

DESCRIPTION: Cerussite, an important ore of lead, is found in the upper oxidized portion of hydrothermal lead deposits. It is usually associated with such minerals as galena, sphalerite, pyromorphite, smithsonite, anglesite, malachite, and wulfenite. Cerussite often forms as an alteration product of galena or anglesite. The high specific gravity, the white color, and the high luster serve to identify it. Massive aragonite and strontianite resemble cerussite, but cerussite is heavier than both. Brucite may also look like cerussite, but again cerussite is heavier. Cerussite is distinguished from similar appearing anglesite by crystal form and effervescence in nitric acid. Cerussite is found in many places, the most notable of which are Broken Hill, Australia, Tsumeb, South West Africa, and Leadville, Colorado. Being an ore of lead and having been derived from galena, cerussite often carries a little silver which had been present in the galena.

Fig. M-54: Granular to massive aggregates of white cerussite from Coeur d'Alene, Idaho, showing vitreous luster.

NAME: **Malachite**

FORMULA: $Cu_2CO_3(OH)_2$

Color: Bright green
Streak: Pale green
Luster: Silky or dull　　　　　　　　　　　　*Cleavage:* Good pinacoidal
Hardness: 3.5—4.0　　　　　　　　　　　　　　　　　*S.G.:* 4.05

CRYSTALLOGRAPHY: Monoclinic. Slender prismatic crystal aggregates, or reniform. Also stalactitic; granular.

DESCRIPTION: This pretty, green mineral serves as an ore of copper and is a fairly widespread secondary mineral of copper deposits. It is often associated with the much less common mineral azurite (of azure-blue color) and also with cuprite, limonite, native copper, and chrysocolla. Some malachite forms by the alteration of azurite and cuprite. Malachite effervesces in acids, has a beautiful green color, and occurs in long prismatic crystals. These three properties are useful in identification. Malachite resembles brochantite, antlerite, and garnierite, but these three minerals do not effervesce in acid because they are not carbonates like malachite. Good crystals of malachite are rare; most of the malachite found is in botryoidal masses and often is banded. Malachite is found at Tsumeb, South West Africa, in Siberia, and in Arizona in the United States. Copper sheeting often develops a coating of green malachite as a result of reaction with carbon dioxide and water.

Fig. M-55: Bright green malachite from the Congo in reniform (botryoidal) layers. Malachite usually forms by alteration of other copper minerals. (Photo courtesy Ward's Natural Science Establishment.)

NAME: **Soda niter**

FORMULA: NaNO₃

Color: Colorless to white
Streak: White
Luster: Vitreous   *Cleavage:* Perfect rhombohedral
Hardness: 1.5—2.0   *S.G.:* 2.25

CRYSTALLOGRAPHY: Hexagonal. Usually massive or as incrustations. Often in beds; rarely in good crystals.

DESCRIPTION: Soda niter is a water-soluble salt and is found as incrustations in arid regions. It is associated with other salts such as niter, halite, gypsum, and glauber salt. Soda niter has a nice cooling taste. It absorbs water easily from the air, producing a loose, powdery, puffy mass on its surface (deliquescence). The cool taste and the deliquescence are useful in identifying soda niter. Niter may resemble soda niter, but niter occurs in orthorhombic crystals. Halite also resembles soda niter, but halite's taste is salty, not cooling. Soda niter is found in very large quantities in northern Chile over an area about 450 miles long by 10 to 50 miles wide. The soda niter occurs interstratified with sand and other salts such as halite, gypsum, and polyhalite. Some is found in California and Nevada. It is used in the making of nitric acid, in certain explosives, and in fertilizers. Natural nitrates now compete with nitrates "fixed" artificially from the atmosphere.

Fig. M-56: Granular masses of soda niter from Terapaca, Chile. This soft white soluble mineral has a vitreous luster and a cool taste.

NAME: **Borax**

FORMULA: $Na_2B_4O_7 \cdot 10H_2O$

Color: Colorless to white
Streak: White
Luster: Vitreous
Hardness: 2.0—2.5

Cleavage: Perfect prismatic
S.G.: 1.71

CRYSTALLOGRAPHY: Monoclinic. Usually short prismatic crystals in masses. Often as incrustations.

DESCRIPTION: Borax is an evaporite mineral and is the most widespread of the borate minerals. It is found in dried lakes and playas in arid regions. Borax is commonly associated with such minerals as ulexite, halite, gypsum, colemanite, kernite, calcite, thenardite, and soda niter. Borax is difficult to identify from other salts. The crystal form and the sweet alkaline taste can be used with great care in identification. Kernite resembles borax, but kernite is usually in long prismatic crystals. Similar appearing colemanite is harder than borax. Ulexite looks like borax, but ulexite is fibrous and silky. Borax crystals effloresce and alter to white puffy, powdery tincalconite ($Na_2B_4O_7 \cdot 5H_2O$). Borax and other borate minerals are mined commercially at Kramer and Searles Lake, California. Borax is used as a cleaning agent, in medicines, and in glass. It also serves as an ore for the metal boron. Boron metal is used to make boron carbide, an abrasive harder than corundum. This metal is also used in motor fuels and rocket fuels.

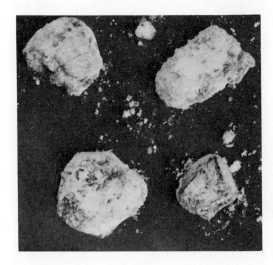

Fig. M-57: Small crystals of borax (about one-half inch across) coated with white fluffy tincalconite.

NAME: **Barite**

FORMULA: $BaSO_4$

Color: White, brown, red
Streak: White
Luster: Vitreous   *Cleavage:* Perfect pinacoidal
Hardness: 3.0—3.5   *S.G.:* 4.5

CRYSTALLOGRAPHY: Orthorhombic. Tabular crystals in divergent groups. Also granular; earthy.

DESCRIPTION: Barite occurs mainly in hydrothermal deposits and is also found in limestones as veins or residual masses. Barite is commonly associated with siderite, fluorite, dolomite, calcite, galena, stibnite, and quartz in the various deposits. Barite often alters to witherite, $BaCO_3$. The high specific gravity, cleavage, crystal form, and luster are good diagnostic properties. Celestite resembles barite very closely, but the two can be distinguished by flame tests, celestite yielding a crimson colored flame and barite giving a yellowish-green flame. Calcite is rather similar but has a lower specific gravity, and barite does not effervesce in hydrochloric acid as calcite does. This important ore mineral of barium is used for other purposes also, such as in drilling muds, pigments, and paper filler. It is found in California, Missouri, Arkansas, and Oklahoma.

Fig. M-58: Large tabular crystals of white to tan barite arranged in divergent masses. Barite is fairly heavy, and is sometimes called "heavy spar."

NAME: **Celestite**

FORMULA: SrSO$_4$

Color: White, often tinted pale blue
Streak: White
Luster: Vitreous   *Cleavage:* Perfect pinacoidal
Hardness: 3.0—3.5   *S.G.:* 3.97

CRYSTALLOGRAPHY: Orthorhombic. Aggregates of tabular crystals. Often fibrous or granular.

DESCRIPTION: Celestite occurs chiefly in sedimentary rocks, often scattered in limestones or sandstones, and occasionally in celestite beds. It is commonly associated with dolomite, fluorite, calcite, strontianite, gypsum, anhydrite, and also with sulfur. Celestite often alters to strontianite. The very common pale blue tint and the specific gravity are helpful in identifying celestite, but they are not conclusive. This mineral is rather hard to identify. Barite resembles celestite rather closely, and these two are distinguished by chemical tests and by flame tests, celestite giving a crimson flame and barite giving a yellowish-green flame. Also, barite has somewhat higher specific gravity than celestite. Good celestite crystals have been found at Put-in-Bay, Lake Erie, at Clay Center, Ohio, and at Lampasas, Texas. Strontium is used in sugar refining and in signal flares. In signal flares the strontium burns with a bright crimson flame.

Fig. M-59: White celestite showing elongated prismatic crystals and vitreous luster.

NAME: **Anglesite**
FORMULA: PbSO$_4$

Color: White to gray
Streak: White
Luster: Adamantine *Cleavage:* Good pinacoidal
Hardness: 3.0 *S.G.:* 6.3

CRYSTALLOGRAPHY: Orthorhombic. Aggregates of tabular crystals. Also massive; in concentric layers.

DESCRIPTION: This lead mineral is found commonly as a secondary mineral of lead deposits of hydrothermal origin. Anglesite is usually associated with galena, cerussite, sulfur, gypsum, wulfenite, smithsonite, and pyromorphite. It sometimes forms from galena, and in turn, anglesite alters to cerussite. The adamantine luster, the high specific gravity, and the close association with galena and cerussite are its most useful distinguishing properties. Anglesite may resemble cerussite, but cerussite effervesces in nitric acid. Barite may also look like anglesite, but anglesite is usually whiter and heavier. Anglesite is not common in the United States, but it is found at Coeur d'Alene, Idaho, and in the Tintic district of Utah. Anglesite is used as an ore of lead. Anglesite is found also at Derbyshire, England, at Broken Hill, New South Wales, in South Africa, in Tasmania, and in Sardinia.

Fig. M-60: White to gray earthy to resinous anglesite from Dividend, Utah, Note the weak concentric layering.

NAME: **Anhydrite**
FORMULA: CaSO₄

Color: White to gray
Streak: White to gray
Luster: Vitreous   *Cleavage:* Perfect prismatic and good pinacoidal
Hardness: 3.5   *S.G.:* 2.96

CRYSTALLOGRAPHY: Orthorhombic. Usually massive or fine grained. Sometimes coarsely fibrous.

DESCRIPTION: Anhydrite is chiefly a sedimentary mineral occurring in beds with gypsum, limestone, and salt. Anhydrite is often associated with quartz, celestite, and dolomite in various deposits. Anhydrite frequently hydrates to gypsum. Its very good cleavage, its hardness, and its specific gravity are usable but not conclusive properties for identification. Anhydrite in massive form is usually difficult to identify. It is distinguished from gypsum by being harder. Calcite may resemble anhydrite, but calcite has good three-directional rhombohedral cleavage, while anhydrite has good prismatic and pinacoidal cleavage. In addition, calcite effervesces in hydrochloric acid. Celestite may also resemble anhydrite, but celestite is somewhat heavier. Anhydrite is found in many localities, notably in New York, New Jersey, Tennessee, and in Texas. It is also found in the Stassfurt salt deposits in Germany. Anhydrite is sometimes used as a source of sulfate for making ammonium sulfate.

Fig. M-61: Crystalline mass of white anhydrite from Mound House, Nevada, showing vitreous luster and granular form.

NAME: **Gypsum**

FORMULA: $CaSO_4 \cdot 2H_2O$

Color: Colorless to white
Streak: White
Luster: Vitreous   *Cleavage:* Excellent prismatic of two types
Hardness: 2.0   *S.G.:* 2.32

CRYSTALLOGRAPHY: Monoclinic. Prismatic crystals in coarse clusters. Often fine-granular; also fibrous.

DESCRIPTION: This very common mineral is found chiefly as sedimentary beds, often interstratified with limestones and shales, commonly taking the form of fibrous "satin spar." Gypsum is also found in volcanic regions and occurs in metallic ore veins where it may be associated with many sulfide minerals, as well as with anhydrite, calcite, dolomite, aragonite, and quartz. The most diagnostic properties for identification are the low hardness, the good cleavage, the prismatic crystals, and the vitreous to pearly luster. Anhydrite may resemble gypsum, but anhydrite is harder. Gypsum often forms by the hydration of anhydrite. Common varieties of gypsum are selenite (transparent), satin spar (fibrous), and alabaster (fine-grained massive). Selenite and alabaster often look like calcite and dolomite, but the latter two minerals are both harder than gypsum. Gypsum is used extensively in the building trade.

Fig. M-62: Transparent vitreous gypsum in cleavable mass. This soft mineral is monoclinic.

NAME: **Alunite**

FORMULA: $KAl_3(SO_4)_2(OH)_6$

Color: White, grayish-white, yellowish-white
Streak: White
Luster: Vitreous to earthy   *Cleavage:* Distinct pinacoidal
Hardness: 3.5—4.0   *S.G.:* 2.6—2.9

CRYSTALLOGRAPHY: Hexagonal. Usually in coarse crystals which resemble cubes. Also massive.

DESCRIPTION: Alunite is found in surface rocks of volcanic regions or where sulfuric acid waters act on potassium feldspars. Alunite is a fairly widespread mineral and is often associated with orthoclase feldspar from which it was derived. Alunite in massive form resembles limestone, dolomite, anhydrite, and magnesite, and X-ray diffraction studies or chemical tests are needed to tell the difference. Kaolinite clay resembles earthy alunite, but kaolinite is softer. Soda niter may resemble alunite, but alunite is much harder and has an acid taste. Brucite also resembles alunite somewhat, but brucite has a waxy to pearly luster. Alunite is found in altered volcanic rocks of the United States, such as at Marysvale, Utah, in Mineral County, Colorado, and in the Goldfield district, Nevada. Alunite finds commercial use in the production of alum and as a filler in fertilizer.

Fig. M-63: Massive white alunite showing granular to dense form and dull luster.

NAME: **Wolframite**
FORMULA: $(Fe,Mn)WO_4$

Color: Black to brownish-black
Streak: Brownish-black to black
Luster: Submetallic
Hardness: 4.0—4.5

Cleavage: Perfect prismatic
S.G.: 7.0—7.5

CRYSTALLOGRAPHY: Monoclinic. Usually in bladed or columnar aggregates. Also massive granular.

DESCRIPTION: Wolframite is found in pegmatites and high-temperature quartz veins. It is commonly associated with cassiterite, molybdenite, topaz, tourmaline, pyrite, apatite, quartz, chalcopyrite, and scheelite. Wolframite is distinguished from others on the basis of its cleavage, specific gravity, and dark color. Cassiterite may resemble wolframite, but stubby crystals of cassiterite differ from bladed crystals of wolframite. Wolframite also has a much darker streak. Columbite resembles wolframite very closely and is very difficult to distinguish from wolframite. However, columbite has a hardness of 6.0 to 6.5, which may help. Flame tests are more decisive. Wolframite is the chief ore mineral of tungsten, and rich deposits are in China, Malaya, and Burma. Wolframite is not abundant in the United States, but it is found in Boulder County, Colorado, in the Black Hills, in Nevada, and in other scattered localities in western United States.

Fig. M-64: Brownish-black wolframite from La Paz, Bolivia, showing bladed crystal habit. Note submetallic luster.

159

NAME: **Scheelite**

FORMULA: $CaWO_4$

Color: White to pale yellow
Streak: White
Luster: Vitreous   *Cleavage:* Good pyramidal
Hardness: 4.5—5.0   *S.G.:* 6.1

CRYSTALLOGRAPHY: Tetragonal. Commonly in good coarse bipyramidal crystals. Sometimes massive.

DESCRIPTION: This tungsten mineral is often associated with wolframite in high-temperature ore deposits of contact metamorphic origin and in deposits of hydro-thermal origin. Scheelite is often associated with other high-temperature minerals such as garnet, quartz, tremolite, cassiterite, wolframite, topaz, apatite, and molyb-denite. Scheelite often alters to the yellow oxide tungstite. The light color, the high specific gravity, and the crystal form usually are sufficient for identification. Certain forms of quartz may resemble scheelite, but scheelite is much heavier and softer. Some crystals of apatite may look a lot like scheelite, but scheelite is heavier. Light-colored feldspars appear similar to scheelite, but crystal forms are very different: scheelite in dipyramids, and feldspars in short blocky prismatic crystals. Scheelite is also softer but heavier than feldspar. Scheelite, an important ore of tungsten, is produced in California and Nevada.

Fig. M-65: Massive and crystalline aggregates of white to tan scheelite from Porcupine, Ontario, showing high vitreous luster.

NAME: **Wulfenite**

FORMULA: PbMoO$_4$

Color: Yellow to orange-yellow
Streak: White
Luster: Resinous to adamantine   *Cleavage:* Good dipyramidal
Hardness: 3.0   *S.G.:* 6.8

CRYSTALLOGRAPHY: Tetragonal. Tabular crystals in coarse aggregates. Also in granular masses.

DESCRIPTION: Wulfenite is a secondary mineral of lead and molybdenum ore deposits of hydrothermal origin, and it is usually associated with pyromorphite, cerussite, calcite, vanadinite, galena, and limonite. Square tabular crystals, yellow color, and high luster are its most diagnostic properties. Crocoite may resemble wulfenite, but the long prismatic crystals of crocoite are very different from the thin tabular crystals of wulfenite. Yellow-colored vanadinite may look a lot like wulfenite, but the tabular crystals of wulfenite are different from the hexagonal prismatic vanadinite crystals. Certain yellow crystals of descloizite may appear similar to wulfenite, but the orthorhombic crystal form of descloizite tells the difference. Very beautiful crystals of wulfenite are found at Red Cloud, Arizona, and also at the Glove Mine, Arizona. There are several other localities in western United States where this molybdenum ore is mined.

Fig. M-66: Square, tabular, thin, fragile crystals of wulfenite with adamantine luster, in roughly radiating clusters.

NAME: **Apatite**

FORMULA: $Ca_5(PO_4)_3(F,Cl,OH)$

Color: Usually some shade of green or brown
Streak: White
Luster: Vitreous
Hardness: 5.0

Cleavage: Poor pinacoidal
S.G.: 3.1—3.2

CRYSTALLOGRAPHY: Hexagonal. Prismatic crystals in coarse aggregates. Also massive to compact.

DESCRIPTION: Apatite is found in many kinds of rocks, including igneous, sedimentary, and metamorphic. It is also found in pegmatites and in hydrothermal deposits. Apatite may be associated with a large variety of minerals. The color, the crystal form, and the hardness are good, usable distinguishing properties. Apatite is distinguished from similar appearing tourmaline by being considerably softer. Beryl may resemble apatite, but apatite is much softer. There are several types of apatite, depending upon which of a few ions may enter the structure, such as fluorine (F), chlorine (Cl), or hydroxyl (OH). Fluorapatite is more common than chlorapatite or hydroxylapatite. Close inspection of apatite often reveals numerous small cracks which make the crystals appear shattered. Impure cryptocrystalline apatite is called collophane and makes up phosphate rock. Phosphate rock is mined in Florida, Tennessee, Idaho, Utah, and Wyoming, and is used for fertilizer.

Fig. M-67: Short prismatic six-sided crystals of light green apatite, showing pyramid terminations, vitreous luster, and shattered appearance.

NAME: **Pyromorphite**
FORMULA: $Pb_5(PO_4)_3Cl$

Color: Green, yellow, or brown
Streak: White
Luster: Resinous   *Cleavage:* None
Hardness: 3.5—4.0   *S.G.:* 6.5—7.1

CRYSTALLOGRAPHY: Hexagonal. Barrel-shaped crystals in groups or masses. Often reinform or fibrous.

DESCRIPTION: Pyromorphite is a secondary mineral found in the oxidized portion of ore deposits which contain galena. Therefore, pyromorphite is usually associated with galena and other lead minerals such as cerussite, as well as with limonite, smithsonite, wulfenite, anglesite, and vanadinite. Actually, much pyromorphite forms by the alteration of galena and cerussite. The good crystal form, the high luster, and the high specific gravity are its best diagnostic properties. Pyromorphite is very similar to mimetite, which is in the same group. Chemical tests are required to distinguish them: mimetite has arsenic; pyromorphite has phosphorus. Vanadinite resembles pyromorphite, but vanadinite is usually ruby red in color. Pyromorphite is found at Coeur d'Alene, Idaho, at Phoenixville, Pennsylvania, and in Davidson County, North Carolina. It is also found in Czechoslovakia, Russia, England, and Mexico. Pyromorphite is sometimes used as an ore of lead.

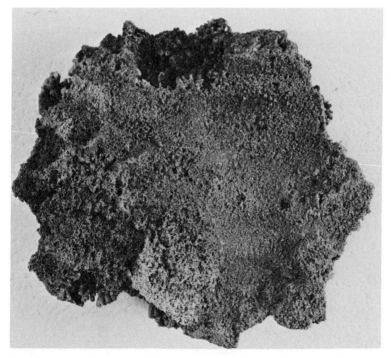

Fig. M-68: Brownish-green mass of porous crusty pyromorphite from Sherman, Idaho. Note a few rounded hollow barrel-shaped crystals.

NAME: **Turquoise**
FORMULA: $CuAl_6(PO_4)_4(OH)_8 \cdot 4H_2O$

Color: Sky blue to bluish-green
Streak: White to pale green
Luster: Waxy to vitreous   *Cleavage:* Perfect basal pinacoidal
Hardness: 6.0   *S.G.:* 2.6—2.8

CRYSTALLOGRAPHY: Triclinic. Usually massive and compact, with dull porcelain-like surface. Crystals rare.

DESCRIPTION: This beautiful blue mineral is of secondary origin, occurring in veins and stringers in volcanic rocks. Turquoise commonly forms by the alteration of rocks rich in aluminum, phosphorus, and copper. The distinctive blue color and the dull porcelain-like luster are its diagnostic features. Turquoise may resemble chrysocolla, but turquoise is much harder. The variety of quartz called chrysoprase closely resembles turquoise, but a chemical test for phosphorus is decisive. Variscite is softer and greener than turquoise, but it appears very similar otherwise. Turquoise is used as a gemstone and has been prized for thousands of years. It has been found at a number of localities in the western United States, such as in Colorado, New Mexico, Arizona, California, and Nevada. Turquoise was mined as early as 5300 B.C. on the Sinai Peninsula of Egypt, and somewhat later in Persia (Iran). Sky-blue, highly prized turquoise is still produced in Persia.

Fig. M-69: Cryptocrystalline compact mass of turquoise from Arizona, showing dull to porcelain luster. (Courtesy Eugene P. Crowell.)

NAME: **Autunite**

FORMULA: $Ca(UO_2)_2(PO_4)_2 \cdot 10\text{-}12H_2O$

Color: Lemon-yellow
Streak: Yellowish
Luster: Vitreous   *Cleavage:* Good basal pinacoidal
Hardness: 2.0—2.5   *S.G.:* 3.1—3.2

CRYSTALLOGRAPHY: Tetragonal. Tabular crystals in coarse or scaly aggregates. Meta-autunite crystals in bundles.

DESCRIPTION: Autunite is usually found as a secondary mineral in the zone of oxidation of uranium-bearing veins and pegmatites. Autunite is rather difficult to distinguish from other yellow minerals, but its lemon-yellow color, its square thin crystals, its strong luminescence in ultraviolet light, and its radioactivity are characteristic. Autunite is commonly associated with other uranium minerals such as gummite, curite, and green torbernite. All of these minerals usually form by the alteration of uraninite. When autunite dries, it changes over, at about 80°C, to meta-autunite whose crystals appear as little bundles. Autunite (or meta-autunite) is found near Spokane, Washington, and at Spruce Pine, North Carolina. Autunite is found at various localities on the Colorado Plateau. Other famous localities are Autun, France, Cornwall, England, in the Katanga district of the Congo, and in South Australia. It is used as an ore of uranium.

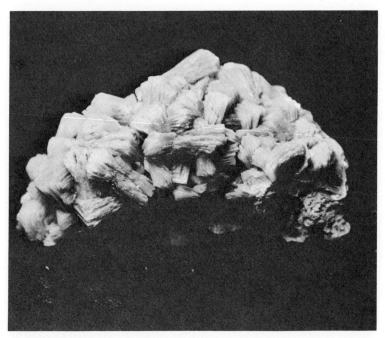

Fig. M-70: Yellow-green sheaflike bundles of autunite (meta-autunite) crystals from near Spokane, Washington.

NAME: **Carnotite**

FORMULA: $K_2(UO_2)_2(VO_4)_2 \cdot 3H_2O$

Color: Bright yellow to lemon-yellow

Streak: Pale yellow

Luster: Dull   *Cleavage:* Good basal pinacoidal

Hardness: 2.0   *S.G.:* 4.0—5.0

CRYSTALLOGRAPHY: Monoclinic. Commonly in loose powdery aggregates. Occasionally compact; rarely in large crystals.

DESCRIPTION: Carnotite is found chiefly in sedimentary rocks, particularly sandstones, brought there by circulating waters carrying both uranium and vanadium. It often becomes associated with tyuyamunite around organic matter such as petrified tree trunks or other vegetable material, or it just becomes disseminated in the sandstones. Carnotite's bright yellow color, its powdery nature, and its radioactivity are good identifying properties. However, many other secondary uranium minerals are similar, especially tyuyamunite, and various optical, chemical, or X-ray diffraction tests are required to make the distinctions. Carnotite serves as an ore for both uranium and vanadium and is found in numerous locations of the Colorado Plateau (Colorado, New Mexico, Arizona, Utah). It is found also in several areas of Wyoming and in the Black Hills of South Dakota. Uranium is currently used in weaponry, but future uses will include power-generating and desalinating nuclear reactors.

Fig. M-71: Powdery yellow carnotite (light at top of specimen) and also disseminated in tan sandstone (dark lower two-thirds of specimen) from Emery County, Utah.

NAME: **Olivine**

FORMULA: $(Mg,Fe)_2SiO_4$

Color: Olive-green
Streak: White
Luster: Vitreous   *Cleavage:* Poor prismatic
Hardness: 6.5   *S.G.:* 3.3—3.4

CRYSTALLOGRAPHY: Orthorhombic. Usually in granular masses. Often scattered in igneous rocks; rarely in large crystals.

DESCRIPTION: Olivine occurs chiefly in basic igneous rocks, often in large proportions. It is usually associated with chromite, corundum, spinel, garnet, magnetite, serpentine, and augite. Olivine often alters to limonite, serpentine, or magnesite. The very glassy luster, the olive-green color, the striated tabular crystals, and the granular nature make olivine rather easy to identify. Epidote resembles olivine, but the pistachio-green color of epidote is different from the green of olivine. Light green diopside may resemble olivine, but diopside is usually in good prismatic crystals and is somewhat lighter green in color. Clear transparent crystals of olivine are called peridot and are used for gemstones. Large peridot crystals come from St. John's Island in the Red Sea and from Snarum, Norway. Peridot has also been found in Burma and in the volcanic rocks around Mt. Vesuvius. An igneous rock composed principally of olivine is called dunite, such as that at Dun Mountain, New Zealand.

Fig. M-72: Olivine crystals from Globe, Arizona, showing dull but translucent nature of this olive-green mineral.

NAME: **Andalusite**

FORMULA: Al_2SiO_5

Color: White, rose-red, brown
Streak: White
Luster: Vitreous to dull   *Cleavage:* Good prismatic
Hardness: 7.5   *S.G.:* 3.15

CRYSTALLOGRAPHY: Orthorhombic. Usually in coarse prismatic crystal aggregates. Sometimes massive.

DESCRIPTION: Andalusite is found chiefly in contact metamorphic shales and in other metamorphic rocks. It is commonly associated with sillimanite, kyanite, garnet, muscovite, cordierite, or tourmaline. Andalusite alters rather easily to sericite (muscovite) which often coats the andalusite crystals. Sometimes andalusite inverts to sillimanite or kyanite if there are big temperature and pressure changes. Andalusite can be identified by its common rose-red color, its hardness, and its prismatic cleavage. Pink tourmaline somewhat resembles andalusite, but square cross-sections of andalusite crystals are enough to tell the difference. One variety of andalusite called chiastolite displays a black cruciform design in cross-section formed of carbonaceous matter. Clear crystals of andalusite are used as gemstones and are found mainly in Ceylon and Brazil. The chiastolite variety is found in California and Massachusetts. Due to its high aluminum content, andalusite is used as a refractory material.

Fig. M-73: Reddish-brown andalusite from Mono County, California, in dull, rough prismatic crystals.

NAME: **Sillimanite**
FORMULA: Al_2SiO_5

Color: White or slightly tinted brown
Streak: White
Luster: Vitreous   *Cleavage:* Good prismatic
Hardness: 7.0   *S.G.:* 3.24

CRYSTALLOGRAPHY: Orthorhombic. Usually in coarse prismatic masses. Often fibrous. Also in compact, tough masses.

DESCRIPTION: Sillimanite is found in high-grade metamorphic rocks such as schists and gneisses. Thus it is often associated with quartz, andalusite, muscovite, corundum, zircon, cordierite, and almandite. Sillimanite alters quite readily to kaolinite, pyrophyllite, and muscovite. When subjected to high pressure, sillimanite converts to kyanite. Sillimanite is usually identifiable by its color and by its long slender crystals revealing prismatic cleavage. Very few minerals are confused with sillimanite. However, tremolite may resemble sillimanite, but sillimanite is harder. Wollastonite also may appear similar to sillimanite, but wollastonite is considerably softer. There are two polymorphs of sillimanite which have the same composition and formula: andalusite and kyanite. Sillimanite is found at Worcester, Massachusetts, where it is in fibrous masses, and it also is found at Norwich, Connecticut, and in New Hampshire. Sillimanite is used principally as a refractory material.

Fig. M-74: Sillimanite from near Dillon, Montana, showing long slender gray-white prismatic crystals in parallel groups.

NAME: **Kyanite**

FORMULA: Al_2SiO_5

Color: Light blue, but patchy
Streak: White
Luster: Vitreous   *Cleavage:* Perfect front pinacoidal
Hardness: 4.0—7.0   *S.G.:* 3.63

CRYSTALLOGRAPHY: Triclinic. Usually long tabular or bladed crystals in divergent masses. Sometimes appears fibrous.

DESCRIPTION: Kyanite is commonly found in medium-grade metamorphic rocks such as schists and gneisses. It is usually associated with corundum, rutile, garnet, and staurolite. This blue mineral alters easily to kaolinite, pyrophyllite, or muscovite. Under high pressure or temperature, kyanite may invert to sillimanite or andalusite, both of which are polymorphs. The blue color, the bladed habit, and the good cleavage serve to distinguish kyanite. Kyanite is not easily confused with other minerals. Kyanite crystals are somewhat flexible, bending a little before rupturing across the long dimension, and they may even be twisted a little. Another unusual feature of kyanite is the variation in the hardness: a hardness of 4.5 along the long direction, and about 7.0 across the long direction. Kyanite is used in the refractories industry and is found in North Carolina and Georgia. Sometimes it is used as a gemstone. Very fine crystals of gem quality are found in Minas Gerais, Brazil.

Fig. M-75: Long tabular bladed crystals of kyanite (patchy blue in color). Note vitreous luster.

NAME: **Staurolite**

FORMULA: $Fe_2Al_9O_6(SiO_4)_4(O,OH)_2$

Color: Brown
Streak: Gray
Luster: Vitreous   *Cleavage:* Good side pinacoidal
Hardness: 7.0   *S.G.:* 3.7—3.8

CRYSTALLOGRAPHY: Monoclinic. Prismatic crystals, singly or in cruciform twins, often in clusters.

DESCRIPTION: Staurolite is found in aluminum-rich gneisses and schists of medium-grade metamorphism. It is usually associated with garnet, muscovite, kyanite, and tourmaline. In fact, staurolite commonly forms from the metamorphism of aluminum-rich rocks. Staurolite alters easily to kaolinite. The brown color, the flat elongated crystal form, and the very common cruciform twins serve to identify staurolite. Tourmaline may resemble staurolite, but tourmaline is hexagonal. Andalusite may appear similar to staurolite, but prism angles are diagnostic; about 89° for andalusite and about 50° for staurolite. Large crystals of staurolite have been found at many places such as at Windham, Maine, and at Grafton County, New Hampshire. Very fine specimens of staurolite, up to one inch in size, come from Fannin County, Georgia, and from Pilar, New Mexico. At Fairfax County, Virginia, well-formed cross twins are called "fairy stones" or "fairy crosses." Transparent crystals are sometimes sold as gems.

Fig. M-76: Brownish-black prismatic crystals of staurolite from Fannin County, Georgia, showing cruciform twins and rough surface.

NAME: **Topaz**
FORMULA: $Al_2SiO_4(OH,F)_2$

Color: Colorless, white, yellow-brown
Streak: White
Luster: Vitreous   *Cleavage:* Perfect basal pinacoidal
Hardness: 8.0   *S.G.:* 3.5—3.6

CRYSTALLOGRAPHY: Orthorhombic. Usually in coarse crystal masses. Often massive. Crystals usually striated.

DESCRIPTION: Topaz is commonly found in pegmatites, in granites, and in high-temperature hydrothermal deposits. It is commonly associated with cassiterite, wolframite, tourmaline, beryl, quartz, molybdenite, apatite, mica, and fluorite. Topaz often alters to sericite or to kaolinite. The crystal habit, the high hardness, the good cleavage, and the yellowish-brown color are its most diagnostic features. Topaz is not often confused with other minerals. Similar appearing quartz and beryl both are not as heavy as topaz. Certain crystals of apatite may resemble topaz, but apatite is much softer and occurs in six-sided crystals, just like quartz and beryl. The great variety of attractive colors and the common transparency allow topaz to be used extensively as a gemstone. Very fine gem quality topaz comes from Brazil, Siberia, Burma, and Ceylon. Topaz crystals also come from Utah, California, Texas, Colorado, and Maine, as well as New Brunswick, Canada.

Fig. M-77: White to light green topaz from Buffalo, Colorado, in long prismatic crystals, showing high vitreous luster.

172

NAME: **Garnet group**

FORMULA: $(Ca,Mg,Fe,Mn)_3(Al,Fe,Cr)_2(SiO_4)_3$

Color: Dark red to brown
Streak: White or pale brown
Luster: Vitreous or resinous   *Cleavage:* None
Hardness: 7.0—7.5   *S.G.:* 3.6—4.3 (depending on composition)

CRYSTALLOGRAPHY: Isometric. In scattered crystals or in coarse granular masses. Crystals usually distinct.

DESCRIPTION: The garnets have great diversity of composition, containing Fe, Mg, Al, Mn, Ca, and Cr in different amounts, along with the SiO_4 ion to produce six species. Garnets are usually of metamorphic origin and are found in schists and gneisses. Garnet may be associated with hornblende, augite, serpentine, spinel, andalusite, kyanite, staurolite, quartz, topaz, idocrase, wollastonite, or epidote, and sometimes even with chromite. The garnets often alter to talc, serpentine, or to chlorite. The isometric crystal form, the color, the hardness, and the vitreous luster are usually sufficient identifying properties. The good dodecahedral habit is very helpful. Green garnet may resemble both jadeite and idocrase, but jadeite is usually in monoclinic fibrous masses and idocrase is commonly in prismatic tetragonal crystals. Clear garnets are used for gemstones; others are used for abrasives because of the high hardness.

Fig. M-78: Fine deep red garnet from Warren County, New York, showing coarse granular form and good vitreous luster.

NAME: **Zircon**

FORMULA: $ZrSiO_4$

Color: Some shade of brown
Streak: White
Luster: Vitreous   *Cleavage:* Poor prismatic
Hardness: 7.5   *S.G.:* 4.6—4.7

CRYSTALLOGRAPHY: Tetragonal. Prismatic crystals with pyramids usually as scattered grains; also in coarse aggregates.

DESCRIPTION: Zircon is commonly found in granitic igneous rocks and in pegmatites. Consequently, it is usually associated with quartz, feldspar, tourmaline, beryl, and many other minerals. Zircon is very resistant to any kind of alteration, and thus it is often found in stream and beach sands, as in Florida. Brown, square prisms with terminating pyramids, the hardness, and the specific gravity are good usable identifying features. Brown-colored idocrase may resemble zircon, but idocrase has a lower specific gravity. Thorite may be mistaken for zircon, but thorite is strongly radioactive. However, zircon may be weakly radioactive and is lighter brown in color. Transparent brown or red zircons are used for gemstones and come mostly from Ceylon and Madagascar. Zircon is also used as an ore of zirconium, a metal which is used in nuclear reactors. Some small crystals are found in North Carolina and in old beach sands in Florida. Very large crystals have been found in Renfrew County, Ontario.

Fig. M-79: Nicely-shaped tetragonal zircon crystals showing dark brown and light brown color, and adamantine luster.

NAME: **Sphene**

FORMULA: $CaTiSiO_5$

Color: Brown, sometimes yellow

Streak: White

Luster: Adamantine　　　　　　　　　　　　*Cleavage:* Good prismatic

Hardness: 5.5　　　　　　　　　　　　　　　　*S.G.:* 3.5

CRYSTALLOGRAPHY: Monoclinic. Usually in coarse, wedge-shaped crystals. Also platy or massive.

DESCRIPTION: Sphene is common as an accessory mineral in igneous rocks as well as in metamorphic rocks. It is often found associated with magnetite, pyroxenes, amphiboles, apatite, quartz, feldspars, scapolite, and zircon. Sphene often alters to rutile or anatase. The brown color, the large flat crystal faces, the wedge-shaped crystals, and the luster are reliable identifying features. Similar appearing staurolite is harder than sphene. Sphalerite may resemble sphene but is much softer. Wedge-shaped crystals of axinite may look like sphene but are much harder. Cassiterite, rutile, and zircon all may resemble sphene, but each is harder and heavier. Sphene is used as an ore for the metal titanium. Good sphene crystals have been found in St. Lawrence County, New York, and near Riverside, California. Large transparent crystals of sphene are used as gemstones; some are found in the Alps, in Switzerland, and in Mexico. The largest Mexican ones are sometimes three to four inches in size.

Fig. M-80: Dark brown platy masses of sphene from Verona, Ontario.

NAME: **Hemimorphite**

FORMULA: $Zn_4Si_2O_7(OH)_2 \cdot H_2O$

Color: White, often tinted brown or blue
Streak: White
Luster: Vitreous   *Cleavage:* Perfect prismatic
Hardness: 5.0   *S.G.:* 3.4—3.5

CRYSTALLOGRAPHY: Orthorhombic. In coarse crystals or in stalactitic form. Also massive.

DESCRIPTION: Hemimorphite is found in the oxidized portions of zinc deposits and is formed secondarily by weathering processes. It is commonly associated with sphalerite, smithsonite, galena, anglesite, and cerussite. Hemimorphite easily forms by the alteration of zinc minerals such as sphalerite. The crystal habit and the prismatic cleavage can be used to identify hemimorphite. Hemimorphite resembles botryoidal smithsonite, but smithsonite effervesces in hydrochloric acid. Similar looking prehnite has a lower specific gravity. Hemimorphite is used as an ore of zinc and is found in numerous places. Some fine specimens come from Sterling Hill, New Jersey. Nice crystals also come from the Leadville district, Colorado. Very nice crystals up to one-half inch across are still obtained from the Mina Ojuela in Durango, Mexico; these are considered as the best available in North America. Hemimorphite also is found in Siberia, Italy, Germany, and Algeria.

Fig. M-81: Small white hemimorphite crystals in multiple divergent groups.

NAME: **Idocrase**

FORMULA: $Ca_{10}Mg_2Al_4(Si_2O_7)_2(SiO_4)_5(OH)_4$

Color: Brown or green
Streak: White
Luster: Vitreous to greasy   *Cleavage:* Poor prismatic
Hardness: 7.0   *S.G.:* 3.3—3.5

CRYSTALLOGRAPHY: Tetragonal. Prismatic crystals in coarse aggregates; crystals usually striated and columnar; massive.

DESCRIPTION: Idocrase is usually found in contact metamorphic limestones. It is commonly associated with diopside, garnet, wollastonite, epidote, sphene, and tourmaline. The common brown color, the striated tetragonal crystals, and the high hardness are good, usable diagnostic features. Zircon may resemble idocrase, but zircon is heavier. Green idocrase resembles jadeite, but jadeite is monoclinic. Epidote looks similar to idocrase, but the pistachio-green color of epidote is distinctive. Brown garnet resembles idocrase, but garnet is isometric and dodecahedral habit is very common. Brown tourmaline looks a lot like idocrase, but tourmaline is hexagonal and in long slender or prismatic crystals. Idocrase is also called "vesuvianite," having been found early at Mount Vesuvius in Italy. Well-formed crystals are rather rare, but some are found in Fresno and Tulare Counties, California. It is also found in New York, New Jersey, and Maine. Green idocrase, called californite, is sold as a gem. Rare massive blue idocrase is called cyprine.

Fig. M-82: Yellowish-brown tetragonal crystals of idocrase showing vitreous luster.

NAME: **Epidote**

FORMULA: $Ca_2(Al,Fe)Al_2O(SiO_4)(Si_2O_7)(OH)$

Color: Pale green to brownish-green
Streak: White
Luster: Vitreous   *Cleavage:* Good basal pinacoidal
Hardness: 7.0   *S.G.:* 3.3—3.6

CRYSTALLOGRAPHY: Monoclinic. Usually in coarse prismatic striated crystals. Sometimes massive.

DESCRIPTION: Epidote is found in low- to medium-grade metamorphic rocks. It is often associated with chlorite, albite, actinolite, scheelite, prehnite, and zeolites. Epidote often forms by the alteration of feldspars, pyroxene, amphibole, and biotite. The color, the hardness, and the cleavage can be used quite successfully in identifying epidote. Epidote is sometimes mistaken for olivine, particularly in color and luster, but coarse columnar crystals of epidote will provide the difference. In addition, epidote nearly always has a pistachio-green color which is very different from the olive-green of olivine. Large green epidote crystals have been found in Austria and Alaska. It also is found associated with garnet in California. Epidote occurs in fair-sized crystals in Adams County, Idaho. Clear epidote crystals are sometimes used as gemstones. Large crystals of epidote occur in pegmatites of Baja California.

Fig. M-83: Elongated striated prismatic crystals of dark green epidote from Butte, Montana.

NAME: **Prehnite**

FORMULA: $Ca_2Al(AlSi_3O_{10})(OH)$

Color: Pale green
Streak: White
Luster: Vitreous   *Cleavage:* Good basal pinacoidal
Hardness: 6.5   *S.G.:* 2.90—2.93

CRYSTALLOGRAPHY: Orthorhombic. Tabular crystals in radiating groups. Also reniform, stalactitic.

DESCRIPTION: Prehnite is found chiefly in cavities in basalts as a secondarily formed mineral. It is also found in minor quantities in metamorphic rocks. It is usually associated with calcite, datolite, zeolites, and pectolite. The green color, the crystal form, and the luster are good usable identifying features. Hemimorphite resembles prehnite, but is of higher specific gravity and is softer than prehnite. Smithsonite may resemble prehnite, but it is softer and effervesces in hydrochloric acid. Good prehnite crystals occur at Farmington, Connecticut, at Paterson, New Jersey, and at Bergen Hill, New Jersey. Some has been found associated with native copper in the Keweenaw area of Lake Superior. It has also been found at Coopersburg, Pennsylvania. Very fine crystals of prehnite are found at Motta Naira, Switzerland. The finest domestic specimens come from Paterson, New Jersey.

Fig. M-84: Rounded groups of tabular light green crystals of prehnite from Colorado in cavities in basalt.

NAME: **Beryl**
FORMULA: $Be_3Al_2Si_6O_{18}$

Color: Pale green to white
Streak: White
Luster: Vitreous   *Cleavage:* Poor basal pinacoidal
Hardness: 8.0   *S.G.:* 2.66—2.92

CRYSTALLOGRAPHY: Hexagonal. Prismatic coarse crystalline masses. Crystal faces usually rough.

DESCRIPTION: Beryl is found in pegmatites, usually as large crystals, and occasionally in granites as small crystals. Beryl is usually associated with spodumene, amblygonite, columbite, mica, tourmaline, lepidolite, feldspar, and quartz. Beryl often alters to muscovite or to kaolinite. The hexagonal crystals, the pale green color, and the hardness can be used to distinguish beryl. Beryl is distinguished from similar appearing apatite by having much greater hardness. Milky quartz may resemble beryl, but beryl is heavier. Beryl is the chief ore for the metal beryllium. Transparent varieties of beryl are used as gemstones, and these are emerald (deep green), aquamarine (light bluish-green), morganite (pink), and heliodor (yellow). Emerald is the most prized of all gemstones, the finest ones being found in Colombia, South America, where they are present in calcite veins. Other beryl crystals are found in pegmatites of the Black Hills, South Dakota, and in San Diego County, California.

Fig. M-85: Prismatic blue-green hexagonal crystal of beryl from Colorado, showing vitreous luster and mottled appearance. (Courtesy Eugene P. Crowell.)

NAME: **Cordierite**

FORMULA: $(Mg,Fe)_2Al_3(AlSi_5O_{18})$

Color: Pale to dark blue
Streak: White
Luster: Vitreous   *Cleavage:* Poor prismatic
Hardness: 7.0   *S.G.:* 2.55—2.75

CRYSTALLOGRAPHY: Orthorhombic. Usually massive or fine granular. Also as scattered small grains.

DESCRIPTION: Cordierite is usually formed by medium- to high-grade metamorphism, and it is found in gneisses, hornfels, and schists. Sometimes it is found as an accessory mineral in granite. Cordierite is commonly associated with garnet, andalusite, mica, quartz, and feldspar. It alters quite easily to mica, chlorite, or talc. The blue color and hardness are its best distinguishing features. Quartz very much resembles cordierite, and the blue color must be used with caution because some smoky quartz may appear blue. Corundum may resemble cordierite, but corundum is considerably harder. Fairly good crystals of cordierite are found in Bavaria, Finland, Madagascar, and Greenland. Cordierite of gem quality comes from Ceylon. Domestic sources are in Connecticut and New Hampshire. Wyoming and the Yellowknife district of the Northwest Territories of Canada also yield quite good specimens occasionally. The transparent cordierite which is used for gems is called *saphir d'eau.*

Fig. M-86: Light to medium blue crystalline mass of cordierite showing vitreous luster. Somewhat resembles quartz.

NAME: **Tourmaline**

FORMULA: $Na(Mg,Fe)_3Al_6(BO_3)_3(Si_6O_{18})(OH)_4$

Color: Black, sometimes brown
Streak: White
Luster: Vitreous   *Cleavage:* Poor rhombohedral
Hardness: 7.5   *S.G.:* 3.0—3.2

CRYSTALLOGRAPHY: Hexagonal. Striated prismatic crystals in radiating groups. Also massive.

DESCRIPTION: Tourmaline is usually found in granitic pegmatites and sometimes in metamorphic rocks. It is commonly associated with quartz, feldspar, mica, topaz, beryl, and apatite. Tourmaline itself often alters to mica or chlorite. The color, the triangular cross-section, and the striated crystals serve to identify it. Apatite may resemble brown tourmaline but is softer. Hornblende resembles tourmaline, but tourmaline has very poor cleavage. Similar appearing green beryl has a lower specific gravity and does not reveal deep striations on crystals, so characteristic of tourmaline. Several varieties of tourmaline exist such as schorlite (black), rubellite (rose), verdelite (green), and dravite (brown). Much tourmaline is used as gemstones, and the better specimens are usually obtained from various pegmatites in Maine. Some is obtained from Mozambique, from Russia, and from the Island of Elba. Very large crystals come from Madagascar.

Fig. M-87: Vertically striated elongated prismatic crystal of black tourmaline from Custer, South Dakota.

NAME: **Tremolite-actinolite**

FORMULA: $Ca_2(Mg,Fe)_5Si_8O_{22}(OH)_2$

Color: White-green
Streak: White
Luster: Vitreous   *Cleavage:* Good prismatic
Hardness: 5.0—6.0   *S.G.:* 2.98—3.46

CRYSTALLOGRAPHY: Monoclinic. Long prismatic crystals in fibrous aggregates. Also coarse or fine granular.

DESCRIPTION: Tremolite and actinolite are found in metamorphic rocks of several types. These two minerals are considered as one; when the Fe content is greater than 2 percent, it is called actinolite rather than tremolite. They are apt to be associated with garnet, wollastonite, epidote, diopside, and even calcite. Tremolite often alters to talc. Actinolite commonly forms by the alteration of pyroxenes, and actinolite itself often alters to serpentine, epidote, or chlorite. The crystals in long slender radiating groups, good cleavage, and color are good distinguishing properties. White tremolite may resemble wollastonite, but the cleavage angle in tremolite (56°) and in wollastonite (84°) may assist in distinguishing them. Tough compact massive tremolite or actinolite is called nephrite and is one of the jade minerals. The other jade mineral is jadeite, a pyroxene. Nephrite jade comes from Turkestan and also from Wyoming.

Fig. M-88: Fibrous aggregates of long, prismatic white crystals of tremolite from South Australia. Note vitreous to silky luster.

NAME: **Hornblende**

FORMULA: $NaCa_2(Mg,Fe,Al)_5(Si,Al)_8O_{22}(OH)_2$

Color: Dark green to black
Streak: White to grayish-white
Luster: Vitreous   Cleavage: Good prismatic
Hardness: 6.0   S.G.: 3.0—3.4

CRYSTALLOGRAPHY: Monoclinic. Coarse columnar crystals or fine granular masses. Also bladed, fibrous, compact.

DESCRIPTION: This very common mineral is found in igneous and metamorphic rocks. It is usually associated with quartz, feldspar, biotite, muscovite, pyroxenes, chlorite, epidote, and calcite. Hornblende sometimes forms by the alteration of pyroxenes, and alters itself into chlorite, biotite, and epidote. The cleavage, the color, and the crystal habit are good distinguishing properties. Hornblende is distinguished from dark pyroxenes by cleavage angles (56° in hornblende and 87° in pyroxenes). Similar appearing black tourmaline lacks cleavage. As a member of the amphibole group, hornblende is the most common one and is distinguished from the others most easily by its black or greenish-black color. Hornblende varies quite a lot in composition, and subspecies are recognized. Black hornblende is found in Lanark County, Ontario. Light-colored hornblende occurs in Argentueil County, Quebec, and at Franklin, New Jersey.

Fig. M-89: Elongated prismatic crystal of black hornblende from Colorado. Very difficult to distinguish from augite.

NAME: **Enstatite-hypersthene**

FORMULA: $Mg_2Si_2O_6$-$(Mg,Fe)_2Si_2O_6$

Color: White to black
Streak: White to gray
Luster: Vitreous   *Cleavage:* Good prismatic
Hardness: 6.0   *S.G.:* 3.2—3.9

CRYSTALLOGRAPHY: Orthorhombic. Usually in massive or fibrous-like forms. Also lamellar.

DESCRIPTION: Enstatite and hypersthene are both essentially the same mineral. If iron exceeds 13 percent, the name hypersthene is used. They occur chiefly in basic igneous rocks and also in metamorphic rocks. These minerals become associated with olivine, augite, and garnet, as well as with micas and feldspars. The color, the luster, and the cleavage are useful identifying properties. Similar appearing hornblende has 56° cleavage, while enstatite has 87° cleavage. Black varieties are hard to distinguish from augite, and optical tests must be made to tell them apart. These minerals are fairly common in the rocks and are found in some of the meteorites as well. In the United States, enstatite is found near Brewster, New York, near Baltimore, Maryland, and near Webster, North Carolina. Hypersthene is found near Cortland, New York. Hypersthene in Labrador was once cut into gemstones called *paulite*.

Fig. M-90: Massive gray enstatite from Norway showing vitreous to silky luster, with tendency to show prismatic crystals.

NAME: **Augite**

FORMULA: $Ca(Mg,Fe,Al)(Al,Si)_2O_6$

Color: Green to black
Streak: White to gray
Luster: Vitreous
Hardness: 6.0

Cleavage: Good prismatic
S.G.: 3.25—3.55

CRYSTALLOGRAPHY: Monoclinic. Usually coarse granular, with short prismatic crystals. Also lamellar.

DESCRIPTION: Augite is the most abundant ferromagnesian mineral in igneous rocks, particularly in basic types. It is commonly associated with olivine, feldspar, magnetite, and hornblende. Augite readily alters to epidote, hornblende, chlorite, serpentine, or talc. The crystal form, the cleavage, and the color are the most diagnostic properties. Augite appears similar to dark hypersthene, and optical tests are needed for distinguishing them. Hornblende looks very much like augite, and the cleavage is the only reliable diagnostic feature: 87° in augite, 56° in hornblende. Augite is the most abundant of the pyroxenes and is important in many kinds of rocks. However, well-formed crystals are not too common, the augite usually occurring as irregular masses in the rocks. Occasionally clear varieties are used for gemstones. Very well formed crystals of augite come from the lavas around Mt. Vesuvius and at Trentino, Italy. In the United States augite occurs in good crystals in Colorado and Oregon.

Fig. M-91: Massive black augite from Kragero, Norway. Augite is very hard to distinguish from hornblende.

NAME: **Spodumene**

FORMULA: $LiAlSi_2O_6$

Color: White to grayish-white
Streak: White
Luster: Vitreous
Hardness: 6.5

Cleavage: Perfect prismatic
S.G.: 3.1—3.2

CRYSTALLOGRAPHY: Monoclinic. Large, elongated, vertically striated, flattened crystals. Also in cleavable masses.

DESCRIPTION: Spodumene occurs chiefly in pegmatite dikes as quite large crystals. Associated minerals include beryl, quartz, lepidolite, amblygonite, tourmaline, muscovite, and feldspar. Spodumene decomposes to muscovite, clay, or albite. The good prismatic cleavage, the large cleavable crystals, and the color serve to distinguish spodumene. Plagioclase feldspar may resemble spodumene, but spodumene often has a wood-like surface alteration on the faces. Spodumene often forms in crystals measurable in several feet in length. Crystals up to 10 feet and occasionally up to 40 feet have been reported. This mineral is the chief ore of the metal lithium, but there are also gem varieties: kunzite (pink) and hiddenite (green). Spodumene is a relatively rare mineral, being found nearly exclusively in lithium-rich pegmatites. Giant crystals have been found in the pegmatites of South Dakota. It is found also in North Carolina, Colorado, and Maine.

Fig. M-92: Coarse crystal of buff colored spodumene from Keystone, South Dakota, showing striated prisms with steep terminations.

NAME: **Wollastonite**

FORMULA: CaSiO$_3$

Color: White
Streak: White
Luster: Vitreous
Hardness: 5.0

Cleavage: Good basal pinacoidal

S.G.: 2.9

CRYSTALLOGRAPHY: Triclinic. Usually in tabular cleavable masses. Also fibrous or compact.

DESCRIPTION: Wollastonite occurs in contact metamorphic limestones and also in other high-grade metamorphic rocks. Wollastonite commonly is associated with garnet, tremolite, epidote, calcite, diopside, idocrase, and calcium feldspars. The very good cleavage, the white color, and the tabular or fibrous habit can be used for identification. White tremolite may resemble wollastonite, but the 84° cleavage of wollastonite differs considerably from the 56° cleavage in tremolite. Fibrous gypsum may resemble wollastonite, but gypsum is much softer. Aragonite may appear similar to wollastonite, but aragonite is softer and effervesces readily in hydrochloric acid. In some localities where it is plentiful, wollastonite is mined for use in the manufacture of tile and in paint pigments. Such large masses are at Willsboro, New York, and also in California. Some good crystals of wollastonite are found at Franklin, New Jersey. It is found also at Chiapas, Mexico.

Fig. M-93: White cleavable fibrous masses of wollastonite showing elongated crystals and vitreous luster.

NAME: **Rhodonite**
FORMULA: $MnSiO_3$

Color: Pink to rose-red
Streak: White
Luster: Vitreous   *Cleavage:* Good prismatic
Hardness: 6.0   *S.G.:* 3.5—3.7

CRYSTALLOGRAPHY: Triclinic. Usually massive or coarse granular. Often coated with black manganese oxides.

DESCRIPTION: Rhodonite is found in hydrothermal deposits and as metamorphosed manganese ores. Rhodonite is usually associated with franklinite, zincite, and willemite at Franklin, New Jersey. Rhodonite alters quite easily to black manganese oxides, especially pyrolusite (MnO_2). The good rose-red color, the prismatic cleavage, and the hardness of 6.0 can be distinguishing features. Rhodochrosite resembles rhodonite, but rhodochrosite is softer, has good rhombohedral cleavage, and effervesces in hydrochloric acid. Rhodonite is insoluble in this acid. Large masses of rhodonite are found in the Ural Mountains of Russia where it is used as ornamental stone. It is also found at Broken Hill, Australia. Other localities where rhodonite is mined are Sweden and Brazil. Very few localities exist in the United States where rhodonite is found in large masses. It usually occurs as scattered crystals in the rocks. Sometimes rhodonite is used as a gemstone.

Fig. M-94: Massive to very finely-granular pink rhodonite with black alteration zone. Specimen from Plainfield, Massachusetts.

NAME: **Kaolinite**

FORMULA: $Al_4Si_4O_{10}(OH)_8$

Color: White
Streak: White
Luster: Dull or earthy   *Cleavage:* Good basal pinacoid
Hardness: 2.0   *S.G.:* 2.6

CRYSTALLOGRAPHY: Triclinic. In clay-like masses, compact or friable. Crystals of large size are rare.

DESCRIPTION: Kaolinite is a very widespread mineral, as it forms chiefly by alteration of aluminum silicates and often is present as a soil constituent. Kaolinite is usually associated with the minerals which are undergoing the decomposition, particularly feldspars. Its clay-like character is its most notable diagnostic feature, but kaolinite is hard to distinguish from other clay minerals. Detailed laboratory tests such as X-ray analysis, electron microscopy, and differential thermal analysis may be required. Because clays can be molded when wet, high-grade clays, mostly kaolinite, are used for refractory products and ceramics. White kaolinite is also extensively used as a filler in paper; about one-third the weight of good-grade paper is kaolinite. It is used abundantly in rubber, in paint, and in plastics. One interesting refractory use for kaolinite is in the brick which makes up the rocket launch pads at Cape Kennedy and in California.

Fig. M-95: Compact clay-like mass of white kaolinite showing dull luster.

NAME: **Serpentine**

FORMULA: $Mg_6Si_4O_{10}(OH)_8$

Color: Green, but variable
Streak: White
Luster: Waxy, greasy, or silky   *Cleavage:* None
Hardness: 4.0—6.0   *S.G.:* 2.2—2.6

CRYSTALLOGRAPHY: Monoclinic. Platy or fibrous masses, or massive. Crystals of serpentine are unknown.

DESCRIPTION: Serpentine usually forms as an alteration product of magnesium silicates such as pyroxenes, olivine, and amphiboles under low-grade metamorphic conditions. It is thus found chiefly in metamorphic rocks, and is usually associated with magnesite, chromite, garnet, spinel, and magnetite. Oftentimes serpentine makes up large masses of rock. Its green color, its waxy to greasy luster, and its frequent fibrous nature are its best distinguishing properties. However, massive serpentine may resemble nephrite jade, but jade is harder. Massive chlorite resembles serpentine, but chlorite is much softer. Fibrous green actinolite very much resembles fibrous serpentine, and X-ray study is recommended for distinguishing them. Positive calcium and iron chemical tests would indicate actinolite. Massive serpentine (verde antique) is often used as ornamental stone. Fibrous serpentine (chrysotile) is used for asbestos. Large deposits are mined in Quebec, Canada.

Fig. M-96: Fibrous serpentine from Thetford, Quebec, showing delicate nature of asbestos fibers.

NAME: **Talc**

FORMULA: $Mg_3Si_4O_{10}(OH)_2$

Color: Pale green, pink, white, or gray
Streak: White
Luster: Pearly or greasy   *Cleavage:* Good basal pinacoidal
Hardness: 1.0   *S.G.:* 2.82

CRYSTALLOGRAPHY: Monoclinic. Foliated masses or fine-grained aggregates. Tabular crystals rare.

DESCRIPTION: Talc is a secondary mineral formed from other magnesium silicates, such as olivine, amphiboles, and pyroxenes. It is found chiefly in metamorphic rocks, and thus is associated with chlorite and serpentine. The most diagnostic properties are the low hardness, the foliated nature, the cleavage, and the greasy feel. However, pyrophyllite may resemble talc, but pyrophyllite usually occurs in radiating aggregates; and a positive test for aluminum (Al) confirms pyrophyllite. Chlorite may resemble light green talc, but chlorite is harder. Serpentine also resembles green talc, but serpentine is considerably harder. Large masses of talc are mined in New York and other eastern states. It is produced in California, Texas, and Georgia, also. In powdered form, talc is used in paint and ceramics, and still in talcum powder. In slabs, talc is used for laboratory table tops, sanitary appliances, and electric switchboards, because talc is acid resistant, fire resistant, and has low electrical conductivity.

Fig. M-97: Green talc from Holly Springs, Georgia, showing fine pearly luster and foliated structure.

NAME: **Montmorillonite**

FORMULA: $Al_2Si_4O_{10}(OH)_2 \cdot nH_2O$

Color: Gray or greenish-gray
Streak: White
Luster: Greasy or dull   *Cleavage:* Unknown; crystals not distinguishable
Hardness: 2.0—2.5   *S.G.:* 2.0—2.7

CRYSTALLOGRAPHY: Monoclinic. Always in earthy masses. Crystals not seen even under high magnification.

DESCRIPTION: Montmorillonite, a clay mineral, forms by the alteration of volcanic ash beds. Clay beds so formed are referred to as bentonite. The clay minerals are difficult to distinguish from one another, but montmorillonite can sometimes be distinguished from the others by its color and by its swelling properties in water. But X-ray analyses are needed, along with electron microscopy, and differential thermal analysis, to differentiate the clay minerals with certainty. Bentonite is a sodium-rich clay and can absorb very large amounts of water, swelling considerably in so doing. This swelling property allows bentonite to be used as circulation mud in drill holes. It lubricates the bit, plasters openings in the walls of the hole, and helps lift rock cuttings up from the hole. Certain types of montmorillonite act as absorbents of impurities from oils, fats, and waxes. Commercial montmorillonite is mined in Wyoming, Texas, and Mississippi.

Fig. M-98: Yellowish-white montmorillonite from Clay Spur, Wyoming, showing the dull luster so common among the clay minerals.

NAME: **Vermiculite**
FORMULA: $Mg_3Si_4O_{10}(OH)_2 \cdot nH_2O$

Color: Yellow to brown
Streak: White
Luster: Pearly   *Cleavage:* Good basal pinacoidal
Hardness: 1.5   *S.G.:* 2.4

CRYSTALLOGRAPHY: Monoclinic. Coarse crystal aggregates with biotite. Crystals may appear crackled.

DESCRIPTION: Vermiculite forms as an alteration mineral from biotite and phlogopite. Its good cleavage, its association with biotite, and its expansion when heated are good usable diagnostic properties. Vermiculite closely resembles biotite, and must be heated to detect characteristic expansion. It has commercial value as a result of this expansion property. Heating vermiculite for a few seconds at 900–1000°C causes expansion up to ten to eighteen times as steam is generated between the mica-like layers. Vermiculite is chemically inert, lightweight, and fairly heat-resistant. So it finds utility as loose-fill insulation for walls and ceilings, in refrigerators, etc., and it is also used as packing material. Much domestic vermiculite comes from Libby, Montana, and from Macon, North Carolina. The United States is the world's leading producer, some 200,000 tons being mined each year.

Fig. M-99: Brown vermiculite, a mica-like mineral, showing good basal cleavage and vitreous luster, from Libby, Montana.

NAME: **Phlogopite**

FORMULA: $KMg_3(AlSi_3O_{10})(OH)_2$

Color: Pale yellow to brown
Streak: White
Luster: Vitreous   *Cleavage:* Perfect basal pinacoidal
Hardness: 2.5   *S.G.:* 2.8—3.0

CRYSTALLOGRAPHY: Monoclinic. Coarse, platy crystal aggregates. Often in foliated masses.

DESCRIPTION: Phlogopite occurs in ultrabasic igneous rocks, in some pegmatites, and in some metamorphic magnesian limestones. It is commonly associated with pyroxenes, amphiboles, and serpentines, and also with dolomite and calcite. The yellow-brown color and micaceous cleavage usually are enough to identify phlogopite. It is distinguished from similar looking biotite by being lighter in color, but this is oftentimes hard to tell. Vermiculite very much resembles phlogopite, and it must be heated for a few seconds to see if it swells greatly. If it expands, it is vermiculite. Phlogopite is found in large amounts in Finland, Sweden, Switzerland, Ceylon, and Madagascar. It is also found in Ontario and in New York State. Phlogopite is used chiefly as an electrical insulator. It has good wearing properties and is often preferred over other types. Phlogopite has good heat resistance, withstanding temperatures up to 1000°C. Synthetic phlogopite is now available, but it is still quite expensive.

Fig. M-100: Phlogopite from Norway showing perfect basal cleavage. Bronzy-black color is not apparent.

NAME: **Muscovite**

FORMULA: $KAl_3(AlSi_3O_{10})(OH)_2$

Color: Colorless to pale green
Streak: White
Luster: Vitreous   *Cleavage:* Perfect basal pinacoidal
Hardness: 2.5   *S.G.:* 2.8—2.9

CRYSTALLOGRAPHY: Monoclinic. Tabular crystals in coarse or fine aggregates. Often in foliated masses.

DESCRIPTION: Muscovite is a common mineral in igneous rocks, in metamorphic rocks, and even in sedimentary rocks. It is commonly associated with orthoclase, microcline, biotite, topaz, tourmaline, beryl, spodumene, quartz, and hornblende. Minute, scaly, fibrous-like muscovite is called "sericite" and is commonly formed by the alteration of feldspar. Muscovite also forms by the alteration of topaz, kyanite, and andalusite. But it frequently is a primary mineral in igneous rocks of granitic types. The perfect cleavage, the clear color, and the flexible and elastic sheet crystals are enough to identify muscovite. Phlogopite may resemble muscovite, but phlogopite is usually some tone of brown color. Lepidolite also resembles muscovite if lepidolite is colorless instead of lilac. Margarite, another mica similar to lepidolite, also may resemble muscovite, but it is usually associated with corundum or emery. Sheet muscovite is used in electrical capacitors and is commonly called *white mica.*

Fig. M-101: Colorless thin sheets of muscovite from Royal Gorge, Colorado, revealing perfect basal cleavage.

NAME: **Biotite**
FORMULA: $K(Mg,Fe)_3(AlSi_3O_{10})(OH)_2$

Color: Dark green, brown, or black
Streak: White to gray
Luster: Vitreous
Hardness: 2.5

Cleavage: Perfect basal pinacoidal
S.G.: 2.8—3.2

CRYSTALLOGRAPHY: Monoclinic. Irregular foliated or scaly masses. Good crystals rare.

DESCRIPTION: Biotite, called *black mica*, is found in igneous rocks of intermediate to acidic composition. It is also found frequently in metamorphic rocks such as schists, gneisses, or hornfels. Biotite is commonly associated with orthoclase, amphiboles, quartz, muscovite, beryl, spodumene, tourmaline, garnet, and even with pyroxenes. Biotite forms by the alteration of amphiboles and pyroxenes, and in turn biotite may alter to chlorite, vermiculite, or epidote. The dark color, the perfect cleavage, and its associates help to identify biotite. Brown phlogopite resembles biotite, but phlogopite is usually bronzy-brown in color. Vermiculite may look very similar to biotite, but heating the specimen for a few seconds can help. If it expands a great deal it is vermiculite. Cleavable green chlorite may resemble some green biotite, but chlorite's flakes are nonelastic. Biotite is of little economic importance.

Fig. M-102: Foliated mass of biotite from Bancroft, Ontario, showing perfect basal cleavage.

NAME: **Chlorite**

FORMULA: $(Mg,Fe,Al)_6(Al,Si)_4O_{10}(OH)_8$

Color: Green
Streak: White to pale green
Luster: Vitreous   *Cleavage:* Perfect basal pinacoidal
Hardness: 2.5   *S.G.:* 2.6—3.3

CRYSTALLOGRAPHY: Monoclinic. Usually in foliated or scaly masses. Often compact. Crystals are rare.

DESCRIPTION: Chlorite is found chiefly in metamorphic rocks such as schists. It is also found in igneous rocks as an alteration product derived from biotite. Chlorite usually becomes associated with garnet, serpentine, amphiboles, pyroxenes, chromite, magnetite, epidote, and apatite. Chlorite often forms by the alteration of garnet, biotite, pyroxene, and amphibole. The chlorite group consists of prochlorite, clinochlore, and penninite, and all are very similar. The micaceous cleavage, the common green color, and the nonelastic flakes are good identifying properties. Chlorite often resembles the micas, but chlorite is neither brittle nor elastic. Massive serpentine may resemble massive chlorite, but serpentine is harder. Green talc resembles chlorite, but talc is much softer. Greenish-colored pyrophyllite may resemble clorite, but pyrophyllite is softer and has a more pearly to waxy luster. Good cleavage plates of chlorite are found in Renfrew County, Ontario, and in Lancaster County, Pennsylvania.

Fig. M-103: Chlorite from Chester, Vermont, in foliated massive form. Note the vitreous luster of this green mica-like mineral.

NAME: **Quartz**
FORMULA: SiO_2

Color: Colorless to white, but often tinted
Streak: White
Luster: Vitreous   *Cleavage:* None
Hardness: 7.0   *S.G.:* 2.65

CRYSTALLOGRAPHY: Hexagonal. Fine or coarse prismatic crystal aggregates; massive. Also microcrystalline.

DESCRIPTION: Quartz is very abundant and is found in igneous, sedimentary, and metamorphic rocks. It is very stable and very resistant. Crystal form, high hardness, lack of cleavage, vitreous luster, and conchoidal fracture are good diagnostic features. Milky quartz is distinguished from white beryl by being softer. Amethyst quartz may be mistaken for fluorite, but octahedral cleavage in fluorite is unmistakable. Clear calcite is much softer than quartz. Cordierite resembles quartz, but cordierite is blue. Scheelite is heavier and softer than quartz. Similar appearing white feldspar has good cleavage. Nepheline looks like quartz but has a greasy luster and is softer. Varieties of quartz, some crystalline, some cryptocrystalline, are amethyst, rose quartz, smoky quartz, milky quartz, chalcedony, chrysoprase, agate, onyx, heliotrope, flint, chert, jasper, and prase. Some are used for gemstones. Clear quartz crystals are often used in optical instruments and in electronic tubes.

Fig. M-104: Well-shaped colorless hexagonal crystals of quartz from Arkansas showing vitreous luster and prismatic form.

NAME: **Opal**

FORMULA: $SiO_2 \cdot nH_2O$

Color: Colorless to white, gray, brown
Streak: White
Luster: Vitreous to waxy   *Cleavage:* None
Hardness: 5.5—6.5   *S.G.:* 2.0—2.2

CRYSTALLOGRAPHY: Not crystalline. Usually in rounded, reniform forms, made up of fine granular masses.

DESCRIPTION: Opal is deposited at low temperatures from waters containing silica which move through cavities and fissures. Opal is associated with many environments and does not become associated with any particular minerals. It is deposited frequently by hot springs such as at Yellowstone National Park. The form of opal, the mode of occurrence, and the low density are its best identifying properties. Chalcedony may resemble opal, but it is harder and has a duller luster. Milky quartz may also resemble opal, but quartz is slightly harder. There are many varieties of opal such as precious opal (with internal colors), fire opal (orange-red colors), hydrophane (white), wood opal (in petrified wood), geyserite (hot springs deposits), and diatomite (opal secreted by microorganisms). Opal with good internal play of colors is used as gemstone, some being found in Nevada. Diatomite in very thick beds occurs near Lompoc, California, and is used as an abrasive, as filtration medium, and as insulation.

Fig. M-105: Reniform (rounded) mass of white opal showing greasy to vitreous luster, conchoidal fracture, and porcelain-like surface.

NAME: **Orthoclase**
FORMULA: $KAlSi_3O_8$

Color: White to pink
Streak: White
Luster: Vitreous   *Cleavage:* Good basal pinacoidal and prismatic
Hardness: 6.0   *S.G.:* 2.56

CRYSTALLOGRAPHY: Monoclinic. Coarsely crystalline masses or irregular crystals. Also fine-grained to massive.

DESCRIPTION: A very common mineral, orthoclase occurs chiefly in igneous rocks of granitic types, also in pegmatites, and in metamorphic rocks, as well as in sedimentary rocks. Orthoclase is commonly associated with muscovite, quartz, and biotite. This feldspar usually alters to sericite or to kaolinite. The color, the hardness, and the good cleavage are usable distinguishing features. Quartz may resemble orthoclase, but quartz lacks cleavage. Plagioclase may appear similar to orthoclase, but plagioclase nearly always exhibits twinning striations. Amblygonite looks a lot like orthoclase, but cleavage angles are different: 90° for orthoclase, 75° for amblygonite. Amblygonite has a lower luster also. Some forms of spodumene may appear similar to orthoclase, but spodumene usually has a wood-like surface alteration on the crystals, and the crystal faces are usually striated and steeply terminated. Orthoclase is hard to tell from microcline, but association is helpful: orthoclase in igneous rocks, microcline in pegmatites and hydrothermal veins.

Fig. M-106: Short prismatic grayish-white crystals of orthoclase from Climax, Colorado, showing common twinning.

NAME: **Microcline**
FORMULA: $KAlSi_3O_8$

Color: White to pink to green
Streak: White
Luster: Vitreous   *Cleavage:* Good basal pinacoidal and prismatic
Hardness: 6.0   *S.G.:* 2.56

CRYSTALLOGRAPHY: Triclinic. Coarsely crystalline masses or in irregular crystals, often revealing graphic intergrowths.

DESCRIPTION: Microcline is found chiefly in pegmatites and hydrothermal veins. Much that actually is microcline is mistakenly called orthoclase. Microcline is commonly associated with quartz, orthoclase, muscovite, beryl, tourmaline, spodumene, and many sulfides such as galena and sphalerite. In pegmatites, microcline is often intimately intergrown with quartz to form "graphic granite" or with albite to form "perthite." The color, the intergrown texture, and the large crystal sizes are identifying properties. Microcline is very difficult to tell from orthoclase without detailed optical examination. But green microcline is distinctive. Plagioclase feldspars appear similar to microcline, but plagioclase is usually striated. Microcline alters rather easily to muscovite and kaolinite, much the same as orthoclase does. Microcline is found in many pegmatites often as quite large crystals. It is mined at Amelia Court House, Virginia, in green masses called *amazon stone,* which is used for ornamental stone.

Fig. M-107: Large crystal of pale green microcline showing good two-directional cleavage and vitreous luster.

NAME: **Plagioclase feldspar**
FORMULA: $(Ca,Na)(Al,Si)AlSi_2O_8$ (ideal)

Color: White or gray
Streak: White
Luster: Vitreous   *Cleavage:* Perfect basal pinacoidal and prismatic
Hardness: 6.0   *S.G.:* 2.62—2.76

CRYSTALLOGRAPHY: Triclinic. Usually in twinned cleavable masses. Also as irregular grains in igneous rocks.

DESCRIPTION: The plagioclase feldspars are a continuous series from albite $(NaAlSi_3O_8)$ to anorthite $(CaAl_2Si_2O_8)$. Intermediate members are oligoclase, andesine, labradorite, and bytownite, with calcium substituting for sodium. The plagioclase feldspars are found chiefly in igneous rocks, but they are present also in sedimentary and metamorphic rocks. They are the most abundant rock-forming minerals and thus are associated with a great many other minerals. Very fine twinning in these minerals is displayed by fine striations on cleavage surfaces. This feature is very diagnostic and distinguishes plagioclase from orthoclase and microcline, which would appear similar otherwise. Amblygonite may resemble plagioclase but has a duller luster. Certain masses of spodumene may appear similar to plagioclase, but spodumene commonly has a peculiar wood-like alteration on the surface. Massive white quartz may look like white plagioclase, but quartz lacks cleavage and is harder.

Fig. M-108: White oligoclase (plagioclase feldspar) from Mitchell County, North Carolina, showing vitreous luster and good two-directional cleavage.

NAME: **Scapolite**

FORMULA: $(Na,Ca)_4[(Al,Si)_4O_8]_3(Cl,CO_3)$

Color: White to gray to pale brown
Streak: White
Luster: Vitreous to pearly   *Cleavage:* Good basal pinacoidal and prismatic
Hardness: 6.0   *S.G.:* 2.55—2.77

CRYSTALLOGRAPHY: Tetragonal. Coarse crystals in aggregates, or compact massive. Crystals may appear partially fused.

DESCRIPTION: Scapolite occurs chiefly in metamorphosed limestones and in schists and gneisses. It usually is associated with garnet, biotite, apatite, zircon, diopside, and amphiboles. In a few unusual pegmatites, scapolites may occur as gem-quality material. Scapolite commonly alters to mica, kaolinite, epidote, talc, zeolites, or feldspar. Scapolite has good cleavage, good crystal form, and often appears fibrous; these are its distinguishing features. Scapolite may resemble the feldspars, but it appears more fibrous and lacks the same cleavage of feldspar. Some masses of spodumene may appear somewhat similar to scapolite, but scapolite looks more woody. Large crystals of scapolite have been found in St. Lawrence County, New York. Yellow-colored scapolite of gem quality is found in Madagascar, Brazil, and Switzerland. Some scapolite is also found in numerous localities in Canada, and also in Burma.

Fig. M-109: Scapolite in massive form showing white to gray color, vitreous to dull luster.

NAME: **Nepheline**

FORMULA: $NaAlSiO_4$

Color: White to gray
Streak: White
Luster: Vitreous to greasy   *Cleavage:* Poor prismatic
Hardness: 6.0   *S.G.:* 2.60—2.63

CRYSTALLOGRAPHY: Hexagonal. As scattered shapeless grains in igneous rocks. Usually in massive form.

DESCRIPTION: Nepheline occurs in both intrusive and extrusive igneous rocks and is also associated with pegmatites of certain types. Nepheline becomes associated with such minerals as sodalite, cancrinite, biotite, feldspar, corundum, and zircon. Nepheline easily alters to sericite, zeolites, and kaolinite. The greasy luster is its most distinguishing property. Quartz resembles nepheline, but quartz has a vitreous luster and is harder. Nepheline is distinguished from feldspar by having poor cleavage. Similar appearing apatite is softer. Some masses of cordierite appear similar to nepheline, but cordierite is harder and usually has a blue color. Nepheline in large masses is found in Russia where it is used in ceramics, in textiles, and other industries. It is also found in Ontario, Canada. Nepheline is also found at Magnet Cove, Arkansas, and in New Jersey and Maine. Nepheline occurring as rather small, glassy crystals is found in the volcanic rocks around Mt. Vesuvius.

Fig. M-110: Massive nepheline, light gray in color, showing greasy luster.

NAME: **Heulandite**

FORMULA: $CaAl_2Si_7O_{18} \cdot 6H_2O$

Color: White
Streak: White
Luster: Vitreous   *Cleavage:* Good prismatic
Hardness: 4.0   *S.G.:* 2.2

CRYSTALLOGRAPHY: Monoclinic. Elongated tabular crystals in coarse aggregates. Crystals coffin-shaped.

DESCRIPTION: This mineral is usually found in cavities in basaltic igneous rocks, formed secondarily from other silicates. Heulandite is most commonly associated with calcite and other zeolites, such as stilbite and chabazite, and also with prehnite, datolite, and quartz. The crystal form, cleavage, and vitreous luster serve to distinguish heulandite. Heulandite is rather difficult to distinguish from other members of the zeolite groups, but the coffin-shaped crystals are diagnostic. Crystal forms of zeolites are usually the most helpful features in identifying them: chabazite is pseudocubic, natrolite in radiating needles, stilbite in sheaf-like crystals, analcite in trapezohedral crystals. Very good crystals are found at Paterson, New Jersey. Heulandite has also been found in Nova Scotia, Germany, Austria, India, Iceland, and Brazil. Heulandite and other zeolites are used in the filtration of hard water and also in the refining of oil.

Fig. M-111: Crystal aggregates of white heulandite from Paterson, New Jersey, showing vitreous to pearly luster and translucency.

NAME: **Stilbite**

FORMULA: $CaAl_2Si_7O_{18} \cdot 7H_2O$

Color: White to cream
Streak: White
Luster: Vitreous   *Cleavage:* Good side pinacoidal
Hardness: 4.0   *S.G.:* 2.2

CRYSTALLOGRAPHY: Monoclinic. In coarse individual crystals. Crystals in sheaf-like bundles.

DESCRIPTION: Stilbite is usually found in cavities in basalts, formed by secondary processes. Stilbite is commonly associated with heulandite and chabazite, as well as with apophyllite, prehnite, and calcite. The sheaf-like form of crystals, the luster, and the good cleavage are usable diagnostic properties. Crystal form is used to distinguish stilbite from other very similar appearing zeolites: heulandite in coffin-shaped crystals, chabazite in pseudocubic crystals, natrolite in radiating needles, and analcite in trapezohedral crystals. Stilbite is found rather abundantly at Paterson, New Jersey, with other zeolites. It is found also in India, Scotland, Iceland, and Nova Scotia. Like other zeolites, stilbite is used to filter hard water. The zeolites are minerals with water molecules, and they have a rather open atomic structure. Water molecules are easily detached and can move about in the open structure without disrupting the structure.

Fig. M-112: Tabular crystals of white stilbite from Brazil showing good cleavage, vitreous luster, and translucency.

NAME: **Chabazite**

FORMULA: $CaAl_2Si_4O_{12} \cdot 6H_2O$

Color: Colorless to white
Streak: White
Luster: Vitreous   *Cleavage:* Poor rhombohedral
Hardness: 4.0   *S.G.:* 2.05—2.15

CRYSTALLOGRAPHY: Hexagonal. In coarse crystalline aggregates. Rhombohedral crystals appear pseudocubic.

DESCRIPTION: Chabazite, a mineral of secondary origin, is usually found in cavities in basalts and other volcanic rocks. Chabazite is usually associated with other zeolites such as heulandite and stilbite, but also with calcite, apophyllite, prehnite, and pectolite. Good rhombohedral crystals which appear pseudocubic, the good vitreous luster, and the color identify chabazite. Chabazite is distinguished from other similar appearing zeolites by crystal form: heulandite in coffin-shaped crystals, natrolite in radiating needles, stilbite in sheaf-like bundles, and analcite in trapezohedral crystals. Some calcite may resemble chabazite, but calcite is softer and effervesces in hydrochloric acid. The same may be said for magnesite when in good crystals. Apophyllite may resemble chabazite, but chabazite lacks pearly cleavage surfaces. Good crystals of chabazite are found with other zeolites at Paterson, New Jersey. It is also found in Ireland, Italy, and Germany.

Fig. M-113: White chabazite crystals showing rhombohedral shapes with nearly cubic angles. Note vitreous luster of this specimen from Nova Scotia.

NAME: **Analcite**

FORMULA: $NaAlSi_2O_6 \cdot H_2O$

Color: Colorless to white
Streak: White
Luster: Vitreous
Hardness: 5.0

Cleavage: None
S.G.: 2.3

CRYSTALLOGRAPHY: Isometric. In coarse crystalline aggregates. Nearly always in trapezohedral crystals. Sometimes massive.

DESCRIPTION: Analcite (analcime) is a secondary mineral formed in cavities in basaltic igneous rocks. It is sometimes present in sedimentary rocks. Analcite is commonly associated with other zeolites, particularly stilbite, heulandite, and chabazite, as well as with calcite and prehnite. The good vitreous luster, the trapezohedral crystal form, and the lack of cleavage are usable diagnostic features. Analcite is hard to tell from other members of the zeolite group, and crystal forms are most helpful: heulandite in coffin-shaped crystals, chabazite in pseudocubic crystals, natrolite in radiating needles, and stilbite in sheaf-like bundles. Leucite may resemble analcite because it has trapezohedral form and is white, but leucite is usually embedded in rock rather than in cavities, and leucite has a duller luster. Pale garnets may resemble analcite, but garnets are much harder. Analcite is found with other zeolites at Paterson, New Jersey, in the Michigan copper deposits, and in Colorado.

Fig. M-114: White crystalline aggregates of analcite from Italy showing vitreous luster and similarity to other zeolites.

Chapter Four

Nature of Igneous Rocks

The solid part of the earth is called the *lithosphere*, a term that means a "ball of rock." The "ball of rock" consists of a number of different kinds of rocks which form under a great many differing circumstances. The study of the nature of rocks and their manners of formation is called *petrology*. Petrology deals with the way rocks are distributed in the Earth's crust, how they form, and how they are related in the historical sense. As we shall see later, rocks do not have definite compositions which can be written in a formula. Usually even the constituent minerals or mineraloids vary in composition.

Many of the rocks formed in the crust of the Earth have developed below the surface, or in offshore areas, or deep in areas where mountains are under construction; hence, their formation is not observable. However, the processes of their formation can be estimated fairly accurately by long and careful field study of associated rocks and also by various laboratory studies. Many of the rocks formed at some depth below the surface have become exposed at the surface after a long period of geologic time during which they are uplifted and eroded.

Based on their manner of formation, rocks are divisible into three very broad groups:

(1) Igneous rocks, derived from the cooling of a hot molten mass of rock material called *magma*.

(2) Sedimentary rocks, derived from the products of weathering and of organic materials.

(3) Metamorphic rocks, derived from previous igneous and sedimentary rocks by the action of heat, pressure, and chemical solutions.

Igneous Processes and the Formation of Igneous Rocks

A large number of rocks of the Earth's crust are of igneous origin, that is, they were derived by the cooling of a hot molten magma. This cooling has taken place either at depth or at the surface, depending on where magma had become disposed while it was in the hot molten condition.

Nature of Magma

Once the magma has been formed as a result of heating and melting, it occupies a certain amount of space called a *magma chamber*. In the magma chamber, the magma behaves essentially as a solution. It tends to become uniform in composition by mixing of ions; but the rather high viscosity of the magma makes this action proceed very slowly. The temperatures of different magmas are estimated at 500—1300°C. However, little evidence exists that magmas below the surface are hotter than about 1100°C. Early in its existence, a magma may assimilate (melt) or "stope" the walls with which it is in contact, and it may be forcibly intruded (moved into surrounding rocks). If intruded at considerable depth, the volatile constituents (vapors) do not have any opportunity to escape because they remain trapped. But if the intrusion is shallow, then vaporous materials may move to areas of low pressure and may escape from the magma to form special rocks. They may also react with the adjacent wall rock or may even be vented into the atomosphere. In a broad sense, magma consists predominantly of only nine of the common chemical elements: oxygen, silicon, aluminum, iron, calcium, sodium, potassium, magnesium, and titanium. These nine elements practically make up over 99 percent of the total magma. The remaining 83 elements, including very rare ones, have been detected in igneous rocks in varying small quantities in one case or another, and hence they are present in many magmas. These remaining 83 elements thus make up the remaining 1 percent of magma if considered in broad aspects. Some of these rarer elements are phosphorus, hydrogen, manganese, sulfur, barium, chlorine, chromium, carbon, lead, copper, zinc, and fluorine. Some of these, such as hydrogen, fluorine, chlorine, and sulfur are volatile (i.e., in vapor state) while in the magma, and even though they are present in small quantities, they have a significant role to play in the crystallization of the rocks and in the development of valuable mineral deposits. These volatiles have a strong tendency to escape from the magma both before and during the solidification processes.

Since only nine common elements make up over 99 percent of the total magma, then these nine elements are involved to a very large extent in the formation of minerals of which igneous rocks are composed. Again, these nine elements, in decreasing order, are oxygen, silicon, aluminum, iron, calcium, sodium, potassium, magnesium, and titanium. While the magma is hot and fluid, these elements are combined into various molecules. Very commonly silicon and oxygen are linked together as a strong unit to which the other atoms become attached to form the minerals.

In a large magma chamber of 100 to 200 miles across, the magma probably is more or less under turbulent conditions as a result of changes of temperature, movements of solutions and vapors, new melting, chemical reactions, etc. Because magma is believed to be generally viscous, the turbulence would necessarily have to be generally sluggish, but it could be accelerated locally where conditions may be proper.

Cooling of Magma, Crystallization, and Composition of Major Rock Types

A relatively large proportion of igneous rocks are formed at the surface, such as around active volcanoes, where their solidification and crystallization can be observed. But a larger proportion of the igneous rocks are formed at considerable depth below the surface where their manners of formation cannot be observed. Consequently, the true nature of crystallization is not completely known, and knowledge of the development, particularly of deep-seated igneous rocks, is partly in the realm of conjecture rather than reality.

If we consider a typical "basic" magma in a magma chamber, we can follow the crystallization processes. The term "basic" here refers to a magma which has MgO, FeO, and CaO dominant over SiO_2. Although large amounts of SiO_2 are present, the other three ingredients together are more abundant. As cooling begins, certain minerals will begin to crystallize and may accumulate to form rocks. Others of these early formed minerals may be redissolved or may react with the magma and form new minerals as cooling proceeds. To explain such processes, Bowen (1922) has established a listing of igneous minerals which diagrammatically gives us conclusions about the various stages of magmatic crystallization, as shown in Fig. 4-1. The reader is urged to note that these minerals are all silicates and contain Si-O in some proportion, along with other ingredients.

Arranged in this order, mineral development from a basic magma is summarized. Minerals in the upper part of the series crystallize early, and the magma changes in composition as the various minerals are removed. Particularly notable is that SiO_2 gradually increases in the remaining magma, even though some of this ingredient becomes involved in mineral formation. Once removed, some of the early formed minerals may collect to form masses of igneous rock having a particular composition according to the mineralogy. For example, olivine, crystallizing very early, may accumulate with pyroxene (notably augite) to form a rock called *peridotite*. Other crystals of olivine may react with the magma for a period of time and be converted to pyroxene, a

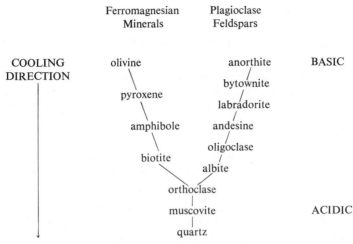

Fig. 4-1: Bowen's reaction series

new, more stable mineral composed of the same elements as the original olivine. The pyroxene may exist for a time, but it becomes unstable as the temperature falls. It then reacts with the solution to form hornblende, and so on.

The important early formed plagioclase feldspar is usually bytownite, as anorthite is rarely abundant. This bytownite, along with pyroxene, may accumulate to form a rock called *gabbro*. But not all of the bytownite and pyroxene will be involved in rock formation. As cooling proceeds, these two early formed minerals will react with the magma and be changed to more stable minerals lower in the series. In some cases, labradorite (or andesine) may become separated to form a mass of rock called *anorthosite*.

This crystallization may be accompanied by settling of the heavier crystals into the magma and floating of the lighter crystals toward upper portions of the magma. There is usually a sequence of minerals formed during this main stage of cooling and crystallization, and the magmatic processes of melting, mixing, assimilation, and escape of volatiles may continue all through this period. The magma continues to increase in SiO_2 during all these phases.

With progressive cooling of the magma, sodium-bearing feldspars begin to form, and along with lesser amounts of orthoclase and quartz, they may accumulate into a rock called *quartz diorite*. At succeedingly lower temperatures, plagioclase and orthoclase may accumulate in nearly equal amounts, together with a little quartz, to form the rock *quartz monzonite*. When the temperature falls to the point where orthoclase feldspar crystallizes in large amounts along with lesser amounts of sodium plagioclase, then quartz and biotite also are produced rather extensively, and a rock called *granite* will develop.

Mineralogical Aspects of Igneous Rocks

It is to be noted that the minerals listed by Bowen are the seven minerals or mineral groups commonly present in *major* amounts in igneous rocks, namely, olivine, pyroxene (both Mg- and Ca-bearing), hornblende, biotite, feldspars, feldspathoids, and quartz. In rarer alkali-rich magmas which are deficient in SiO_2, the feldspathoids will form in place of the feldspars. The feldspathoids are sodium and potassium aluminosilicates, with low Si-O content. These important igneous-rock-making minerals or mineral groups are usually classed as *essential* constituents. A few other minerals may be present in very small amounts, namely, magnetite, ilmenite, apatite, muscovite, corundum, fluorite, zircon, pyrite, and pyrrhotite; thus they are called *accessory* minerals.

The left-hand series consists of minerals known as the *ferromagnesian group* because they contain iron (Fe) and magnesium (Mg), and their series is discontinuous, i.e., each one is a different mineral with a different crystal structure. The right-hand series is made of the various species of the very important plagioclase feldspar, and their general composition is from calcic (Ca) to calc-sodic (Ca-Na) to sodic-calcic (Na-Ca) to sodic (Na) feldspar. Each is a variety of the isomorphous plagioclases, and there is no change of crystal structure. Thus the righthand series is a continuous one, i.e., having continuous structure.

The quantities of the essential minerals in different igneous rocks can vary fairly widely. In a few cases, an igneous rock may be composed nearly completely of just one mineral, e.g., such rocks as anorthosite (plagioclase mineral), dunite (olivine mineral), or pyroxenite (pyroxene mineral). But most igneous rocks contain two or three of the *essential* minerals, particularly feldspars, and two or three of the *accessory* minerals.

As a result of many factors acting on the magma during the cooling and crystallization history, igneous rocks developing from any particular magma are varied. There is less variation of igneous rocks produced from an acidic magma than from a basic magma because, in a basic magma, the complete reaction series theoretically can proceed to completion or at least to a greater degree.

The minerals which make up the bulk of igneous rocks are formed by separation of ingredients in any group or by chemical reactions which combine them. The most predominant molecule in the magma which is involved in the various reactions is silica (SiO_2), and thus it is present in the essential minerals.

The composition of an *average* igneous rock is given in Table 4-1 and it is expressed in percentages of the most abundant elements (column 1) and in oxides (column 2). The table shows that oxygen and silicon are the most abundant elements in magmas. During crystallization then, the minerals that form have silicon and oxygen as a basic part and thus are called silicates.

Table 4-1: COMPOSITION OF AVERAGE IGNEOUS ROCK (percent by weight)

1 Elements		2 Oxides	
Oxygen	46.59	SiO_2	59.12
Silicon	27.72	Al_2O_3	15.34
Aluminum	8.13	Fe_2O_3	3.08
Iron	5.01	FeO	3.80
Calcium	3.63	CaO	5.08
Sodium	2.85	Na_2O	3.84
Potassium	2.60	K_2O	3.13
Magnesium	2.09	MgO	3.49
Titanium	0.63	H_2O	1.15
		TiO_2	1.05

Data from Tyrrell, 1929

The various silicate minerals are generally made up of SiO_4 in groups to which are attached cations such as Al, K, Ca, Na, Fe, and Mg in various proportions and arrangements. In addition to the silicates, a few oxides may be present in the rocks in significant quantities, such as magnetite, Fe_3O_4, rutile, TiO_2, and ilmenite, $FeTiO_3$.

Table 4-1, giving the composition of an average igneous rock, is not a definite analysis; it is an abstract analysis merely to show the general composition of magma. There is a great variety of igneous rocks, but every variety was not produced from a magma which had the composition shown by that variety. Geologic evidence reveals that the large number of igneous rock types are produced from a limited number of magma types by a series of differentiations (crystallizations and separations). In fact, it may be possible to trace these few types of magma ultimately to only *two* major types. These two so-called primary magmas can account for all the known varieties of igneous rocks as far as the compositions are concerned. Briefly, these two types of primary magma are *basalt* and *granite*. However, many geologists believe that all types of igneous rocks can be developed from just one type, the basaltic type, of magma.

Textures of Igneous Rocks

The texture of an igneous rock includes all the small-scale features such as the degree of crystallinity, the shapes, the sizes, and the arrangements of the minerals which make up the rock. It also includes noncrystalline features. Depending on the sizes of crystals of the component minerals, texture may be rather easily observed by eye if crystals are fairly large, or with the aid of a hand lens if smaller, or with the aid of a microscope if of very fine size. Texture is best studied in "thin-sections" under the microscope, that is, very thin slabs of the rock through which light is transmitted. Texture is important to determine because it is related to the cooling and crystallization history of the rock, and it also assists in detecting physical and chemical environment at the time of consolidation.

Crystallinity

The term crystallinity refers to the proportions of crystallized and non-crystallized material of an igneous rock. If an igneous rock consists entirely of crystals of the constituent mineral matter, even though there may be some variation in sizes and shapes of crystals, the texture is called *holocrystalline*. When crystals have very good outlines and shapes, the term *euhedral* is applied to them. Those with euhedral form usually have developed early in the crystallization sequence, and their growths were not interfered with by any neighboring crystal. Crystals with moderately to poorly developed faces are said to be *subhedral*, and those with very poorly developed faces or no faces at all are said to be *anhedral*. The latter two terms, subhedral and anhedral, imply that the crystals were formed late and that crystal growth was interfered with by the presence of early formed crystals which were in the way (Fig. 4-2).

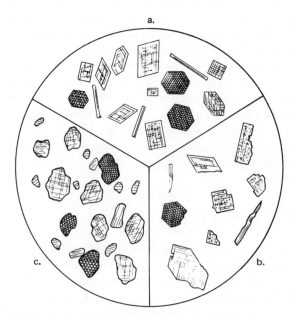

Fig. 4-2: Degree of face development in crystals: (a) euhedral; (b) subhedral; (c) anhedral.

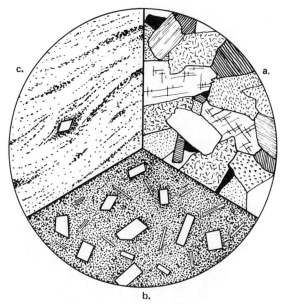

Fig. 4-3: Degree of crystallinity: (a) holocrystalline; (b) merocrystalline; (c) holo-hyaline.

In some cases, a rock may be composed entirely of glass, with no crystals at all. Such a rock has a *holohyaline (or glassy) texture.* Holohyaline rocks frequently occur as crusts or chill zones of larger igneous rock bodies; they also occur as lavas, as dikes, and as sills. The holohyaline texture is developed as a result of very rapid cooling, such that crystals did not have time to develop. However, this inference is slightly misleading, because in most natural glasses there are a number of very minute bodies of various shapes which represent the beginnings of crystallization. Such embryo crystals are called *crystallites* or *microlites.* The term *crystocrystalline* is often applied to these crystallites and microlites. The term *merocrystalline* actually is more appropriate. Merocrystalline means that the rock is composed partly of crystals and partly of glass, and the term applies to the texture whether the crystals are microscopic in size or can be easily seen by the eye. Degrees of crystallinity of igneous rocks are illustrated in Fig. 4-3.

Sizes of Crystals

In the case of holocrystalline texture, the sizes of crystals are quite variable, and a method of describing the grain size has been used by geologists for some time. The following subdivisions of grain sizes are employed by some geologists:

less than 1.0 mm	fine-grained
from 1.0 to 5.0 mm	medium-grained
greater than 5.0 mm	coarse-grained

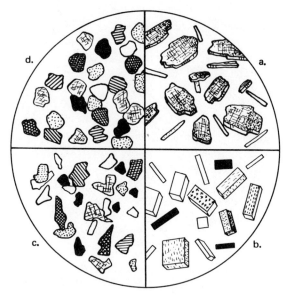

Fig. 4-4: Crystal shapes: (a) tabular; (b) prismatic; (c) irregular; (d) equidimensional.

If crystals are large enough to be seen easily by eye or with the aid of a simple pocket lens, the rock texture is *phaneritic*. If crystals are so small that they can be seen (as individuals) only with a microscope, the rock's texture is said to be *microcrystalline*. However, the terms microcrystalline, merocrystalline, cryptocrystalline, and holohyaline are grouped as *aphanitic* textures.

Shapes of Crystals

In addition to the three terms *euhedral, subhedral,* and *anhedral* which describe the degree of crystallinity and development of faces on crystals, there are other terms which are used to describe shapes of crystals. Crystal shapes are often described in accordance with the relative dimensions in the three directions. Crystals which are fairly uniform (or equal) in the three directions are *equidimensional*. Those which are extended in two directions but not in the third are *tabular*. When only one dimension is elongated, the crystals are *prismatic*. Another term, *irregular*, includes such shapes as wisps, shreds, ragged patches, veins, and skeletons (Fig. 4-4).

Mutual Arrangements of Crystals

The texture of an igneous rock also includes the pattern made by the constituent mineral material. It is the relative sizes, and not the absolute sizes, together with their arrangement, which develops the pattern. There are four main textures dependent upon arrangement of crystals and any glassy

material which may be present. These are as follows: equigranular, inequigranular, directive, and intergrown (Fig. 4-5).

Equigranular textures are developed when the constituent minerals all have
approximately the same size, whether they be fine-grained or coarse-grained,
and regardless of whether they are euhedral, subhedral, or anhedral.

Inequigranular textures are developed when the constituent minerals have
significant differences in sizes. There are commonly two dominating sizes of
the crystals, and few or no crystals of intermediate size are present. There are
two important subtypes of inequigranular texture: *porphyritic* and *poikilitic*.
A porphyritic texture is produced when individual large crystals, called *phenocrysts*, are surrounded by a groundmass made up of phaneritic, aphanitic, or
glassy material. A porphyritic texture is usually developed when the phenocrysts form in the magma at great depth, under slow cooling, and under high
pressure. Later when the magma is transferred or injected to a higher, cooler
level in the crust, or extruded at the surface where cooling becomes more
rapid, the loss of volatiles, the increase in viscosity, and the higher rate of
cooling cause the remaining magma to develop small crystals (or glass) as
the groundmass. A porphyritic texture may also be developed by large growth
of crystals which are relatively insoluble in the magma. In still another way,
the large phenocrysts and smaller-sized groundmass material may crystallize
at the same time. The phenocrysts grow large because of more favorable
ability due to their higher molecular concentration in the magma. Por-

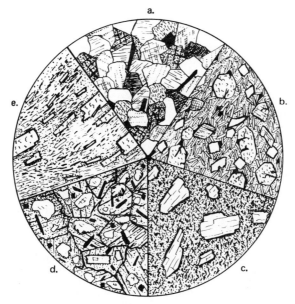

Fig. 4-5: Igneous rock textures: (a) equigranular; (b) inequigranular; (c) porphyritic; (d) poikilitic; (e) directive.

phyritic texture is most often found in volcanic rocks or in rocks of intermediate depth. Occasionally though, this texture is also found in deep-seated intrusive masses such as porphyritic granite. A *polkilitic texture* is produced when small-sized crystals are enclosed within large-sized crystals. There must be enough randomly oriented small-sized crystals to produce a distinctive pattern. Poikilitic texture is most commonly found in rocks called *syenites* and *monzonites*, in which orthoclase crystals grow large and act as the surrounding host for the smaller crystals.

During crystallization, the newly formed crystals, microlites, or crystallites may be carried along by the magma so that the long dimensions of the solid particles become lined up in parallel or subparallel bands. These bands follow stream lines of the magma. When the magma crystallizes completely, these bands remain. Such a texture produced by flow in a magma during crystallization is called a *directive texture*. Textures of this type are often seen in lavas which contain feldspar crystals. Feldspar crystals are usually lath-like, and these often become orientated in parallel position by flow to produce a *trachytic* (directive) texture.

Graphic texture is produced when two different minerals crystallize at the same time from the same magmatic solution and consequently grow very intimately together (Fig. 4-6). *Granophyric texture* arises often between quartz and feldspar when irregular patches, shreds, and blebs of quartz are seen within feldspar crystals. However, if the feldspar albite is present in irregular patches within orthoclase crystals, the special term *perthite texture* is used (Fig. 4-7).

Classification of Igneous Rocks

There are several different classifications of igneous rocks which have been used at different times in the past. In practically all cases, each classification was founded on only two simple principal factors: texture and mineralogical composition. Even so, no general agreement exists among geologists, particularly petrologists, as to the details of the classifications. Because there is a fairly great variety of igneous rocks, both in chemical and mineralogical makeup, indefinite boundaries exist within the various ranges of composition. It should be pointed out further that, with the finely crystalline rocks, the high-power microscope is necessary in order to make mineral identification with certainty. And with this technique, a greater amount of subdivision can be made in many cases.

For a classification which has good utility, without the aid of a high-power microscope or other sophisticated equipment, the one herein presented is recommended. However, a simple pocket lens of 10X magnification is recommended for assistance. Since only the human eyes and a simple pocket lens are used for determining the name of an igneous rock, this classification

Fig. 4-6: Pegmatite rock from Bedford, New York, showing coarse graphic texture between feldspar (light) and vitreous to smoky quartz (medium gray).

Fig. 4-7: Perthite texture formed by microcline (gray) and multiple small veinlets of albite (white).

becomes a *megascopic classification*, based on textural and mineralogical characteristics determinable in hand-sized specimens. Consequently, the subdivisions are comparatively broad.

The most important factor in the following classification is the fact that the rock's texture may be phaneritic, porphyritic, or aphanitic, in which crystals may be absent in some cases to present a *glassy* texture. Except in glassy textured rocks and in very-fine-grained rocks, the minerals can be identified and their amounts measured or estimated. Such a *textural* and *mineralogical* classification not only accounts for the composition but also leads to knowledge of the cooling history of the rock.

For the study and determination of igneous rocks, it is of utmost importance to obtain as fresh and unaltered specimens as possible. Rocks exposed at the Earth's surface become altered by weathering processes rather rapidly, and the presence of secondarily produced minerals and partially or completely destroyed minerals makes the correct assignment very difficult. The minerals present in igneous rocks are usually classed as *essential, accessory*, and *secondary*. The essential and accessory minerals are said to be primary or original with the rock at the time of formation. The secondary minerals are those formed by weathering, metamorphism, or circulating solutions. But the rocks are really broadly classified only on the basis of the essential minerals because these make up the larger proportion of the rock. Accessory minerals may be present in small amounts, but these are not considered as being necessary for classifying the rock.

The texture of the rock is very important in indicating the general conditions under which the cooling took place; thus it yields information on the rock's mode of occurrence. Among igneous rocks, there are two principal modes of occurrence: *plutonic* (intrusive) and *volcanic* (extrusive). When crystallizing under deep-seated plutonic conditions, igneous rocks form by a slow cooling process. Pressures are high, and volatile (vaporous) constituents are retained in the magma. Under these conditions, any crystals which form are generally coarse-grained, and all of the magma develops crystals, thus forming rock which is holocrystalline and coarse. In the case of magma crystallizing under very shallow conditions or out on the surface under volcanic conditions, igneous rocks form by a rapid cooling process. Pressures are low, and volatile constituents are easily lost. Under these conditions, any crystals which form must necessarily be very fine-grained, or in many parts of the rock mass, no crystals at all may form because cooling is far too rapid. Textures so developed are thus merocrystalline or glassy and may even exhibit vesicular structure or flow structure. Rocks formed under such conditions often have a porphyritic texture.

Some (but not all) geologists also utilize a *third* mode of occurrence and development of igneous rocks. The third mode is called *hypabyssal* (intermediate between plutonic and volcanic), the rocks in this group occupying an intermediate position in the Earth's crust between the deep-seated plutonic and surficial volcanic lava flows. Rocks crystallizing from magmas in such intermediate positions usually develop merocrystalline to holocrystalline textures and take the form of sills, dikes, laccoliths, ring dikes, cone-sheets, etc. If magma is suddenly transferred from deep to upper levels of the crust, the rock which finally forms will have a porphyritic texture.

Classification Scheme

The classification of igneous rocks set forth below does not require the use of a microscope or other complicated equipment. However, a pocket lens is recommended. Three main textures are used: phaneritic, porphyritic, and aphanitic, in the broadest sense.

Phaneritic: The rock is completely crystalline: more than half of the crystals are large enough to be seen with the naked eye and identified, with or without the aid of a hand lens. To identify the remaining crystals would require the aid of a microscope.

Porphyritic: The rock contains phenocrysts (large crystals) embedded in a finer groundmass. The phenocrysts are large enough to be identified by the naked eye, with or without the hand lens. The remaining finer groundmass, if coarse-grained, may also be readily identified. But if the groundmass is very-fine-grained, microscopic determinations would be required.

Aphanitic: The rock consists entirely of very-fine-grained crystals (microcrystalline), entirely of glass, or is made up of both very-fine-grained crystals and glass. None of the fine crystals can with certainty be identified by the naked eye, even with the aid of a hand lens. A microscope would need to be used to identify the crystals and/or glassy portion.

Three very broad compositional divisions are used in association with the three main textures: acidic, intermediate, and basaltic. A fourth division sometimes used, ultrabasic, includes fairly uncommon rocks. These first three are based on the predominant feldspar present; the fourth division contains no feldspar. The general scheme places the most acidic rocks on the left and the most basic on the right, with intermediates placed between. This does not mean that the rocks are uniformly more basic from left to right, because this trend is not strictly so for some intermediate points. The compositional terms are defined briefly as follows:

Acidic: Orthoclase feldspar makes up more than two-thirds of the total feldspar; quartz is significant.

Intermediate: Orthoclase feldspar and plagioclase feldspar are in nearly equal proportions; quartz is less significant than in acidic.

Basic: Plagioclase feldspar makes up more than two-thirds of the total feldspar; quartz is in very minor quantities, but ferromagnesian minerals become more significant.

Ultrabasic: No feldspars and no quartz are present; pyroxenes and olivine become important, and magnetite, ilmenite, chromite, and pentlandite are somewhat significant.

Major Groups of Igneous Rocks

The general scheme of classification, using three principal textures and three principal compositions, is presented in Table 4-2. The major crystalline groups, shown in the vertical columns, are: granites and rhyolites, syenites and trachytes, quartz monzonites and quartz latites, monzonites and latites, quartz diorites and dacites, diorites and andesites, gabbros and basalts, and peridotites-pyroxenites. Each of these groups includes those with porphyritic textures. Glassy textured rocks are presented in the lower portion of the table.

Pegmatites

As the bulk of igneous rock-making minerals crystallize into masses of major rock types, the remaining magma becomes progressively richer in silica and alumina along with volatile components (vapors). The main volatiles in the remaining magma are H_2O, HF, H_2CO_3, B, P, Be, and Li, and these have a strong tendency to keep the magma very fluid. Because of this, the volatiles can move easily to sites of crystal growth and permit large crystals to develop. As a result, very-coarse-grained igneous rocks are produced. This late stage in the crystallization, after the bulk of the igneous rocks are formed, is called the pegmatite stage, and the coarse-grained igneous masses are called *pegmatites*. Crystals of quartz, feldspar, mica and many other minerals often reach sizes of 1 inch up to several inches, and frequently even larger.

Pegmatites usually become crystallized near the borders of large intrusive masses, mainly outside but also inside the borders. They take the form of small sheets, dikes, and veins (Fig. 4-8). Pegmatites are more often associated with granites than with other types of igneous rocks, but they make up only a small percentage of the total igneous rock with which they are associated, probably only about 1 percent. Most pegmatites are simple in mineralogy, being made up of minerals also commonly found in the main granite mass, notably quartz and alkali-feldspars, since most pegmatites are associated with granites. Muscovite and biotite frequently are present in substantial amounts. Minerals rich in volatile components which frequently are present are tour-

Table 4-2: CLASSIFICATION OF IGNEOUS ROCKS

		Composition						
ESSENTIAL MINERALS		Acidic		Intermediate		Basic		Ultrabasic
		Potash Feldspar > 2/3 of Total Feldspar		Potash Feldspar and Plagioclase Feldspar about equal		Plagioclase Feldspar > 2/3 of Total Feldspar		Pyroxene and Olivine
						Sodic Plagioclase		Calcic Plagioclase	
TEXTURE		Quartz	No Quartz	Quartz	No Quartz	Quartz	No Quartz	Calcic Plagioclase	
PHANERITIC		granite	syenite	quartz monzonite	monzonite	quartz diorite	diorite	gabbro	peridotite pyroxenite
PORPHYRITIC		granite porphyry	syenite porphyry	quartz monzonite porphyry	monzonite porphyry	quartz diorite porphyry	diorite porphyry	gabbro porphyry	
		rhyolite porphyry	trachyte porphyry	quartz latite porphyry	latite porphyry	dacite porphyry	andesite porphyry	basalt porphyry	
APHANITIC		rhyolite	trachyte	quartz latite	latite	dacite	andesite	basalt	

GLASSY: obsidian and its varieties: pitchstone, perlite, pumice; pyroclastic varieties: tuff, agglomerate, scoria, basalt glass, volcanic breccia

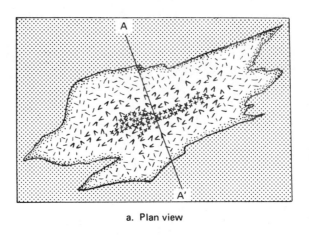

a. Plan view

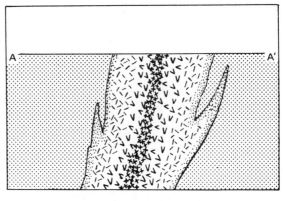

b. Cross-section

Fig. 4-8: Typical complex pegmatite.

maline, fluorite, apatite, beryl, topaz, and lepidolite. However, some pegmatites are rather complex in mineralogy, are often structurally zoned, and often contain numerous valuable ore and gem minerals. In addition to the common minerals found in simple pegmatites, minerals which often occur in complex pegmatites include cassiterite, amblygonite, wolframite, spodumene, uraninite, zircon, and columbite, as well as minerals containing molybdenum, copper, radium, bismuth, and arsenic.

In pegmatites, quartz and feldspar often become intimately intergrown to form so-called "graphic granite." And in some cases, potash feldspar becomes closely intergrown with albite to form "perthite."

Pegmatites are predominantly granitic in composition and are commonly referred to as *granite pegmatites.* However, *syenite pegmatite, monzonite*

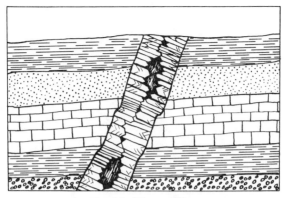

Fig. 4-9: Fissure filling.

pegmatite, diorite pegmatite, and *gabbro pegmatite* are also known, but these latter four types are uncommon.

Hydrothermal Deposits

As pegmatites continue to develop as characteristic minerals become precipitated, any remaining magmatic solutions become richer in water and silica. Such solutions are called *hydrothermal solutions* and develop later than pegmatites. These warm-water solutions carry large quantities of sulfur in solution, often with a fair amount of metals such as copper, iron, lead, zinc, mercury, tin, tungsten, molybdenum, gold, silver, arsenic, and bismuth. Other elements often are present also. Mineral deposits formed by these solutions are called *hydrothermal deposits.* Hydrothermal deposits commonly take the form of sulfide veins, formed by the filling of fractures and fissures in country rock adjacent to igneous rock bodies (Fig. 4-9). However, many develop as irregular masses partially or completely replacing other rock (Fig. 4-10). A third type occurs as widely *scattered* (disseminated) *minerals*

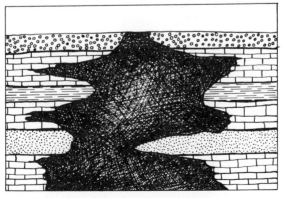

Fig. 4-10: Replacement deposit.

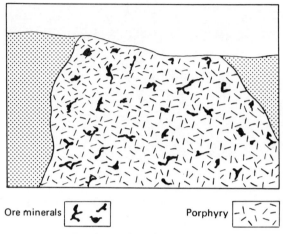

Ore minerals | Porphyry

Fig. 4-11: Disseminated deposit.

within a large mass of igneous rock (Fig. 4-11). Hydrothermal deposits often contain large quantities of important ore minerals (mostly sulfides); consequently, they are of high commercial value. Hydrothermal deposits support the mining industry of the western United States. However, numerous hydrothermal deposits are actually barren of valuable minerals.

The shape and general physical nature of a sulfide vein depends upon the type of fissure into which the solutions moved. Most deposits are tabular, i.e., long, sheet-like, but thin. Veins may occur singly, in groups, or in multiple interlacing bodies. The walls of a vein may be sharply defined or gradational with the enclosing rock, depending upon the amount of reaction that has occurred between them. Hydrothermal deposits of greatest economic value are deposited from fairly warm ascending magmatic solutions, charged with water, derived as residual solutions after the cooling and crystallization of magma. These ascending solutions deposit their mineral content as temperature and pressure decrease even further.

Hydrothermal deposits are classified into three types, according to temperature and pressure conditions: hypothermal, mesothermal, and epithermal. *Hypothermal* deposits are formed at great depth of several miles and at temperatures in the range of 500–300°C. Common valuable ore minerals of hypothermal deposits include cassiterite, wolframite, scheelite, molybdenite, gold, chalcopyrite, chalcocite, galena, pyrrhotite, pyrite, arsenopyrite, bismuthinite, and magnetite. *Mesothermal* deposits are formed at intermediate depths at moderate pressures and at temperatures in the range of 300–200°C. Common valuable ore minerals are pyrite, chalcopyrite, galena, arsenopyrite, sphlerite, and gold. *Epithermal* deposits are formed at shallow depth under moderate pressure and at temperatures in the range of 200–50°C.

Common ore minerals of value are gold, silver, sphalerite, cinnabar, and stibnite. In the various deposits, certain valueless minerals called *gangue* may be present. These include quartz, fluorite, tourmaline, topaz, calcite, siderite, rhodochrosite, barite, and opal.

Deposits of Fumaroles and Hot Springs

Hydrothermal solutions and vapors occasionally reach the surface through various types of openings. Liquids exiting at the surface are called *hot springs*, while gases reaching the surface are called *fumaroles*. Hot springs and fumaroles usually occur together in active volcanic areas and are generally very dilute, as far as mineral content is concerned, since they have been mixed with groundwater. Consequently, mineral deposits of most *hot springs* consist only of opaline silica (siliceous sinter) and perhaps small amounts of sulfur and sulfides. Common fumarolic minerals deposited in volcanic areas include sulfur and chlorides. But others in minor quantities may include hematite, magnetite, pyrite, molybdenite, realgar, galena, sphalerite, and various borates.

Forms and Structures of Igneous Rocks

Igneous rocks can be seen in volcanic areas where magma has issued forth at the surface and solidified, and magma itself can be observed around active volcanoes. Magma poured out on the surface is called lava and is said to be *extrusive*. However, lava flowing at the surface in a volcanic area is simply a small geologic event in comparison to the much greater igneous activity taking place at deeper levels below the surface. The magma at these deeper levels is said to be *intrusive*, and igneous rocks formed from deep-seated magma can be seen only after they have been exposed through subsequent weathering and erosion or through earth movement.

Forms

Extrusions

The quiet eruption of magma at the surface in the form of a *lava flow* may come from individual volcanic cones or from a series of fissures. Lava flows are usuallly tabular in shape, and they often become spread over fairly extensive areas as extrusive sheets. They are quite thin and are usually elongated in the main direction of flow (Fig. 4-12). Lava flows may vary from a few inches to several hundred feet thick, and they may cover just a few acres or several hundred square miles of area. Fissure eruptions commonly flood over large areas. On the other hand, central eruptions are much more local and produce

Fig. 4-12: Lava flow.

simple volcanic cones. Lava cones constructed of very fluid lava during eruption are broad and low and are referred to as *shield volcanoes.*

Extrusions from numerous fissures often result in piles of superposed flows many thousands of feet thick. The fluidity of a flow has a controlling effect on the general form which develops, but this is directly related to composition and temperature of eruption. Basic lavas such as basalt are quite mobile and thus can spread or flow fairly easily. Acidic lavas such as rhyolite, on the other hand, are more viscous and sluggish and thus are more apt to get heaped up around the volcanic vent.

Volcanic eruptions sometimes are explosive, producing fragmental types of igneous rocks called *pyroclastics.* Fragments are produced from hard rocky material in the throat of the volcano which is blown to pieces or from liquid magma which is blown up through the vent. They range in size from chunks weighing several hundred tons to the very finest dust sizes. The ejected material is usually classified as follows:

$$
\begin{array}{ll}
> 32\ mm & \left\{\begin{array}{l}\text{bombs if rounded} \\ \text{blocks if angular}\end{array}\right. \\
32 - 4\ mm & \text{lapilli or cinder} \\
< 4\ mm & \text{ash}
\end{array}
$$

The large fragments usually accumulate in and around the vent, and they commonly are progressively smaller in size as the distance from the vent becomes greater. If most of the material is cinder-sized, the cones are called cinder cones. Any single accumulation of coarse fragments is apt to be thickest near the vent and becomes progressively thinner out away from the vent.

The dust-sized material is commonly deposited out away from the vent also, as it may be carried by wind currents for many miles over the country-

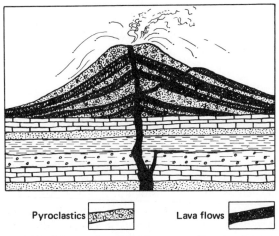

Pyroclastics   Lava flows

Fig. 4-13: Cross-section through a composite volcanic cone, showing alternating pyroclastics and lava flows.

side. Such accumulations of ash will be generally pretty thin, but may be spread over a great many square miles, depending on the amount of material and where it was transported by the wind.

Volcanic cones constructed chiefly of pyroclastic material are called pyroclastic cones. In some cases, a cone may be built of both pyroclastics and lava flows in alternating layers. These are called composite cones (Fig. 4-13). Volcanic cones vary in size from tens of feet to several thousand feet high, covering many square miles.

Freshly erupted pyroclastics, such as the volcanic bomb in (Fig. 4-14),

Fig. 4-14: Volcanic bomb, seven inches long, formed by solidification of lava during flight.

may lie around on the surface for some time. But by cementation and compaction, these materials are converted to rock. Coarse, rounded fragments become agglomerate, while angular fragments become volcanic breccia. Lapilli are converted to lapilli tuff, and fine-sized ash becomes tuff.

Intrusions

Masses of molten rock material which have been injected between layers and blocks of the Earth's crust take on a large variety of forms. The forms depend upon geologic structures and associated subordinate features. The geologic structures and features are, for the most part, openings of various kinds below the surface where magma has moved and solidified. Intrusions are usually classed as *major* intrusions or *minor* intrusions.

Major intrusions are *discordant* if they do not conform with the rocks into which they have been intruded; they are *concordant* if they do conform to the bedding or layers of crustal rock. Major discordant intrusions are called *batholiths, stocks,* or *bosses,* depending on their sizes. *Batholiths* are irregular bodies of igneous rock of large area, great depth, and great volume. Lengths and breadths measurable in hundreds of miles and depths of several miles are often observed. Batholiths occur on all continents and are associated with large mountain chain belts. Stocks and bosses commonly are associated with the larger batholiths and are usually considered as smaller offshoots (Fig. 4-15). They have cylindrical shapes with a domed top and somewhat circular outcrop area.

The contact between a batholith and the adjacent country rock may be sharp or may be gradational, depending upon degree of melting, movement of magma, and other similar factors at the time of association. Most batholiths are granitic in composition, but basic ones such as gabbro, as well as intermediate ones such as diorites, are also known.

Major concordant intrusions are classed as laccoliths, lopoliths, phacoliths, and sills, depending upon structural relationships and shapes of the igneous rock bodies.

The *laccolith* is described as a lenticular mass having a dome-shaped upper surface and a relatively flat base. The magma apparently was fed through a central channelway. Because of the high viscosity together with the pressure of the rocks above, the magma does not spread laterally very far, but generally tends to heap itself up around the feederway. This causes a lifting of the strata above it into a general domed shape. The outcrop area of a laccolith is generally circular or elliptical in plan and often somewhat elongated. Small-sized sills may extend from the margins, and small dikes may be produced in fractures in the stretched domed-up sediments above (Fig. 4-16).

A *lopolith* is described as a basin-shaped mass of igneous rock, generally concordant, lenticular in gross shape, intruded into sediments (Fig. 4-17).

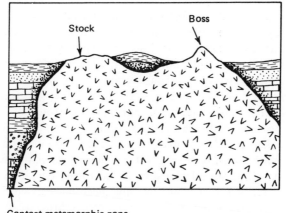

Contact metamorphic zone

Fig. 4-15: Diagram of batholith, stock, and boss.

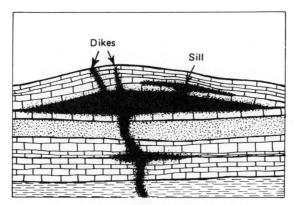

Fig. 4-16: Laccolith with dikes and sill.

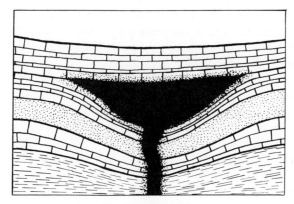

Fig. 4-17: Lopolith.

The downsagged structure probably requires tectonic forces, resulting in removal of support from beneath an area, development of space, and later infilling by intrusive magma. Lopoliths are generally larger than laccoliths, and they have different forms. Both lopoliths and laccoliths may show layering of dense minerals which have sunk through the magma after being formed.

A *phacolith* is described as a mass of igneous rock intruded into folded sediments. During folding, crests of anticlines and troughs of synclines become weakened due to stretching and tension. Magma tends to find its way into these weakened and fractured crests and troughs, and under pressure it may spread out between beds (Fig. 4-18). Thus, a phacolith develops a lens-like form. Phacoliths are relatively small intrusives, similar to laccoliths.

A *sill* is described as a relatively thin sheet-like mass of igneous rock, intruded along nearly horizontal strata, or along schistosity, or along layers of igneous rock. The shape of a sill is broadly lenticular, but upper and lower surfaces may be parallel for fairly long distances (Fig. 4-19). Thicknesses of sills vary from a few inches to hundreds of feet. During sill emplacement, the magma can spread only to a distance which is dependent on the force of injection, temperature, degree of fluidity, and the weight of rock above it which must be lifted by the incoming magma. Most larger sills are basic in composition and hence quite fluid at the time of injection.

Minor intrusions also may be discordant or concordant with the rocks into which they have intruded. Minor discordant intrusions are classes as dikes, veins, cone sheets, ring dikes, and volcanic necks.

A *dike* is described as a mass of igneous rock intruded into near-vertical fissures (cracks) in the surrounding rocks. Consequently, a dike develops a narrow, elongated, and parallel-sided form (Fig. 4-20). Dikes vary in thickness from inches to hundreds of feet, but generally they are just a few feet thick. Their lengths vary from a few yards to many miles. Dikes tend to occur in swarms or systems, lying parallel to one another or in a radial pattern around a central point. Dikes are usually indicative of regional tension around an area of igneous activity. The intrusive magma may force open existing cracks, thus relieving some of the tension by pushing aside blocks of crustal material which otherwise would be held under tension. Emplacement of dikes may take place rather rapidly.

A *vein* is an irregular tongue of igneous rock, located along the margins of large intrusions. Veins often occupy tensional fractures in the country rock and often develop sinuous courses as a result of the manner of fracturing of the rock as shown in Fig. 4-9. Veins commonly are small, like dikes, but usually not as numerous.

A *cone-sheet* is a single mass of igneous rock having an arcuate structure. Cone-sheets usually occur in series, all dipping toward a common center. Each cone of igneous rock is separated from others by intervening screens of

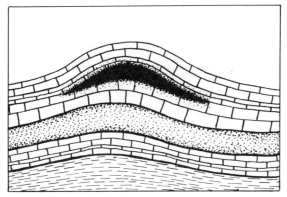

Fig. 4-18: Phacolith.

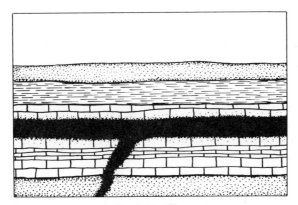

Fig. 4-19: Sill.

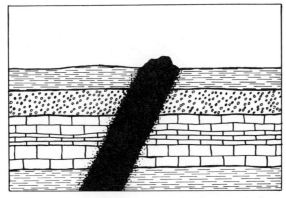

Fig. 4-20: Basalt dike cutting across sedimentary beds.

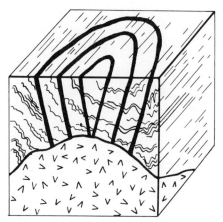

Fig. 4-21: Cone sheets.

country rock. Such structures form as a result of pressure from below in a small area. Magma then fills a series of conical fractures to produce the cone-sheets. (Fig. 4-21).

A *ring-dike*, too, is a conical-shaped mass of igneous rock having an arcuate structure. Although ring-dikes usually occur in series, they all dip away from a common center. Each ring-dike is separated from others by a screen of country rock. Such structures develop as a result of subsidence of a large block of crustal rock into a magma chamber. The magma then is forced upward along the margins of the sinking block to produce ring-dikes (Fig. 4-22).

A *volcanic neck* is described as a mass of igneous rock sealing up the vent of an old volcano, frozen there when extrusion of lava ceased. A volcanic

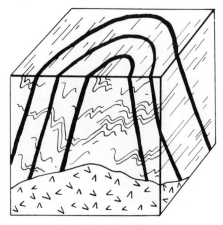

Fig. 4-22: Ring dikes.

Fig. 4-23: Volcanic neck and cinder bed.

neck may occupy the cylindrical vent or may also be partly intrusive into the volcanic rock around the vent. There may even be "offshoot" dikes and sills into the surrounding area (Fig. 4-23).

Structures

The term *structure* refers to large-scale features of the rocks, as these features are seen in the field. Also included in the term structure are a few small-scale features.

Vesicular and Amygdaloidal Structures

Lavas extruding at the surface release various gases rapidly, with the production of cavities or bubbles whose shapes vary from spherical to irregular. Such cavities or bubbles are called vesicles, and the structure is termed *vesicular.* When vesicles are later filled by secondary minerals, the materials so formed are called amygdules, producing an *amygdaloidal structure.* The minerals of amygdules are often calcite, silica, zeolites, or hydrated ferromagnesian minerals (Fig. 4-24).

Blocky and Ropy Lava

The surface of a lava flow may be covered with a mass of rough, jagged, angular blocks of hard rock of various sizes which were carried along by flowing lava. The blocks were formed early by crusting of lava by cooling, and then later broken up as gases continue to escape from below the encrustation, often violently. Such a structure is called *blocky lava.* Highly mobile lavas usually solidify with smooth surfaces which exhibit wrinkled, ropy, or

Fig. 4-24: Amygdaloidal basalt showing vesicles filled in with secondary minerals such as calcite and zeolites. Rest of rock contains fine-grained augite and labradorite, with minor hematite to yield reddish-brown color.

corded appearance. There may also be lava blisters and low humps of material on the surface. Such a structure is called *ropy lava* (Fig. 4-25).

Ropy lavas issue at higher temperatures than do blocky lavas, but there are lesser amounts of gas. This gas escapes quickly and the lava freezes quickly. A much larger gas content in blocky lavas causes the lava to be fairly mobile, and solidification starts early and proceeds quickly. Escape of gases later becomes very rapid and often violent, but the magma still is gas-saturated and solidification continues.

Pillow Structure

Sometimes lava exhibits the appearance of piles of small masses, with vesicular crusts, and often with glassy glaze on the outer surface. These small lump-like masses resemble pillows or cushions, and they often appear to indent adjacent masses like indentations in soft cushions or pillows. Many irregularly shaped pillows exist, ranging from rounded to elongated to ropy. Such *pillow structure* develops as a result of very rapid cooling where water is present, such as on the floor of the ocean, in very moist land areas, beneath ice sheets, or into water-logged sediments. In these situations, the hot lava

Fig. 4-25: Rough ropy lava of Kilauea volcano, Hawaii. (Photo courtesy Jennie A. Matthews.)

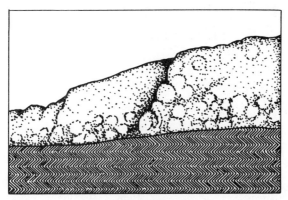

Fig. 4-26: Pillow lava.

enters cold water and develops large spheroidal masses. The exterior becomes rapidly chilled and gases trapped in the lava help keep the rounded bulbous shape. Steam generated by the hot lava causes a certain buoyancy which permits different spheroidal masses to roll forward on one another for a time (Fig. 4-26).

Flow Structure

Lava flowing at the surface is never completely homogeneous. Contained in the lava are patches and masses of constituents which are more concentrated than elsewhere as a result of the presence of contained gases, of different viscosity, or of different degree of crystallization. During flow, these patches, masses, and sometimes layers become drawn out into parallel lenses, streaks, bands, and lines by the cooling flowing lava (Fig. 4-27). Such *flow structures* may be detected by differences in colors, textures, or crystal sizes, arranged in parallel to subparallel alignment.

Fig. 4-27: Basalt showing crude flow structure produced by fluid basic magma and partially solidified material. (Courtesy James B. Stevens.)

Fig. 4-28: Columnar jointing in igneous rock of Devil's Tower, Wyoming. (Photo courtesy Jennie A. Matthews.)

Joints, Sheeting, Platy Structures

Joints (or fractures) are found in all types of igneous rocks from granites to basaltic lavas. Three sets of joints are often present, two vertical and one horizontal. When the horizontal joints are quite closely spaced, they are referred to as *sheeting*. The jointing may be curved or sinuous and oftentimes takes on very arcuate curvatures. Very closely spaced jointing may be developed in such a way that a *platy structure* is produced.

Joints are caused in some cases by tensile stress due to contraction of the rock mass after cooling. In other cases, jointing is produced by tectonic factors which impart tensional, compressional, or torsional stresses on the rocks.

Columnar and Prismatic Structures

When cooling and contraction due to cooling are fairly uniform, the jointing becomes quite regular. The resultant blocks of rock between joints develop a *columnar* or *prismatic* form of four-, five-, or six-sided prisms (Fig. 4-28). Such blocks develop perpendicularly to the cooling surface. Consequently, in a sill or lava flow the blocks are vertical, but in a dike they are horizontal.

Fig. 4-29: Granite from Texas containing pink orthoclase, vitreous quartz, and black hornblende.

Fig. 4-30: Hornblende granite from Rockport, Massachusetts, showing light colored feldspars, glassy quartz, and substantial amounts of black hornblende.

The Crystalline Igneous Rocks

Granites and Rhyolites

Granites have phaneritic textures and are usually evenly granular. They consist of predominant euhedral orthoclase (or microcline), a little plagioclase (albite or oligoclase), and irregularly shaped colorless or white quartz. The quartz usually fills in the spaces between other crystals. Of the total feldspar content, more than two-thirds is orthoclase. Usually a small amount of dark ferromagnesian minerals such as biotite, hornblende, or occasionally augite is present (Fig. 4-29). Other occasional accessories are magnetite, hematite, tourmaline, and pyrite. Biotite seems to be more common than the others. When biotite is the only dark mineral present, the granite may be called *biotite granite* (or *normal granite*); hornblende granite is fairly common also (Fig. 4-30). Granites containing abundant phenocrysts, particularly feldspar (and quartz) phenocrysts, are called *granite porphyries* (Fig. 4-31). Other varieties are *tourmaline granite* and *muscovite granite*.

Very coarsely crystalline aggregates of feldspar may occasionally be

Fig. 4-31: Granite porphyry with large phenocrysts of pale pink orthoclase in a groundmass of vitreous (glassy) quartz, black hornblende, biotite, and a little gray feldspar.

intimately intergrown (often with quartz) so that a cuniform texture pattern develops to produce a rock called *graphic granite* (see Fig. 4-6). Sometimes orthoclase feldspar and albite feldspar may be intimately intermixed to produce an intergrowth texture called *perthite*, consisting predominantly of orthoclase with streaks of albite (see Fig. 4-7).

The colors of granites vary from white to pink to red to gray, and they occasionally have tints of blue. The color of any particular granite is due to the feldspars which are present, orthoclase being white or pink, microcline being light blue, plagioclases being light gray. Quartz is usually colorless and transparent and therefore imparts little color. Slightly gray tones are imparted by small quantities of biotite, hornblende, or augite. But when those dark accessory minerals are more abundant, the rock becomes darker.

The average granite contains about 50 percent potash feldspar (orthoclase or microcline), about 20 percent plagioclase (albite or oligoclase), more than 20 percent quartz, and 3 to 15 percent biotite. These values may range significantly without transgressing the boundaries of "granite." In a few granites, minerals have been arranged in irregular bands formed when the magma was forced to flow during solidification processes.

The fine-grained rocks of this group, the *rhyolites*, are generally light-colored, since they contain very siliceous minerals such as quartz, orthoclase, and plagioclase. The average rhyolite contains about 50 percent potash feldspar, about 20 percent plagioclase, a significant amount of quartz, and some biotite. Colors of rhyolites range from light gray to pink to red and to brown (Fig. 4-32). When the texture is very-fine-grained throughout, it is very diffi-

Fig. 4-32: Brownish-pink rhyolite showing fine granular texture and fairly dense appearance.

Fig. 4-33: Rhyolite porphyry from Chaffee County, Colorado, containing white feldspar phenocrysts in a brownish-red aphanitic matrix.

cult to tell rhyolites from trachytes, quartz latites, latites, and dacites in hand specimens, and microscopic study is usually required. The rhyolites sometimes are slightly porphyritic, and when a few phenocrysts are developed, the rock can more easily be determined. The presence of quartz or orthoclase phenocrysts usually indicate rhyolitic composition. A few dark ferromagnesian minerals such as hornblende and biotite may also be present and may even be phenocrysts.

When phenocrysts of orthoclase and quartz are abundant (one-tenth to one-half of the rock), *rhyolite porphyry* is the name given to the rock (Fig. 4-33). The groundmass of these porphyritic rocks is usually very-fine-grained and sometimes even glassy. If no phenocrysts are present at all and the rock is composed of only dense glass, the rock is called *obsidian* rather than rhyolite.

Fragmental material cast out of volcanoes by explosive eruptions often form extensive deposits after falling. If the fragments are all of small pieces (ash size) of volcanic glass and minerals of rhyolite rocks, the rock so accumulated is called *rhyolite tuff*. But when the fragments are composed of larger pieces of rock and volcanic bombs, the rock is called *rhyolite breccia*.

Syenites and Trachytes

The rocks of this group all have light-colored feldspars, orthoclase and plagioclase, but quartz is lacking. *Syenites*, those with phaneritic texture, are quite similar to granites, but granites contain large quantities of quartz. The quartz content of syenites is very much in the minority or is lacking alto-

Fig. 4-34: Nepheline syenite from near Little Rock, Arkansas, showing medium gray color due to feldspars and nepheline, and minor amounts of black ferromagnesian minerals.

gether. The predominant feldspars typically are alkali types, such as orthoclase, but a little sodic plagioclase (albite or oligoclase) may be present. Orthoclase usually makes up more than two-thirds of the total feldspar. Syenites often contain a little hornblende and some biotite as accessories. Other accessory minerals usually include magnetite, apatite, and zircon. *Biotite syenite* and *hornblende syenite* are varieties. Another variety is *nepheline syenite*, shown in Fig. 4-34.

In those rocks in which orthoclase feldspar phenocrysts are present, the texture becomes porphyritic. If phenocrysts in a groundmass of feldspar grains are abundant, the rock is *syenite porphyry* (Fig. 4-35).

The average syenite contains about 80 percent feldspar of which more than two-thirds is orthoclase; the remainder of the feldspar is plagioclase, usually albite or oligoclase. The balance of the rock, about 20 percent, consists of dark minerals such as hornblende, biotite, or even occasionally pyroxene. The colors of syenites are similar to granites: white, gray, pink, red, and occasionally blue. These colors arise from the predominance of the same minerals as in granites. Syenites thus are closely related to granites, but syenites contain little or no quartz. Syenite is a much rarer rock than granite.

Fig. 4-35: Syenite porphyry from near Wausaw, Wisconsin, showing light pink orthoclase, gray plagioclase feldspars phenocrysts and minor amounts of black hornblende. Quartz absent.

Fig. 4-36: Trachyte from Teller County, Colorado, showing fine grained nature. Rock is light gray in color, and has a few phenocrysts of feldspars. Large dark area is weathered portion.

The fine-grained rocks of this group, called *trachytes*, are usually light-colored, containing light-colored feldspars as the predominant ones. Common colors are white, gray, pink, and red. As in syenites, quartz fails. On the average, trachytes contain about 80 percent feldspar, predominantly orthoclase. The fine-grained rocks sometimes are slightly porphyritic, containing feldspar phenocrysts (Fig. 4-36). Biotite and hornblende may be present in small quantities. *Biotite trachyte* and *hornblende trachyte* are varieties when these minerals are in fair amounts. When the texture is very-fine-grained throughout, it is difficult to tell trachyte from rhyolite and quartz latite. But when a few phenocrysts are present, the identification becomes easier. Since quartz is absent, phenocrysts are usually orthoclase feldspar which, according to their composition, aid in naming the rock. When phenocrysts are abundant, *trachyte porphyry* is the name given to the rock (Fig. 4-37).

The groundmass of these porphyritic rocks is usually very-fine-grained. If no phenocrysts are present at all and the rock is wholly dense glass, the rock is called *obsidian*.

Fig. 4-37: Trachyte porphyry containing white to light pink orthoclase in a reddish-brown aphanitic groundmass.

Fig. 4-38: Quartz monzonite from Bishop, California, showing uniform granular texture of feldspar, hornblende and quartz.

Fragmental material of trachyte composition deposited as volcanic ejecta, consisting of volcanic glass, are called *trachyte tuffs*. But these are quite rare in America, and they resemble rhyolite tuffs, except for possible quartz content of the latter.

Quartz Monzonites and Quartz Latites

The *quartz monzonites*, the phaneritic rocks of this group, have light-colored minerals, quartz, orthoclase, and plagioclase as essential minerals. The orthoclase and plagioclase are in nearly equal amounts, each ranging from one-third to two-thirds of the total feldspar content (Fig. 4-38). Common dark-colored minerals are biotite, hornblende, or augite, generally as accessories, but others such as apatite, pyrite, or magnetite may also be present. If phenocrysts of feldspars are abundant, the rock becomes *quartz monzonite porphyry* (Fig. 4-39).

Quartz monzonites, as an average, contain about 20 percent quartz, about

Fig. 4-39: Quartz monzonite porphyry containing plagioclase and orthoclase phenocrysts in a finer grained gray matrix.

Fig. 4-40: Quartz latite with a few phenocrysts of feldspar and aphanitic light gray matrix.

60 percent feldspar, both orthoclase and plagioclase being in nearly equal amounts, and 10 to 20 percent dark silicates. Colors of quartz monzonites vary from pink, white, to light gray.

The fine-grained rocks of this group are called *quartz latites*. These rocks contain quartz, orthoclase, and plagioclase. The orthoclase and plagioclase each make up between one-third and two-thirds of the total feldspar (Fig. 4-40). The quartz latites somewhat resemble rhyolites and dacites; so the actual quantities of the two feldspars are needed for accurate classifying. This requires microscopic study.

Quartz latites usually contain 15 to 20 percent quartz, about 60 percent feldspar, and 10 to 20 percent of the dark silicates biotite or hornblende. Other dark accessories such as apatite, pyrite, or magnetite may be present.

The presence of abundant phenocrysts of quartz or feldspar permits the name *quartz latite porphyry* (Fig. 4-41). The groundmass of these porphyritic rocks is usually very-fine-grained or sometimes glassy. If no phenocrysts are present and the rock is wholly dense glass, it may be called *obsidian.* Cellular lavas (pumice) of *quartz latitic* composition are not common.

Fig. 4-41: Quartz latite porphyry with numerous white feldspar phenocrysts in an aphanitic pink groundmass. Minor amounts of fine grained quartz are present.

Fig. 4-42: Monzonite from southern California showing predominance of feldspar, both plagioclase and orthoclase, with fair amount of black hornblende. Quartz absent.

Fragmental material deposited as volcanic ejecta and consisting of quartz latitic composition are *quartz latite tuffs*. They resemble tuffs of rhyolitic or dacitic composition.

Monzonites and Latites

Monzonites, the phaneritic rocks of this group, contain orthoclase and plagioclase in nearly equal amounts. Both types of feldspars range from one-third to two-thirds of the total feldspar. Quartz is lacking or absent altogether. Dark silicates, hornblende, biotite, or sometimes augite are common accessories, but any of the three may be fairly abundant, as shown in Fig. 4-42.

Monzonites contain about 70 percent feldspar, both orthoclase and plagioclase, 15 to 25 percent dark silicates, and a few percent of other accessories such as pyrite, apatite, or magnetite. Monzonites are usually medium to dark gray in color. If phenocrysts of feldspars are abundant, the name *monzonite porphyry* is applied to the rock (Fig. 4-43).

The fine-grained rocks of this group are called *latites*. The latites contain both orthoclase and plagioclase feldspar (andesine or oligoclase) in nearly equal amounts, each ranging from one-third to two-thirds of the total feldspar. The latites somewhat resemble trachytes, dacites, and andesites, and

Fig. 4-43: Monzonite porphyry containing white and pink feldspar phenocrysts. A small amount of quartz and dark silicates are present.

Fig. 4-44: Light gray latite showing aphanitic texture of predominent sodic plagioclase. Note few small black crystals of hornblende. Slightly porphyritic.

microscopic determination of feldspar quantities are usually needed for accurate classification. Dark silicates are usually biotite, hornblende, or augite in small quantities (Fig. 4-44). Other dark accessories such as pyrite, magnetite, and apatite are often present too.

As an average, latites contain about 70 percent feldspar, both orthoclase and plagioclase, 15 to 25 percent dark silicates, and a few percent of other accessories. Latites most commonly are medium to dark gray in color.

When phenocrysts are abundant, the name *latite porphyry* is given to the rock (Fig. 4-45). The groundmass of these porphyritic rocks is usually very-fine-grained but sometimes glassy. If phenocrysts are absent and the rock is wholly glass, the rock is called *obsidian.* Cellular lavas (pumice) of latite composition are not uncommon. Fragmental material of latite composition, deposited as volcanic ejecta, are *tuffs* and may resemble tuffs of trachytic or andesitic composition.

Quartz Diorites and Dacites

The coarse-grained rocks of this group, called *quartz diorites*, have a predominance of the light-colored minerals feldspar and quartz. The chief

Fig. 4-45: Latite porphyry containing large light gray feldspar phenocrysts in a light pinkish-gray aphanitic groundmass.

249

Fig. 4-46: Quartz diorite containing abundant plagioclase, minor large black hornblende and quartz.

feldspar is sodic-plagioclase (oligoclase or andesine) which is present in dominant quantities (Fig. 4-46). The dark silicates biotite or hornblende are usually present in minor amounts. Quartz diorites are somewhat closely related to granites, but quartz diorites contain dominant quantities of plagioclase, whereas granites have dominant orthoclase. However, when orthoclase increases, such a quartz diorite is called *granodiorite* (Fig. 4-47).

Quartz diorites usually contain about 20 percent quartz, about 60 percent feldspar, and 20 percent or less of biotite or hornblende. In addition, magnetite in small quantities often is present, along with apatite, chlorite, epidote, and garnet. When feldspar (or quartz) phenocrysts are abundant, the term *quartz diorite porphyry* is employed.

The fine-grained rocks of this group are called *dacites* and contain light-colored sodic-feldspars and quartz. They resemble rhyolites very strongly and are very hard to distingush from rhyolites. But study reveals that the dacites contain predominant plagioclase feldspar; in rhyolites the predominant feldspar is orthoclase. Dacites usually contain about 15 to 20 percent quartz,

Fig. 4-47: Granodiorite, with plagioclase and substantial orthoclase, with minor amounts of hornblende and quartz. Usually classed as a variety of quartz diorite.

Fig. 4-48: Dacite, an aphanitic volcanic rock, containing predominant plagioclase, with minor quartz and hornblende. Note a few small white feldspar phenocrysts.

about 60 percent feldspar, and 10 to 20 percent biotite or hornblende (Fig. 4-48).

The quantity of potash feldspar (usually orthoclase) may vary somewhat. But in this group, the quantity of orthoclase may be less than or more than 10 percent of the total feldspar. Detailed microscopic study and laborious mineral counts are required to determine the amount of orthoclase. Dark silicates, usually in small quantities, are apt to be biotite, hornblende, or rarely augite. Biotite is the most common. Other accessory minerals, particularly magnetite, apatite, calcite, chlorite, and epidote, may be present, but they are usually inconspicuous. Colors of dacites range from light grays to yellows to pale reds.

When phenocrysts of plagioclase and quartz are quite abundant, the rock is called *dacite porphyry* (Fig. 4-49). The groundmass of these porphyritic rocks is usually very-fine-grained, but it may grade to glassy. If no phenocrysts are present and the rock is wholly dense glass, the rock is called *obsidian.* In addition, cellular lavas (pumice) of dacitic composition are known in volcanic areas.

Fragmental material deposited as volcanic ejecta, consisting predomi-

Fig. 4-49: Mica dacite porphyry from Colorado with plagioclase phenocrysts, minor black mica and quartz.

Fig. 4-50: Diorite from Salem, Massachusetts, composed of feldspar and dark pyroxene, biotite, and hornblende.

nantly of dacitic composition, are referred to as *dacite tuff*. They somewhat resemble other tuff of trachytic or rhyolitic composition.

Diorites and Andesites

The phaneritic rocks of this group, called *diorites,* contain sodic-calcic plagioclase feldspar in large quantities with the sodic type more abundant. Plagioclase dominates over orthoclase. Quartz is lacking, but small quantities of augite may be present. Hornblende is usually more abundant than biotite (Fig. 4-50). Diorites are very similar to gabbros, but rocks are called diorites if the plagioclase feldspar is more acidic (sodic) than labradorite, i.e., in the oligoclase-andesine range. Similar rocks with the more basic (calcic) plagioclase are called gabbros. But microscopic study is necessary to determine the composition of the plagioclase.

When phenocrysts are abundant, the term *diorite porphyry* is applied (Fig. 4-51).

Average diorites consist of about 65 percent plagioclase, 35 percent of the dark silicates biotite or hornblende (or both), and augite in very minor

Fig. 4-51: Diorite porphyry with light gray phenocrysts of sodic plagioclase (oligoclase-andesine) in a dark groundmass of additional phaneritic plagioclase and black hornblende. Quartz absent.

Fig. 4-52: Reddish-brown andesite near Alpine, Texas, showing uniform aphanitic fine-granular texture.

quantities. Accessory minerals of magnetite, quartz, apatite, calcite, chlorite, garnet, epidote, and potash feldspar are not uncommon. A common variety is *hornblende diorite*. Colors of diorites range from light to dark gray or grayish-green.

The fine-grained rocks of this group, called *andesites*, consist of predominant feldspars, most of which are plagioclase; but sodic plagioclase is the main type. Quartz is absent, but the dark silicates hornblende, olivine, and augite are present. Biotite is often present also but usually in smaller quantities than the hornblende and augite. Andesites usually contain about 65 percent plagioclase, and the remainder of the rock is usually biotite and hornblende. If biotite or hornblende is fairly abundant, *biotite andesite* or *hornblende andesite* become variety names. Colors of andesites range from grays to gray-greens, but they often are red or pink (Fig. 4-52). Andesites which are completely fine-grained are difficult to distinguish from dacite, latite, trachyte, or rhyolite. However, most andesites are porphyritic, and when a few phenocrysts are present, the composition can be determined with more certainty. In cases where phenocrysts (usually of feldspars) are very numerous, the rock is called *andesite porphyry* (Fig. 4-53). The groundmass

Fig. 4-53: Andesite porphyry containing phenocrysts of white feldspars. The grayish-green matrix is aphanitic.

Fig. 4-54: Pyroxene gabbro from near Stillwater, Montana, with light gray plagioclase and abundant dark pyroxene.

of these porphyritic rocks is usually very-fine-grained feldspars or even glassy. If no phenocrysts are present and the rock is wholly dense glass, the rock is called *obsidian.* Cellular lavas (pumice) of andesitic composition are known.

Andesitic fragmental material, *tuffs* and *breccias,* are fairly common in the western United States in volcanic areas, but they resemble tuffs and breccias of other compositions.

Gabbros and Basalts

Gabbros, the phaneritic rocks of this basic series, make up a large and variable group. In the gabbros, dark silicates are nearly as abundant as the light-colored ones, but many gabbros exist with variable ratios of dark to light minerals. In addition to large amounts of dark calcic plagioclase (labradorite), the dark silicate minerals in significant amounts are nearly always one of the pyroxenes, usually augite, as in Fig. 4-54, and olivine (Fig. 4-55). Small amounts of hornblende or biotite are only rarely present in certain gabbros. If the pyroxene crystallized in the orthorhombic system, the rock is called *norite* (a variety of gabbro). But microscopic study is required to detect

Fig. 4-55: Olivine gabbro from Stillwater, Montana, containing predominant plagioclase, some dark pyroxene and olivine.

Fig. 4-56: Anorthosite from Stillwater, Montana, containing coarse crystals of light bluish-gray plagioclase feldspar (labradorite).

this. In certain gabbros, quartz, garnet, and corundum may be present in quite small quantities.

Gabbro rock consisting wholly of pure coarsely crystalline labradorite (rarely anorthite) feldspar is called *anorthosite*, a variety of gabbro (Fig. 4-56). Anorthosites are commonly brown or dark gray. Large bodies of anorthosite rock are simply differentiates of large masses of gabbroic magma. When plagioclase, augite, or olivine phenocrysts are abundant, the rocks are called *gabbro porphyry*. Gabbros, on the average, contain about 50 percent plagioclase (calcic), about 30 percent augite, and 10 percent olivine. Magnetite, ilmenite, iron, and copper sulfides are usually present in very small amounts. It must be understood, however, that the gabbro group is a variable group, and these percentages are not constant. Gabbros are known with practically 100 percent feldspar (anorthosites) to down to about 10 percent feldspar. Gabbros may be dark gray, green, or black in color.

The fine-grained rocks of this group, called *basalts*, contain large quantities of dark-colored silicates including dark (usually dark gray) calcic feldspar (Fig. 4-57). Consequently, basalts are dark gray to dark green to brown to

Fig. 4-57: Porphyritic basalt. Note scattered augite crystals (phenocrysts) in dark gray groundmass.

Fig. 4-58: Dark gray vesicular basalt, known as the Rawls basalt, from west Texas. Vesicles are lined with white crusts of opaline material.

black in color. In the basalts, augite and olivine are abundant, but hornblende and biotite are scarce or absent altogether. Magnetite is often present in small quantities. A variety of basalt which contains long feldspar crystals lying at random positions and associated with augite is called *diabase*. Basalts usually contain about 50 percent plagioclase, about 30 percent augite, and 10 percent olivine, but these are variable. Vesicular basalt is shown in Fig. 4-58.

The basalts generally have a very-fine-grained texture and are heavy rocks. When phenocrysts of augite, olivine, or plagioclase are abundant, the term *basalt porphyry* is employed (Fig. 4-59). The groundmass of these porphyritic rocks is very-fine-grained or may be glassy (Fig. 4-60). If phenocrysts are absent and the rock is wholly glass, the rock is called *obsidian*. Cellular basalts (scorias) are very abundant in volcanic regions.

Fragmental materials of basaltic composition, deposited in volcanic regions, are referred to as *basaltic tuff*, *agglomerates*, or *breccias* according to their physical aspect.

Peridotites and Augitites; Pyroxenites and Limburgites

Gabbros grade over to *pyroxenites* and *peridotites* by substantial decrease in plagioclase. In pyroxenites, the rocks increase in pyroxene (usually augite) substantially (Fig. 4-61). The peridotites come in when both pyroxene (usually augite) and olivine increase. These two rocks, the pyroxenites and peridotites, are generally classed as *ultrabasic* rocks and are usually very dark in color, dull green to black. Hornblende, magnetite, pyrrhotite, biotite, and garnet may be present. Peridotites consisting nearly completely of olivine are called

Fig. 4-59: Olivine basalt porphyry from Colorado. Note the large phenocrysts of olivine and plagioclase, with minor augite in dark gray to black fine crystalline groundmass.

Fig. 4-60: Basalt porphyry containing large black augite phenocrysts in a fine grained dark gray matrix.

Fig. 4-61: Pyroxenite from South Africa showing coarse texture produced by dark brown pyroxene crystals.

Fig. 4-62: Green peridotite consisting almost entirely of granular glassy olivine, and thus is a dunite. Note a few small black grains of chromite.

dunites, which are pale green or dark greenish-brown in color (Fig. 4-62). In some peridotites, magnetite is quite abundant, and chromite is often present. A common mineral of both peridotites and pyroxenites is black hornblende. When this mineral is present in large quantities, the rock is called the variety *hornblendite*. Brown biotite is also common in peridotites and pyroxenites. *Porphyritic mica peridotite* (called kimberlite) is rare, but in South Africa this rock contains diamonds.

When phenocrysts are quite abundant in a phaneritic groundmass, the rock is called *peridotite porphyry* or *pyroxenite porphyry*, accordingly.

The fine-grained rocks of these two groups, the *augitites* and *limburgites*, contain large quantities of dark silicates, particularly augite in the former and both augite and olivine in the latter. Plagioclase feldspar is scarce or absent altogether. Hornblende, magnetite, and pyrrhotite may be present. Chromite too may be present. These rocks are usually glass-rich.

It is to be pointed out that augitites and limburgites are very rare rocks. When they are found, they are apt to be called basalt until a great deal of microscope work is completed to make mineral determinations.

When the fine-grained nature of these rocks fails, the rocks become glassy, and such rocks are called *basaltic obsidians* rather than *augititic obsidian* or *limburgitic obsidian*. Cellular glassy rocks of this composition are included in the term *scoria*.

The Glassy Igneous Rocks

A number of igneous rocks have glassy texture. In most of these rocks, no crystals can be seen, but in some cases, very tiny crystallites or microlites of minerals may be present. Such crystallites or microlites could be seen under a high-power microscope. The glassy igneous rocks have glassy (vitreous) or resinous luster.

Fig. 4-63: Black obsidian with excellent conchoidal fracture. This is a glassy volcanic rock.

Obsidian and Its Varieties

Obsidian is volcanic glass with a bright vitreous luster, and it has a good conchoidal fracture. Obsidian is commonly black, but other colors such as red, brick red, brown, and gray have been reported. The colors give no clue to the composition (Fig. 4-63). Colors are most likely due to dust-like particles of magnetite or hematite. Most obsidian has a specific gravity of about 2.4 and is somewhat harder than window glass. Obsidian may form as a result of very rapid cooling of extrusive magma, preventing crystal growth to occur. But obsidian may also form as a result of a very viscous magma which was simply too rigid for crystal growth (Fig. 4-64).

Obsidian often contains crystallites and microlites, but it also often contains spherulites and stone bubbles. Spherulites are rounded, layered masses ranging from microscopic to pea size. Crystals of feldspar often radiate from the centers of the spherulites. Stone bubbles (also called lithophysae) are also rounded masses of varying sizes composed of concentric layers containing aggregates of fragile crystals. There is often a central cavity lined with tiny crystals of quartz, feldspar, tourmaline, or others.

Actually, obsidian may have a composition comparable to any of the phaneritic rocks. Since the glassy rock cannot be identified by eye due to absence of crystals, a chemical analysis may be required in order to determine

Fig. 4-64: Dark brown and black obsidian from Oregon, showing flow structure and good conchoidal fracture.

Fig. 4-65: Pitchstone, a variety of obsidian. Note the dull pitchy luster.

its exact composition. Although most obsidian compares to granite in composition, other compositions are common. But the name *obsidian* is sufficient for general purposes and nomenclature.

A fairly common variety of obsidian is *pitchstone*. This variety has a dull pitchy (tar-like) luster due to the presence of microscopic water bubbles (Fig. 4-65). Pitchstone is commonly black or gray, but brown and green ones are also reported.

Another variety of obsidian, called *perlite*, is a highly fractured obsidian, the fractures having developed by contraction during cooling. The fractures are nearly always concentric and produce onion-like masses. Perlites commonly are gray or bluish-gray, and they contain 3 or 4 percent water. Perlite is shown in Fig. 4-66.

Pumice is a variety of obsidian having a light glassy frothy texture. The rock is composed of silky glass fibers and has tiny pores or vesicles between fibers. Because the fibers enclose empty sealed bubbles, pumice usually is so light that it floats on water. This type of obsidian is commonly produced in

Fig. 4-66: Perlite, a variety of obsidian, containing numerous tiny curving cracks.

Fig. 4-67: White pumice, a volcanic rock, showing fine, frothy nature. Note tiny pores produced by escaping vapors. Rock is composed of silky glass fibrous material, and is lighter than water.

the lava in the throat of a volcano or in the top part of a lava flow where pressure is relieved and vapors and gases are permitted to escape. Fragments blown out by exploding volcanoes are often pumice. The escaping vapors and gases tend to puff up the magma into spongy form (Fig. 4-67). Pumices range in color from white to light yellow to gray-brown. Most pumice has a composition similar to granite, but a chemical analysis is necessary to make the determination.

When basaltic lavas develop vesicular form as a result of expansion of gases, they are called *scoria*. This variety of obsidian resembles cindery furnace slag, and it usually has a stony or glassy texture. The scoriaceous nature of this rock is developed when water and other gases bubble up through fairly stiff lava, with the production of holes that are much larger than those in pumice. Scoria ranges in color from reddish brown to dark gray to black and commonly has a basaltic composition. Scoria from Oregon is shown in Fig. 4-68.

Fig. 4-68: Scoria from Oregon, a cellular volcanic rock of brownish-black and reddish-brown color. Cellular structure results from separation of dissolved vapors.

Basaltic glass, a variety of obsidian, is a rather rare rock, but it does occur in small quantities as thin crusts on lava flows and as fragmental material such as bombs, cinders, or scoria thrown out by basaltic volcanoes. The texture of basaltic glass may be vesicular or spherulitic, and a few crystals of feldspar or olivine may be present. Basaltic glass, usually called *tachylyte*, is usually black in color, but very dark brown ones are also reported.

Another common type of obsidian is *vitrophyre*. Vitrophyre is a porphyritic obsidian, containing small phenocrysts of feldspar or biotite or quartz. The phenocrysts may be few or many. Vitrophyre is usually named according to the phenocrysts contained in them, such as *feldspar vitrophyre*, *biotite vitrophyre*, or *quartz vitrophyre*.

Fragmental Varieties

Fragmental igneous rocks blown out from volcanoes are termed *pyroclastic material*. During violent explosive eruptions, a great deal of material may be deposited around in the volcanic area as well as over adjacent areas. Explosive eruptions are due to release of gases under pressure in the rising magma as it approaches the surface. The escaping gases carry the liquid magma up to great heights. The magma hardens while aloft and then falls as spongy solid fragments. Often mixed with these new spongy pyroclastics are solid pieces of older lava torn from the volcanic vent after having been solidified there during a previous eruption.

Fragments of lava, including pieces of old lava and solidified pieces of newer lava thrown out during violent eruptions, range in size from very fine dust to large masses several feet across. Particles smaller than 4 millimeters include *volcanic ash* and *volcanic dust*, the latter being the very finest portion. Fragments ranging between 4 and 32 millimeters across are called *lapilli*, and those larger than 32 millimeters are called *bombs* or *blocks*. Bombs, often but not always rounded to subrounded in shape, are formed by solidification of liquid magma in flight, while blocks are merely solid, usually angular pieces of rock tossed out during the eruption. Needless to say, bombs may take on different shapes according to the shape of the magma while it was solidifying during eruption. Such varieties are *ribbon bombs* and *spindle bombs*, while *bread crust bombs* have cracked outer crusts as a result of the expansion of interior gases after the exterior crust had hardened.

Fig. 4-69: Pink to tan colored volcanic agglomerate, showing scoriaceous and rounded nature of this rock from Los Alamos County, New Mexico.

Fig. 4-70: Light tan colored tuff from west Texas showing fine-grained gritty-like rough volcanic ash of which it is composed.

Broken rock material which is blown out of volcanoes may become mixed as they fall to the ground. The fragmented material making up this mixture may be obsidian, scoria, lava, ash, lapilli, and bombs. If the material has a dominance of rounded fragments, the rock deposit is called *agglomerate* (Fig. 4-69). If the fragmented material is chiefly angular in shape, the rock deposit is called *volcanic breccia.* Loose fragmented rock deposits often become cemented together by volcanic ash or by lime or silica from surface waters.

Agglomerate and volcanic breccia vary in color from browns to reds to even yellows and grays, depending upon the color of the main fragmental material and the cementing ingredients. Agglomerates and volcanic breccias commonly are cemented by tuff and may contain angular lapilli. Often interlayered with this volcanic debris can be fragments of old limestones, shales, sandstones, granites, gneisses, or schists, rocks which may have served as the subvolcanic foundation rocks which yielded to explosive volcanism.

A fine-grained volcanic pyroclastic rock with a low specific gravity is called *tuff.* Tuff is volcanic ash and lapilli which have become cemented rather firmly by mineral material deposited by surface waters. Tuff may resemble other light-colored volcanic rock and often may be mistaken for a fine-grained sediment because the ash which makes up a tuff may fall in layers on the ground or in lakes. Tuff ranges from white to yellow to pink to light brown to light gray (Fig. 4-70). Rarely are they dark in color. They are rough to the feel because of the presence of sharp-cornered dust particles. In most tuffs, the volcanic material ranges from coarse (lapilli) to fine (ash), and thus they are poorly sorted. The glassy ash grains may be nearly any shape, ranging from elongated, curved, to spicule-shaped. For the most part, the glassy particles are colorless. Tuff may even contain fossils of vegetation if it falls on land, or fossils of marine-type animals if it falls in shallow lakes and along coastlines.

A tuff, then, is a type of stratified sandstone, sedimentary in nature, containing volcanic debris of glass, crystals, and igneous rocks. Layers are produced by ash falls and by deposition by surface waters. If normal sedimentary material is dominant over the volcanic debris, the rock may be called *tuffaceous sandstone.* On the other hand, if volcanic material exceeds the sedimentary material, the name *sandy tuff* or *clayey tuff* may be used.

Chapter Five

Nature of Sedimentary Rocks

The second great subdivision of rocks is known as *sedimentary rocks*. Sedimentary rocks consist of material that was derived from preexisting rocks by various weathering processes, then was transported, and later deposited under a wide variety of environments. However, all sedimentary environments are considered as being within normal temperatures, i.e., approximately 0°C to 100°C, and under low pressures of 1 atmosphere. Such environments are located on the surface or very high in the outer crust of the Earth. The material of which sedimentary rocks are composed was obtained from rocks of all classes of igneous, sedimentary, and metamorphic types. Rocks of various types, located on the land masses, are subjected to weathering processes which develop decay products of two main types: mechanical breakdown and chemical decomposition.

The products of weathering are then moved along and deposited chiefly by running water. In some environments, however, the wind moves and deposits them, while in other environments, moving ice has caused the material to be moved and deposited. However, material that has been moved along and deposited by the action of water makes up a very large part of the sedimentary material. Material moved and deposited by the wind is far less abundant, and that moved and deposited by ice is still less abundant.

Since the larger proportion of sedimentary rock material is carried in and deposited by water, a little more discussion of the factors concerned is advisable at this time. Material derived from the landmass undergoing weathering is made up of fragmental particles, derived mechanically, and also of chemical matter, derived chemically. These two types of material, fragmental and chemical, are moved by water to new locations where they accumulate. The fragmental material accumulates as conglomerates, sandstones, and clays. The chemically derived material is carried in solution in the water and then later is precipitated from those water solutions by various chemical processes or by organic processes. In many cases, some of the chemically precipitated material may become mixed with the fragmental material, as both may be deposited in the same area. In some cases, organically deposited material may be mixed with fragmental or chemical deposits. It may also happen that in areas where chemically precipitated deposits are forming, varying quantities of fragmental material can become intermixed. On the other hand, large masses of fragmental material and large masses of chemical material may be deposited in adjacent areas with little or no mixing by the other. Thus, in a broad sense, there are two classes of sedimentary rocks:

(1) Clastic (or fragmental), produced chiefly by mechanical processes;
(2) Nonclastic (or chemical), produced by chemical reactions of two types.

Even though there are two basic classes, it is sometimes quite difficult to draw a clear line between them.

It is common practice to speak of material which has been deposited by sedimentary processes as *sediment*. But the term sediment usually is applied to the material when it is in a soft and unconsolidated condition. Sediment becomes firm sedimentary rock only through a series of lithification processes called *diagenesis* (discussed more fully in a later section). Even after sediment becomes sedimentary rock, the term sediment is often carelessly applied to the rock.

Weathering, Sedimentary Processes, and the Formation of Sedimentary Rocks

Whenever rocks are exposed to the atmosphere by being lifted up by earth forces or by removal of overburden by erosion, they undergo a gradual decay or alteration. Such decay brought on by the action of air, water, and various water solutions is generally called *weathering*. Weathering is partly a *physical* breakdown and partly a *chemical* breakdown of rock at the surface. The physical disintegration (or mechanical breakdown) results in a reduction in size of the rock and mineral particles without any chemical changes occurring in the fragments. However, as the particles are made smaller by progressive breakdown, they become more subject to chemical decomposition as more surface areas are exposed on the fragments. The chemical decomposition or chemical breakdown results in a change in the chemical characteristics and physical characteristics of the rocks and minerals. As the particles become changed by the chemical processes, they may also be continually acted upon by the mechanical processes. In fact, processes of chemical decomposition seldom proceed alone; they are usually intimately associated with the mechanical processes of disintegration. In many cases, the action of physical breakdown promotes chemical action, and the action of chemical breakdown often permits physical processes to proceed more rapidly. A good example of this interaction of physical and chemical processes can be seen where a sandstone is being weathered. Chemical processes are removing the soluble cement between sand grains; the result is that the cohesion of the sandstone is weakened, and the rock crumbles as the loosened sand grains fall away.

Weathering of rocks proceeds quite gradually, and the total breakdown of rocks is brought about by the interrelated action of several factors. It seems appropriate at this time to discuss separately the different factors which cause the physical disintegration and the chemical decomposition of rocks and minerals undergoing weathering.

Physical Disintegration

The heating and cooling of rocks by daily and seasonal changes of temperature sets up strains in the rock, with resultant breakdowns. These strains can find relief through *granular disintegration*.

Another process of disintegration is known as *exfoliation*. This process occurs when a rock is heated intensely by the sun or by a forest fire. The outer portion of the rock increases in temperature, expands, and becomes weakened. The process of exfoliation occurs also when the temperature falls. Heat in the rock is lost more rapidly from the surface than from the interior. The surface layers tend to contract, and small cracks often develop on the surface allowing the weakened surface layers to spall off in slabs.

Frost wedging occurs if any water is present in the rock pores, cracks, or bedding planes and becomes frozen during extreme cooling; the rock may be shattered as the ice forms. Water expands when it freezes, and it crystallizes with a fairly strong force, forcing apart the rock very suddenly.

Another important and effective agent of mechanical disintegration is the *abrasive action* of rock material while it is being transported. It includes two associated actions called *grinding* and *impact*. All three of these processes tend to reduce the sizes of all particles. Abrasion occurs when rock particles being carried in water, in ice, or in the wind are rubbed against other rock masses over which they move to wear the latter gradually away. The grinding process occurs when small particles are caught between larger particles and become ground down to very fine powder. Impact occurs when two particles or masses of rock are smashed against one another as transportation proceeds, whether it be by water or wind. The result of impact is that flaking and chipping take place; rocks are thus reduced in size. Abrasion increases with increase in the amount of material being transported by air, water, or ice.

Chemical Decomposition

Rocks which are exposed at the surface of the Earth undergo chemical decomposition as well as mechanical disintegration, both sets of processes essentially operating simultaneously. Chemical decomposition is brought about by the action of natural water, various natural acids, oxygen of the air and free oxygen in natural waters, and the carbon dioxide content of waters. Chemical decomposition generally proceeds more rapidly when temperatures are high and when moisture is readily available than when temperatures are low and moisture is not available. These two factors, elevated temperatures and moisture, are important because they promote rapid chemical reactions on the minerals of the rocks. Consequently, chemical weathering is more rapid in warm humid regions than in cool or dry climates.

The most important agent of chemical decomposition of surface rocks is water. Water, consisting of H_2O, is partially dissociated, i.e., a little of the $H+$ ions are separated from the $(OH)-$ ions. This dissociation permits other ions to be dissolved in the water. In fact, water dissolves oxygen and carbon dioxide out of the atmosphere and also easily dissolves sodium and potassium from soil and surface rocks; it thus becomes a powerful reactive agent. Additional dissolved substances in surface waters are derived from decaying vegetation.

Minerals in the surface rocks undergoing chemical decomposition may consist of silicates, oxides, sulfides, carbonates, sulfates, etc. These minerals are changed chemically into hydrated silicates, hydroxides, new carbonates, new oxides, new sulfides, etc., which, for the most part, are more soluble than the ones from which they were derived. This easily dissolved material is then later redissolved and carried away in water of streams and on to the ocean. Once in the ocean, the material may stay dissolved in the water for fairly long periods of time, but eventually these soluble salts will be deposited by various processes.

Because there is a wide variety of minerals present in the various rocks undergoing the weathering processes, there are varying degrees of solubility of the different minerals. Some minerals are very soluble, others are very insoluble, while others fall in between in solubility by weathering solutions. In the area undergoing weathering, the minerals which are very soluble will be removed easily, while those progressively more resistant will be removed more slowly. Those which are the most resistant will be dissolved very little. They will not undergo much decomposition but will remain fairly much as is, even if they are transported to a new location. While all this dissolving and mechanical loosening of material on the rock surfaces are proceeding very slowly, any accumulation of a layer or mantle of such broken and altered rock, developed during weathering, usually is called *soil*.

Development of Soil

Soil actually consists of well-weathered as well as partly weathered rock material. A layer of well-weathered and partly weathered rock is present nearly every place that rocks are exposed, except along ledges, cliffs, or crags

of hills. The soil is usually thicker in areas of gentle slope on flat lands than in areas of steep slope or mountainous areas, because in areas of steep slope, the soil is removed more quickly by water and gravity.

The processes involved in the chemical decomposition of surface rocks and resultant development of soil, in the presence of water, are hydration, oxidation, carbonation, and solution. *Hydration* is a process by which water, in the form of the hydroxyl ion (OH) water or H_2O water, is chemically added to minerals, into the molecular structure, producing a different mineral. Hydration occurs most often in silicates and oxides, but it may occur in others as well. A common example of hydration is given by the reaction:

$$2KAlSi_3O_8 + 2H_2O + CO_2 \longrightarrow Al_2Si_2O_5(OH)_4 + 4SiO_2 + K_2CO_3$$

In this example, orthoclase feldspar is changed to kaolinite, as the kaolinite has taken on four (OH) water molecules to become $Al_2Si_2O_5(OH)_4$, after potassium (K) has joined the CO_2 to become K_2CO_3.

Oxidation is a weathering process which involves the addition of oxygen to minerals. This process is nearly always accompanied by hydration, and it very commonly occurs in minerals containing iron. A simple example of oxidation is shown by the reaction:

$$2FeS_2 + 2H_2O + O_2 \longrightarrow 2FeSO_4 + 2H_2SO_4$$
$$\text{(pyrite)} \quad \text{(water)} \quad \text{(oxygen)} \quad \text{(ferrous} \quad \text{(sulfuric acid)}$$
$$\text{sulfate}$$
$$\text{mineral)}$$

Carbonation is a process whereby carbon dioxide, CO_2, or bicarbonate, HCO_3, ions combine with minerals. Carbonation is a very important and intense process which causes some of the major breakdowns of rocks. Hydration, also a strong process, usually is associated with carbonation. Rocks containing calcium (Ca), magnesium (Mg), and potassium (K) are quite easily decomposed by waters which have absorbed CO_2 from the atmosphere or other places to become weak carbonic acids, shown by the reaction:

$$2KAlSi_3O_8 + H_2CO_3 + H_2O \longrightarrow K_2CO_3 + 4SiO_2 + Al_2Si_2O_5(OH)_4$$
$$\text{(orthoclase)} \quad \text{(carbonic} \quad \text{(water)} \quad \text{(potassium} \quad \text{(silica)} \quad \text{(clay)}$$
$$\text{acid)} \quad \text{carbonate)}$$

The action of CO_2 on other minerals also produces carbonates or bicarbonates which are soluble and easily removed, ultimately to the ocean.

Solution (dissolving) comes into play as a weathering process in many different ways. It involves the dissolving of many minerals chiefly by water. The real importance of solution in weathering comes into play, not so much in the primary dissolving of original constituents of rocks in the weathering area, but in dissolving and removing the secondary products developed by other processes of decomposition. When natural waters dissolve different substances, they may become acid or alkaline, which permits them to be even more efficient in their dissolving powers.

Removal of Soil and Transportation of Sediment

As mentioned earlier in this chapter, the mantle of soil covering most of the solid rocks of the Earth's surface is derived as a result of a series of mechanical and chemical processes of weathering. This soil actually is the material from which a large part of the sedimentary rocks are formed, especially the clastic type. Other sedimentary rocks, especially the chemical type, are composed of some of the soluble mineral matter which was released (by solution) from the rocks in the weathering area. A third type of sedimentary rock is composed in large part of organic matter derived from the land surface or from the oceanic area, and it may be mixed with either or both of the other two types.

Once the soil has formed in an area, it becomes subject to removal chiefly by running water, but also by wind or ice if present. A soil is made up of a variety of materials available to be moved and worked over into sediments, e.g., undecomposed fragments of the original rock, partially decomposed and partially reconstituted products of alteration by weathering solutions, and new solutions formed as a result of decomposition.

Movement and transport of the mechanically loosened material, on its way to deposition as fragmental (clastic) sediment, is accomplished chiefly by running water. Transportation of fragmental material by running water generally involves temporary halts along the course of the stream until it ultimately reaches the sea. The dissolved load, however, usually reaches the sea more rapidly, but it too may be halted along the way temporarily, as it may become precipitated in lakes or similar places. Once the clastic (fragmental) material reaches the sea as sediment, it can easily be moved around by waves and currents. These same waves and currents also can cause coastline erosion of shore rocks, and thus additional material can be added to that supplied by the rivers.

A similar type of mechanical transport of sediment is produced by the wind. In fact, the wind is considered to be a very effective agent of transport and has been responsible for the transport of enormous quantities of dust during duststorms and also at other times. Wind action also plays a significant role in the movement and accumulation of desert sand dunes, as well as of sandy shores.

Transport of fragmental material is also accomplished by moving ice (glaciers). Broken and fragmental rock material of many different sizes may be carried atop the ice, frozen within the ice mass during accumulation, or dragged along beneath the ice.

While fragmental material is being transported by water, it is broken into finer pieces, and the angular corners become rounded off. Wind appears to be more efficient than water in the rounding of particles, and moving ice is very ineffective.

In many cases, fragmental material enroute to a depositional site may pass from the influence of one transporting medium to another. For example, fine silt may be picked up by a glacier and later be dropped out as melting occurs. Then it may be blown around by the wind for a time, and still later it may be picked up and moved by a stream.

Deposition (Sedimentation)

The fragmental and chemical material being transported by water, wind, or ice ultimately reaches the oceans. However, there may be times and conditions when this material is deposited temporarily along the way on the land. These deposits of sedimentary materials may stay on the land for long periods of time, perhaps several geological periods, before they continue on toward the sea.

Sedimentation may occur in a variety of environments on the Earth's surface: some on the land, some along coastal areas, and some out in the sea itself. Consequently, various types of sediments and sedimentary rocks are distinguished as being *terrestrial*, *marginal*, or *marine*, respectively, according to their place of deposition. Deposition may be by mechanical processes or by chemical processes, depending on whether the material was carried as clastic (fragmental) sediment or as chemical (in solution) sediment. Any material carried in suspension by water, wind, or ice will be deposited when the transporting agent becomes overloaded, slowed, or changed physically or chemically, whether it be on land or in the sea. For example, a river emptying into the sea in a deltaic region loses velocity quite rapidly, and the suspended load of sand, silt, and clay is deposited as the river current loses its power to carry.

Considerable portions of rock material are transported as *dissolved load*. The dissolved load (soluble material) also may be deposited as sediment by chemical processes, such as precipitation or evaporation, or possibly by organisms.

Bacterial activity also can cause precipitation of material from solution, as a result of living processes which affect the nature of the soluble material. For example, the deposition of bog iron ore in lakes is brought about by the vital activities of bacteria in the lake water.

Since most sediments are transported to depositional areas by water, they are deposited as bedded or stratified deposits. Strata (or beds) may differ from one another in composition, texture, or both as a result of variations in type of material or size of grains of material, or both factors at the time of deposition.

In most cases, any sediment is a mixture of both mechanical (clastic) material and chemical (including biochemical) material. The actual type deposited is determined by the relative amounts of the two types of sediment, and the balance can be shifted one way or the other by changing the volume of either.

Environments of Deposition

The sedimentary environment includes all the physical, chemical, and biological conditions which exist where a mass of sediment is accumulating. This combination of various types of conditions has a strong influence on the properties of the sediments so accumulating. The various processes operating in a given environment are broadly controlled to a large extent by the physiographic (or geographic) setting, by the relations of land and sea, and by the action of the geologic agents operating in the environment.

Most environments are dominated by a combination of factors, but occasionally only one factor dominates. The variations of conditions in the different environments are brought about by the following factors:

(1) Nature of the medium (water, wind, or ice) from which sediment is deposited.

(2) Depth of water plus size and characteristics of the sedimentary area.

(3) Temperature, pressure, and general climatic factors.

(4) Nature of motion of the medium depositing the sediment.

(5) Types and abundances of organisms present in the environment.

(6) Nature and composition of the sediments being brought in to the sedimentary area.

It becomes apparent that there are a great number of environments of deposition, but it is possible to group them into just a few major types which have similar characteristics.

Terrestrial

In the terrestrial (continental) environments, deposition takes place on the land masses above sea level, except very rarely where a basin is extremely low. Five subtypes of continental environments are observed: alluvial (stream), lacustrine (lake), paludal (swamp), desert, and glacial.

Marginal

In the marginal environments, deposition takes place at or very near sea level, along the margins of land masses. Marginal (transitional) environments are classed as deltaic, lagoonal, and littoral (shoreline), and all involve deposition by water. Numerous subdivisions occur among these, and some involve fresh, brackish, or normal marine water.

Marine

The marine environments include those in which oceanic water is the depositing medium. There are three main marine environments: neritic (now called sublittoral), which extends from low-tide level out to a depth of 600 feet; the bathyal, extending from 600 feet out to 13,500 feet; and the abyssal, which extends from 13,500 feet to 21,000 feet.

Lithification

The nature of a new sedimentary deposit is largely determined by the exact manner in which it is deposited and on the environment under which it came to rest. But after being deposited, the sediment may undergo significant changes while it is being converted into firm rock. All processes that lead to rock consolidation are referred to as *lithification* (also known as diagenesis). Diagenetic changes may be partial or complete. They may occur during deposition, soon after deposition, or a long time afterward.

A simple diagenetic process is *compaction*. As sediment accumulates, the weight of the material increases also, thus causing the particles to be pressed together. Much of the water is squeezed out, open space is reduced, particles become adhered to one another, and the rock gains strength.

Lithification may be accomplished also by *cementation*, a process by which

(1) a cementing material is precipitated in the open spaces, thus bonding the particles together, or

(2) fine-grained muddy material occupies the open spaces, thus producing a bond.

The latter is accomplished partly by compaction of the fine-grained materials in the open spaces. Excluding these fine-grained materials, the two most common cementing agents are calcite, $CaCO_3$, and silica, SiO_2, followed far behind by iron oxide (limonite), $FeO(OH) \cdot nH_2O$, and gypsum, $CaSO_4 \cdot 2H_2O$. Other common cements are dolomite, $CaMg(CO_3)_2$, and siderite, $FeCO_3$. The cementing material, which may be derived from the sediment itself or brought in from outside, may be introduced during or after sedimentation. The sediment thus becomes hardened by the presence of cement.

Diagenesis may be partly due to *authigenesis*, a process whereby new minerals form *in situ* in a rock usually in relatively small quantities. Minerals formed by authigenesis are products of chemical and biochemical action. Common ones are quartz, calcite, dolomite, siderite, feldspar, illite, chlorite, rutile, gypsum, and pyrite.

Secondary enlargement, growth of material around a nucleus of the same material, also occurs in sediment to a large degree. Both authigenesis and secondary enlargement add strength to the rock in limited ways.

Composition of Sedimentary Rocks

General Statement

The composition of a sedimentary rock is usually stated in terms of the minerals present, but sometimes the composition must be expressed in terms of its chemical composition. The chemical composition must be considered in the case of very-fine-grained rocks whose minerals cannot be readily determined. In these cases, some sort of a chemical analysis must be made, and to interpret such an analysis properly requires some knowledge of the compositions of other sedimentary rocks with which to make comparisons.

Mineralogy

As stated before, the sedimentary rocks consist basically of two types of material:

(1) clastic (or detrital) (also called fragmental) material, and

(2) chemical (and organic) material.

In a great many rocks, these two types of material are mixed in varying proportions, but in general, one type is dominant. But both types of materials are derived from the weathering and breakdown of rocks exposed on land surfaces. Such weathering yields three kinds of material which proceed to form the sedimentary rocks. These materials, listed in Table 5-1, may be

Table 5-1: COMMON MINERALS OF SEDIMENTARY ROCKS*

Detrital
  Stable Primary: quartz, feldspar, mica, rock fragments
  Stable Secondary: clay minerals, limonite, goethite, chlorite
Chemical
  Precipitates: chalcedony, silica, calcite, dolomite, aragonite, barite
  Oxidates: limonite, hematite, pyrolusite, psilomelane
  Reduzates: marcasite, pyrite, organic material, sulfur, siderite
  Evaporates: halite, gypsum, anhydrite, sylvite

---

*Modified after Mason and Berry

grouped according to the degree of stability they possess in the weathering complex. These three types of material are:

(1) stable primary minerals of the parent rock which survive weathering processes and are mechanically released as the rock is broken down;

(2) stable minerals which form secondarily by partial chemical decay of unstable minerals and sometimes partial reconstitution; and

(3) dissolved chemical ions from the parent rock which are precipitated by various chemical processes.

The first type, stable primary minerals which survive weathering processes, usually are deposited as sand-sized detritus, but much coarser and much finer-sized material may also be deposited. Since minerals have varying degrees of resistance to weathering, any of them may be moved as detritus, depending upon how incomplete weathering has been. However, only the most resistant minerals survive in most weathering areas, and these will also survive transportation, to be deposited mainly as *sands*. The most common of these "resistate" minerals are quartz, feldspar, micas, some others in very minor amounts, and also rock fragments.

The second type, the stable secondary minerals, formed by partial decomposition of unstable minerals and partial reconstitution, usually are of clay-sized detrital material. They are mostly crystalline, but also some amorphous material is included. During weathering of rocks, unstable silicate minerals are easily decomposed, partially or completely. When partially decomposed, they may be reconstituted to a small degree by the addition of water to yield "hydrolyzates." In their reconstituted condition, these usually can resist destruction to be deposited mainly as *clays*. The most common of these secondarily produced resistates are the *clay minerals*, usually assigned to three groups: kaolinite, illite, and montmorillonite. Also sometimes present is chlorite. Other important hydrolyzates are goethite and limonite, and occasionally bauxite, glauconite, chamosite, and collophane.

The third type, generally called the "chemical" materials because they are all carried as solutions of dissolved chemical ions, usually are deposited by chemical processes according to the environment in the depositional area. However, many chemical factors may influence chemical precipitation at any time. Depending upon the nature of the chemical reaction mechanism, the chemical sediments are classified as precipitates, oxidates, reduzates, and evaporates. The chemical sediments may be deposited alone or mixed with the other two types of material.

Textures of Sedimentary Rocks

When a sedimentary rock is described for purposes of identification and origin, the texture and the chemical composition must be considered. The *texture* refers to characteristics of particles and how the particles are arranged with one another. The *composition* means the mineralogical or chemical makeup of the rock. Sediments usually are made up of both clastic material (sand, silt, or mud) and chemical material (calcite, gypsum, or dolomite). Either of these two types of materials may be deposited alone (which is very rare) or mixed in any proportions, yielding sediments ranging from pure types to mixed intermediate types.

Clastic (or detrital) rocks usually have a *fragmental texture*, as shown in Fig. 5-1, in which the particles are broken, partially abraded, irregular, or

Fig. 5-1: Clastic textures: (a) rounded grains; (b) broken grains; (c) partially abraded grains; (d) irregular grains.

fairly well abraded (rounded). These particles are usually in direct contact with one another, displaying an interlocking pattern. Chemical rocks commonly have a *crystalline texture* consisting of interlocking crystals, many with crystal faces. The texture of a sedimentary rock is made up of three components: the dominant particles, the matrix (finer) material filling in the open spaces between the dominant particles, and the cementing material which acts as a binder of the dominant particles and matrix material. In some rocks, there is not very much distinction between the dominant and matrix material, but the cementing agent usually can be distinguished. This is because the clastic material may all be very nearly the same size, but the cementing material is usually a chemical material precipitated in the interstices of the clastic grains after they were deposited.

Textures of Clastic Rocks

The texture of clastic rocks is made up of four elements: particle size, particle shape, particle surface texture, and particle orientation (the fabric).

Particle Size

One of the most important elements of rock texture is the *particle size*. The size of constituent particles of any clastic rock is related to the energy of transportation as well as the energy in the depositional environment. The actual sizes of particles can be measured by passing the crumbled rock through a nest of sieves; thus the particles are sorted into size groups as determined by the sizes of openings of the sieves. Particle size can also be determined by their settling velocities in water.

Table 5-2: SIZE CLASSIFICATION OF DETRITAL MATERIAL

Material		Diameter in mm
Gravels	boulders	256 or more
	cobbles	256–64
	pebbles	64–2
Sands	sand grains	2–0.05
Silts	silt grains	0.05–0.005
Clays	clay particles	0.005 or less

The constituent particles in a clastic rock are classified according to their diameters measured in millimeters, according to the following arbitrary but precise classification given in Table 5-2. This classification has come into general use in recent years and is a very practical and usable one.

Particle Shape

Another important element contributing to rock texture is the *particle shape*. The shape of a particle involves the maximum and minimum dimensions and also the roundness of the particles. The shape of the particles partly controls how well the particles can be transported, and the roundness (or angularity) reflects the distance the particle has traveled or how intense the transportation conditions were. Consequently, the particle shape is made up of two parts, both of which are considered in studies of sediments.

The shapes of clastic particles vary a great deal, ranging from disk-like to blade-like to roller-like to spheroidal as illustrated in Fig. 5-2.

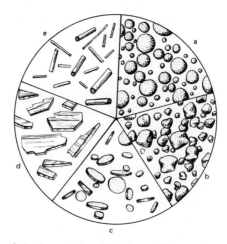

Fig. 5-2: Shapes of clastic particles: (a) spheres; (b) angular grains; (c) discs; (d) blades; (e) rods.

Roundness of particles is concerned with the sharpness of edges and corners and actually is independent of shape. Particles of very different shapes, for example, disks and rods, may have the same roundness, because roundness applies to the degree of rounding of corners and edges and not to the particle shape. The roundness of particles increases in the direction of transport as gradual abrasion of corners and edges continues. Large angular particles become rounder quite rapidly during transport, while small particles become rounded more slowly. The rate of rounding also depends in part on the hardness of the material making up the particles. For example, particles of limestone (hardness 3) become rounded more rapidly than particles of quartz (hardness 7). Consequently, quartz may not become rounded even after great distances of transport.

Surface Texture

Another contributing element to the texture of a sedimentary rock is the detailed *surface texture* of the particles. Surface texture (also called surface roughness) of a clastic particle is the composite appearance of its minute surface features. The surface features may be developed by various abrasive processes during transport or they may be developed by chemical solution on the surface at some time after they were deposited. Surface textures are usually described as follows:

abraded	appearing chipped or ground
lobate	cobbled appearance
corroded	material removed by solution
smooth	no pronounced markings
faceted	displaying crystal planes

In examining a clastic sediment, the particles could be viewed under the microscope if necessary, groups could be counted, and descriptions could be made more detailed.

Particle Orientation

Occasionally when clastic particles are deposited, they tend to come to rest in a fixed *orientation*. When a significant number of similar nonspherical particles are orientated in the same or nearly the same direction, they display a preferred orientation (fabric pattern). The following several examples will serve to indicate preferred orientation: long pebbles may become aligned, shells of mollusks may all be orientated in the same direction, pebbles in glacial deposits may all be aligned in the direction of ice movement, and pebbles rolled on a beach may come to rest with their long dimension parallel to the shore.

Fabrics are of two types: *apposition* and *deformation*. An apposition fabric is a primary fabric which develops at the time the material is deposited, and many sedimentary rocks have an apposition fabric. On the other hand, a deformation fabric develops when an external stress is applied to the rock and some of the particles are rotated or moved under the stress, or in some cases, when the growth of new particles in common orientation is controlled by the stress. The compaction of shales often produces a deformation fabric.

Textures of Nonclastic Rocks

The textures of nonclastic rocks are made up of particles which are aggregated in a vastly different manner than the particles of a clastic sedimentary rock. In the nonclastic rocks, particle roundness is not at all related to abrasion, as in the clastic particles. In fact, in some nonclastic rocks, the constituents may be deposited in rounded shapes, for example, as oolites and as pellet-like particles. The textures of nonclastic (chemical) sediments primarily are the result of crystallization from solutions or from gels or from the recrystallization of microcrystalline or amorphous material. These textures are grouped into four major types:

(1) crystalline,

(2) colloform,

(3) organoform, and

(4) mixed clastic and chemical.

Crystalline Textures

The *crystalline* types may be formed in two ways. If the constituent particles were formed by crystallization from a fluid, then the texture is a primary one and is termed *crystalline granular*, shown in Fig. 5-3. The constituent crystals may be of various shapes, but many are equigranular.

If the texture is developed by secondary processes of diagenesis (after deposition) either by crystallization or recrystallization without the aid of solutions, the texture is a secondary texture and is termed *crystalloblastic*, shown in Fig. 5-3. The crystals may be of nearly any size, but there are few rocks which are coarsely crystalline. Most nonclastic rocks have crystals whose sizes are less than 1 mm or so. If the crystals are larger than 0.75 mm, the term *macrocrystalline* is used; if between 0.75 and 0.20 mm, *mesocrystalline*; if 0.20 to 0.01 mm, *microcrystalline* (a microscope is needed to determine the size); and if less than 0.01 mm, *cryptocrystalline*, as presented in Table 5-3.

During normal crystal growth, a mosaic of anhedral crystals develops as they interfere with one another. If confining pressures develop, some solution of crystal boundaries may occur to produce a suture-like zigzag contact. Many clastic rocks contain variable proportions of nonclastic material, and the resulting textures are mixed (or hybrid). Any rock that has a texture produced predominantly by recrystallization may carry relict textures and

Fig. 5-3: Chemical (nonclastic) textures: (a) crystalline calcite in limestone; (b) crystalloblastic texture produced by dolomite rhombs in dolomite; (c) oolitic texture.

Table 5-3: SIZES OF NONCLASTIC CRYSTALS AND PARTICLES

Size	Size Term
> 0.75 mm	macrocrystalline
0.75 — 0.20 mm	mesocrystalline
0.20 — 0.01 mm	microcrystalline
< 0.01 mm	crystocrystalline

structures of the original rock, i.e., textures and structures which were not completely destroyed during the recrystallization. Such features as bedding, oolites, fossils, or original clastic textures may be retained.

Colloform Textures

A number of rocks and minerals appear to be of colloidal origin, i.e., composed of particles of materials of very small size, ranging from 0.002 mm down to 0.000001 mm. Particles larger than 0.002 mm are said to be in suspension, and those smaller than 0.000001 mm are in true solution. Thus, colloids are between suspensions and solutions. When liquid colloids, called *sols*, coagulate into a gelatinous mass, they are called a *gel*. Such gels later lose considerable portions of water and develop into amorphous minerals. Colloids are easily converted into crystals, usually very small at first, but often later becoming larger. Even after crystallization, the original form of the gel may be retained. Minerals which develop often retain the original concretionary, botryoidal, reniform, mammillary, nodular, oolitic, or pisolitic textures and structures, all of which are concentric features. Other features indicative of colloids are shrinkage cracks and lenticular cracks.

Oolites are small spherical to ellipsoidal masses in a rock, ranging from 0.25 to 2.00 mm in diameter. If they are larger than 2.00 mm, they are termed *pisolites*. Oolites usually are radial or concentric in structure, and growth has probably been outward from a center. In many cases there is a nucleus, such as a sand grain or shell fragment (see Fig. 5-3). Oolites may be composed of calcareous material, aragonite or calcite. But dolomite, silica, pyrite, hematite, and other material compose many of the oolites. Oolitic texture is common among phosphorites, bauxites, and chamositic iron ores.

Colloform granules are products of coagulation of gels. They have no internal structure but still assume roundish, ellipsoidal, or subspherical shapes. Colloform granules compare closely in size to oolites.

Spherulites are similar to oolites in being minute in size and being nearly spherical in shape. However, the surface of spherulites is irregular in contrast to smooth surfaces of oolites. Spherulites have a radial type of internal structure.

Organoform Textures

Various organoform textures are discussed later in this chapter under Carbonate Rocks.

Mixed Clastic and Nonclastic Textures

Mixed clastic and chemical textures are due to various combinations and amounts of these two main types of sedimentary material; they need no further discussion.

Color of Sedimentary Rocks

The most obvious and easily observed feature of sedimentary rock is the *color*. The *color* of a sediment (or sedimentary rock) is defined as the overall hue displayed. It is caused by combination of grain color, surface coating, matrix color, and cement. Sometimes it is quite difficult to describe accurately the color of a rock, so a set of color-comparison charts is advised for those attempting to establish colors of sediments. Color charts are available from several sources, for example: National Research Council (1948); American Association of Petroleum Geologists (1944); U.S. Geological Survey.

Colors of sediments include white and various tones of gray, black, yellow, brown, red, and green. *White* and *light grays* result from the presence of uncolored minerals such as calcite or quartz. *Dark grays* and *blacks* are due principally to the presence of varying proportions of carbonaceous matter, which may be present between or may be coating constituent particles. Blacks and grays can also be due to the presence of black or gray constituent particles such as basalt particles, biotite or hornblende crystals, etc.

Yellows (and *browns*) and *reds* are usually due to the presence of iron oxides, as a cement, as coating, or as disseminations. The two most common iron oxides yielding these colors are hematite (red) and limonite (yellow). *Green* colors are usually due to the presence of green-colored mineral particles such as chlorite, serpentine, epidote, or glauconite, and usually not from the presence of interstitial material or coatings.

In the cases of interstitial material or coatings, only a relatively small percentage of the coloring ingredient is required to produce the predominant color.

Classification of Sedimentary Rocks

The composition of sedimentary rocks is commonly expressed in terms of minerals present or in terms of chemical elements present. The composition is used as a basis for classification, usually being more important than texture in deriving a correct rock name. Many of the sedimentary rocks have fairly uniform mineralogical compositions, but most sedimentary rocks are composed of mixtures of several minerals, and a few have a very great variety of minerals as well as rock fragments. Well over one-hundred different mineral species have been identified in sedimentary rocks; however, many of these minerals are considered as rare. Actually, only about twenty minerals can be considered as common constituents of sedimentary rocks.

The common minerals are usually grouped into two broad classes, *detrital* (clastic) and *nondetrital* (nonclastic). The detrital minerals consist of abraded and fragmental particles deposited by mechanical processes to form the clastic sediments. The most abundant detrital minerals are quartz, clay minerals, and fine-grained micas. Feldspar, coarse-grained micas, and chert are in subordinate amounts. Any heavy minerals present in very small quantities are referred to as accessory minerals, the most common of which are hematite, zircon, tourmaline, epidote, garnet, and hornblende.

The nondetrital (nonclastic) minerals are usually precipitated from solution, as a result of a chemical reaction or by biological effects, and commonly accumulate very close to the location where they are precipitated. The most common minerals of the nonclastic rocks are calcite and dolomite. In minor quantities are chert, secondary quartz, and gypsum (or anhydrite). Accessory minerals also may be present.

The construction of a perfect classification of sedimentary rocks is impossible because of the many gradations between one type and another. Such gradations among the rocks result from variations in the source area, in the composition, in the depositional agent, and in the sedimentary environment. The classification presented in Table 5-4 is a modification of that of Pettijohn, and it is based on two broad modes of formation, clastic (detrital) and chemical (nondetrital), but the boundaries are not clearly defined.

The *detrital* rocks are divided into three main size groups, according to the size of particles making them up, i.e., coarse-grained ($>$ 2.0 mm), medium-grained (2.0–0.05 mm), and fine-grained (0.05–0.005 mm and $<$ 0.005 mm). The student and reader are to note that the composition of the particles does not enter into detrital groups. This part of the classification is thus a *size* grouping. Neither does this part of the classification consider the cementing material which binds the detrital material together.

The remaining part of the classification divides the non-detrital rocks into three subgroups, *chemical*, *biochemical*, and *organic*. Note that no consideration is given to *size* of any of the crystals or organic material which may make up the chemical and organic rocks. Thus this part of the classification is a *composition* grouping. Furthermore, this part of the classification does not consider any detrital material which may be mixed with the chemical and organic constituents.

The Clastic Sedimentary Rocks

Coarse-Grained Clastics

The coarse-grained clastics (conglomerates and breccias) are composed of particles larger than 2 mm across. When this coarse material exists as unconsolidated sediment, it is referred to as *gravel*. Gravel is commonly subdivided into three general size categories:

(1) fragments larger than 256-mm material are called *boulders*;

(2) fragments between 256 and 64 mm are called *cobbles*; and

(3) fragments between 64 and 2 mm are referred to as *pebbles*.

After consolidation into rock, gravels are called conglomerates (or breccias). By definition, then, a conglomerate is a clastic rock containing rounded fragments larger than 2 mm, while a breccia contains angular fragments. The coarse-grained fragments must make up greater than 10 percent of the rock; however, there is no universal agreement on this value (Fig. 5-4). In most cases, the larger fragments are rock fragments rather than mineral fragments; the most common rock fragments of conglomerates (and breccias) are quartzite, vein quartz, chert, rhyolite, granite, and sometimes limestone (Fig. 5-5). The larger fragments are usually located at the base of the rock

Table 5-4: CLASSIFICATION OF SEDIMENTARY ROCKS

| Clastic | | | Nonclastic | | | | |
| Grain Size | | | Chemical | | | Biochemical—Organic | |
Coarse > 2.0 mm	Medium 2.0–0.05 mm	Fine < 0.05 mm	Oxidates	Evaporates	Precipitates	Accretionates	Reduzates
conglomerate	sandstones:	shale	iron oxide rocks	rock gypsum	C A R B O N A T E R O C K S (limestones)		coal
breccia	orthoquartzite graywacke arkose tuff	argillite	manganese oxide rocks	rock salt	chemical types: travertine tufa	accretionary types: organic reef biostromal ls pelagic ls clastic types: coquina lithographic ls altered type: dolomite	black shale bituminous ls bog ores

siliceous rocks:
chert
diatomite
phosphate rocks
iron-bearing rocks

Fig. 5-4: Coarse quartz-pebble conglomerate made up of large pebbles of white rounded quartz surrounded by finer grained sand and clay matrix.

Fig. 5-5: Limestone breccia composed of large angular pebbles ($\frac{1}{2}$—2 inches across) of light gray limestone in dark fine grained carbonaceous limey material.

x

Fig. 5-6: Pebble conglomerate consisting of rounded pebbles of a variety of rocks, varying from black to brown to light gray in color.

mass, and some grading to smaller sizes is often seen toward the top. True bedding of conglomerates (and breccias) may or may not be present. The larger fragments may be all composed of one rock type, such as quartz pebbles, or they may be composed of a wide variety of rock fragments. The latter type is the more common (Fig. 5-6).

The matrix of conglomerates (and breccias), the finer portions of the rock together with a cementing agent, lithifies the previous gravel into a firm mass. The cementing agent may be derived from without or from within the conglomerate rock mass itself, by solution and redeposition of material in the rock, and it commonly consists of calcite, silica, or iron oxide. The matrix may consist of a wide variety of materials such as quartz, feldspar, or clay particles.

Names of conglomerates are applied according to the nature of the large fragments, such as *quartz-cobble conglomerate, quartz-pebble conglomerate, granite-pebble conglomerate, granite-boulder conglomerate,* or *graywacke conglomerate.* Corresponding breccias are somewhat less abundant. The colors of conglomerates are determined mainly by the large fragments and matrix. Quartz-pebble conglomerates are generally light colored, and granite-pebble conglomerates are pink or gray, according to the type of rock making up the pebbles. Other common rock fragments in conglomerates are phyllite, gneiss, schist, basalt, and others. Fossils are rare in conglomerates, and when present, they have a very fragmental nature.

Most gravels are developed under aqueous conditions such as in river beds, in alluvial fans at the foot of rapid mountain streams, and also in very shallow marine waters. Gravels deposited by rivers and streams are generally local in extent and become deposited along river courses where slack water occurs. Those deposited by torrential mountain streams may be somewhat more extensive, especially where several alluvial fans coalesce. Gravels deposited under shallow marine conditions may be rather extensive as a result of transport by longshore currents, but they often develop discontinuous masses.

Many breccias result from tectonic action as crustal forces often crush, shatter, fault, or fold brittle rock, and may even pulverize the rock in some cases.

Some gravels are formed under glacial conditions, deposited by melting ice as a glacier retreats. This material is referred to as *glacial till*, and once it becomes indurated into firm rock it is called *tillite*, a type of conglomerate. Tillite is an uncommon rock. It is generally poorly sorted, unstratified, and variable in composition, as all types of rock may be present.

Rarely a gravel may accumulate under aeolian conditions since wind velocities may be high enough to cause transport and deposition of gravel-sized material.

Medium-Grained Clastics

The medium-grained clastics have grains whose diameters lie between 2 mm and 0.05 mm and thus are of sand size. Sand-sized sediment is lithified to *sandstone* by pressure, by compaction of any clay constituents, or by the introduction of a cementing agent. Sandstones are the most important and widespread of the medium-grained clastics.

In *sandstones*, the particles of sand size make up more than 50 percent of the rock, and these particles are commonly cemented together with finer-grained matrix and cement. The constituent sand grains may be composed of various material such as quartz, chert, feldspar, rock particles, and occasionally even calcite or dolomite. The majority of sandstones are composed primarily of quartz particles, but rarer sandstones composed of other minerals in substantial quantities are known, such as *blacksand* (with magnetite), *greensand* (with glauconite), or *brownsand* (with iron oxide). Sandstones are commonly cemented together by mineral material introduced from outside. But in other cases, sandstones are lithified by bonding produced by the finer-grained clayey matrix usually present in subordinate quantities.

Sandstones are subdivided into three main types on the basis of the most common detrital components, quartz, feldspar, and rock particles. The three main types are thus *orthoquartzite*, *askose*, and *graywacke*, respectively. The limits are artificial and exist only for descriptive purposes. Sandstones grade into one another by change of composition of the constituent particles, but they also grade to conglomerates by increase in particle size, or to siltstones or shales by decrease in particle size. A further gradation to tuffs may occur as the presence of volcanic ash increases.

A sandstone composed primarily of quartz may be called "pure" sandstone if the cementing agent is silica, or as "calcareous" sandstone if cemented by carbonate material, calcite or dolomite. "Argillaceous" sandstones are cemented by fine-grained clayey matrix. The term "quartzose sandstone" is used if quartz is present up to 95 percent, but not cemented by silica. When silica cement is present, the term "quartzitic sandstone" is used.

Orthoquartzites consist very predominantly (80 percent or more) of quartz, usually well-rounded and well-sized (Fig. 5-7). Well-worn grains may appear frosted. Other less well-worn grains may exhibit crystal facets as a result of a secondary overgrowth of quartz on a worn grain. Orthoquartzites may also contain other detrital constituents. If feldspar grains (usually orthoclase or albite) are present to about 10 percent, the rock is called *feldspathic sandstone*. Glauconite, a green iron-silicate mineral, may be present as a minor component in orthoquartzites. Fossil shells and shell fragments of calcareous or phosphatic composition may be present in fairly small amounts. Other detrital components in extremely minor quantities may include tourmaline, zircon, rutile, and garnet. Substantial quantities of

Fig. 5-7: White sandstone (Entrada formation) from near Boulder, Colorado, showing fine granular surface.

Fig. 5-8: Dark reddish-brown micaceous sandstone from Portland, Connecticut, showing granular texture produced by quartz, feldspar, and muscovite.

muscovite may yield the name *micaceous sandstone* (Fig. 5-8). Cement composed of siliceous or calcareous material is usually introduced by ground waters, but it may be formed from seawater at the time of deposition, or from shelly material. The presence of ferric oxides in significant amounts may permit the name *ferruginous sandstone. Orthoquartzites* are usually well-bedded and often cross-bedded. This type of sandstone usually is deposited as widespread blankets of sand, although generally thin.

 Graywackes contain rock fragments (usually fairly large fragments), are generally poorly sorted, and contain substantial quantities (15 percent or more) of fine-grained matrix (Fig. 5-9). They are gritty or sandy. The matrix is usually composed predominantly of clay minerals, chlorite, and sericite, but also hematite and chert. Graywackes usually contain some feldspar and can be called *feldspathic graywackes* if feldspar is more abundant than rock chips. However, if rock chips are more abundant than feldspar, the name *lithic graywacke* may be used. Graywackes are usually medium to dark gray in color, but yellowish-gray, greenish-gray, bluish-gray, or smoky-gray colors also are recorded. In graywackes, the quantity of dark rock fragments is over 20 percent. The dark rock chips are usually of quartzite, basalt, slate, phyllite, rhyolite, greenstone, or schist, but dark minerals such as augite,

Fig. 5-9: Dark brownish-gray graywacke showing granular texture produced by quartz, a few small elongate rock chips, and scattered black grains.

Fig. 5-10: Light gray subgraywacke showing gritty texture. A few rock chips are present among the sand grains. Matrix is fine silty to clayey material.

serpentine, hornblende, and iron ores may also be present. The cementing agent of graywackes is usually silty, muddy, or calcareous material. The main ingredients of graywackes are quartz (20 to 70 percent), feldspars (up to 50 percent), and rock chips (over 20 percent).

Bedding in graywackes is variable. Some are quite massive, particularly the coarse-grained ones, but occasionally thin-bedded finer-grained material is present. Cross-bedding is rare, and ripple marks are nearly always lacking. Fossils are very rare, except for occasional wood fragments.

Subgraywacke, a very common type of sandstone, is an intermediate rock between graywacke and orthoquartzite, but it actually differs from both of these in several aspects. For example, feldspar is scarce (only 1 to 15 percent), quartz particles are better rounded and more abundant than in graywacke, and there is only a minor amount (5 percent) of rock fragments. Subgraywackes are usually better sorted than graywackes, are locally cross-bedded, and locally display ripple marks. Most are lighter in color than graywacke, as shown in Fig. 5-10.

Sandstone containing 25 to 30 percent or more of feldspar derived from acidic igneous rock is called *arkose* (Fig. 5-11). Potash and soda feldspars

Fig. 5-11: Reddish-brown arkose showing sandy texture produced by quartz and subordinate feldspar.

usually dominate over other feldspars. The amount of feldspar in an arkose rarely exceeds 50 percent. Quartz is the most abundant clastic mineral, feldspar is commonly second in abundance, and both muscovite and biotite may be present in minor quantities. Minor amounts of hornblende, apatite, rutile, or garnet may be present. Even a few rock particles may be present. Arkoses are usually cemented together by calcite, iron oxide, or clay minerals, while silica cement is observed only rarely. Arkoses usually have a light pink to pinkish-gray color as a result of the feldspars present. The particles of arkoses often are fairly coarse-grained, angular to subrounded in shape, and moderately well sorted. Bedding in arkoses is generally poor, but in some cases cross-bedding is strongly developed.

Sandstones are formed under marine conditions, particularly in shallow continental shelf areas. Close to shore, sandstone masses are apt to be variable in thickness and distribution as a result of tidal and storm effects. But further out toward the deeper water, sandstones are more uniform in thickness and distribution. Sandstones formed in freshwater lakes, in river courses, and in estuaries are usually of local extent and of variable thicknesses. Sandstones are also deposited beneath and around the margins of glaciers.

Materials ejected from volcanoes may fall on land or into standing water and thus become clastic in nature. That which falls on land is subject to erosion, transportation, and redeposition in water. This volcanic material, deposited originally on land or in water, may be mixed with other sedimentary clastic material to form a sandstone called *tuff* (Fig. 5-12). Although volcanic ejecta range widely in sizes, tuffs are composed of material whose size ranges between 4 mm to $\frac{1}{4}$ mm, or *fine tuffs* if less than $\frac{1}{4}$ mm. If individual volcanic fragments are larger than 4 mm, they may be called cinders, bombs, or blocks, and these will form other rock types such as *agglomerates* or *volcanic breccias*. But tuffs are a type of stratified sandstone whose volcanic constituents are glass, crystals of various minerals, or igneous rock debris (andesite

Fig. 5-12: Light colored tuff from west Texas showing gritty nature produced by volcanic ash which fell and was deposited as sediment.

Fig. 5-13: Gray sandy shale showing layering and slightly gritty nature.

and basalt) accumulated by deposition of volcanic ejecta, coarse or fine ash. Layers are produced by ash falls and by deposition by surface waters. If the amount of normal sedimentary clastic material exceeds the volcanic material, the rock is known as *tuffaceous sandstone*. But if volcanic material is present more abundantly than clastic sedimentary debris, the rock may be called *sandy* or *clayey tuff*.

In most tuffs, the volcanic debris ranges from coarse to very fine, and thus they are generally poorly sorted. The grains of ash, since they are glassy in nature, take on a variety of shapes, ranging from elongate, curved, and spicule-shaped fragments. These glassy particles are usually colorless. The crystal grains are commonly broken fragments of previous lava crystals. Rock fragments in tuffs most commonly are acidic aphanitic igneous rocks, but andesitic and basaltic fragments are also observed in some cases.

Fine-Grained Clastics

Fine-grained clastics consist of grains whose diameters are less than 0.05 mm and thus are of silt to clay size. Silt sizes range from 0.05 to 0.005 mm, and clay sizes are less than 0.005 mm. Silt or clay sized sedimentary material is lithified into *silty shale* and *clay shale* by pressure, by compaction, or rarely by introduction of cement. Shales are more abundant than any other sedimentary rock.

Shale is a thinly laminated or fissile siltstone or claystone (Fig. 5-13). If laminations and fissility (property of being easily split into slabs) are lacking, the term *mudstone* is used. In shales, more than 50 percent of the rock must be made up of silt-sized and/or clay-sized particles. The term *argillite* is used to describe a shale that has undergone a much higher degree of indura- tion than a common shale, perhaps partially hardened by recrystallization. The fine-grained particles are lithified by compaction and very rarely by a cementing agent. The constituent particles of shales are composed predomi- nantly of clay minerals such as kaolinite, illite, or montmorillonite; also

Fig. 5-14: Carbonaceous shale composed mainly of clay minerals and silt, but also containing organic residual material.

hydrated oxides of aluminum and ferric iron, fragments of silicate minerals such as quartz, mica, and feldspar, and others in minor quantities may be present. In some cases, authigenic minerals, i.e., minerals derived in the rock after deposition, may be present, such as chlorite, illite, glauconite, calcite, dolomite, opal, pyrite, rutile, and ferric oxide. Also present may be organic materials derived from animal shells and plant substances (Fig. 5-14). Colors of shales, usually due to a secondary pigmenting agent rather than the clay minerals, range from black (organic) to green (chlorite or glauconite) to red (hematite) to blue and gray (siderite and ankerite).

The constituents of these fine-grained rocks are actually the finest products of silicate debris from weathering processes. They are deposited with large quantities of water admixed among the tiny grains. At the time of initial sedimentation, nearly 75 to 80 percent of the volume is water, so that consolidation takes place by progressive compaction as water is expelled. As compaction increases, water is eventually completely expelled and the grains interlock, and possibly some recrystallization may take place to produce consolidation. Even though shales contain significant quantities of clay minerals, several varieties of shale exist, according to the nature of some of the other ingredients. *Quartzose shale* contains substantial quantities of rounded quartz of silt size. In addition, if a particular mineral such as calcite, glauconite, limonite, or hematite is present in significant quantities, the names *calcareous quartzose shale, glauconitic quartzose shale,* or *ferruginous quartzose shale,* respectively, may be used. And there are still others. *Feldspathic shales* carry more than 10 percent feldspar as silt-sized material. These may grade to other types, depending upon composition of constituents. *Chlorite shales* carry a fair quantity of chlorite. *Micaceous shales* contain large quantities of mica, along with common constituents quartz and feldspar. Micaceous shales are usually gray in color and are commonly well-laminated.

Clays and silts are classified into two major groups according to their origin: residual and transported. Residual clays are formed by weathering processes and are formed *in situ*; thus they are a type of soil. The transported clays and silts contain products of weathering, products of abrasion, and chemical materials added during transport or after deposition.

Black shales are exceptionally rich in organic matter and are very fissile (easily separated into thin beds). Much iron sulfide, usually pyrite, is commonly present as fossil replacement, as nodules, or as scattered grains. Only selected types of fossils are present, and they generally are hardy types which can survive under the unusual conditions of black shale formation.

Clay and silt are commonly transported by running water to sites of deposition such as lakes, mountain valleys, lagoons, estuaries, deltas, continental shelves, and even to the deep sea floor. Deposits in lakes and mountain valleys are generally temporary in nature, while those in marine waters are considered much more permanent. Once consolidated into firm rock, the clay and silt become *shale*.

Loess is fine-grained material, deposited from suspension by air currents. It is commonly poorly consolidated, it lacks bedding, and it usually contains large proportions of quartz fragments. Other fragmented minerals may be feldspar, mica, chlorite, chert, and calcite, as well as still others in very minor quantities. Clay minerals are often present. Loess originates in a glacial environment. Dust-sized outwash material is carried by surface waters and deposited in flood plains where it dries out. During the dry winters, the dust is then carried and deposited by the wind.

The Nonclastic Sedimentary Rocks

Nonclastic rocks are classified according to their composition. The most important nonclastic sedimentary rocks are the carbonate rocks, limestones of two main types called precipitates and accretionates in the classification. Limestones altered to dolomite are also discussed. Other precipitates are relatively uncommon. Another small nonclastic group, called evaporates, includes rock salt, anhydrite, and gypsum. Still less abundant are various organic sediments, including coal, composed of vegetable and/or animal matter, a group known as reduzates. A fourth nonclastic group is called oxidates.

Oxidates

The *oxidates* are deposited from natural waters as a result of oxidation processes. The two most important types are iron oxides and hydroxides and manganese oxides and hydroxides.

Various iron salts are commonly present in natural waters, and they become precipitated when conditions are right. Iron oxides and hydroxides are deposited under oxidizing conditions, i.e., when oxygen is abundant.

Iron silicates, carbonates, and sulfides will form if conditions are reducing, i.e., when oxygen is lacking. Those deposited in seawater are usually fairly extensive and are often economically important. *Iron-bearing rocks* usually contain over 10 percent iron, or in other terms, over 15 percent iron oxide.

Deposition of iron-rich oxidate sediments consisting of ferric hydroxide and ferric oxide may take place in fresh water or in shallow marine water. Those deposited in shallow seas are the most extensive and the most important. Ferric hydroxide, chiefly as goethite, $HFeO_2$, is precipitated as small concretions called oolites, with varying amounts of earthy ferric hydroxide serving as the surrounding matrix material. These deposits thus have an oolitic texture and commonly are brown to yellowish-brown in color. Ferric oxide, chiefly as sedimentary hematite, Fe_2O_3, may also be precipitated as oolites or mixed with and replace broken or crushed fossils, and it may become cemented by a mixture of calcite and hematite (Fig. 5-15). Many deposits of hematite, such as the Clinton oolitic iron ores of Silurian age, are essentially pure and take on the typical dark red color. However, other hematite iron beds are mixed with clay, sand, and silica. In a similar manner, ferric hydroxide also becomes mixed with other sediments to produce *ferruginous sandstone, ferruginous shale,* or *ferruginous limestone.*

Manganese, like iron, is very often deposited as an oxidate mineral. It is

Fig. 5-15: Sedimentary hematite from near Birmingham, Alabama. Note the small oolitic structure of this iron oxide material.

Fig. 5-16: Dark brown manganese oxide pellets, 1 to 7 mm in diameter, from the bottom of Green Bay (Lake Michigan). Note a few clear vitreous quartz grains.

easily dissolved and carried to the sea, and it becomes precipitated when conditions are oxidizing. Manganese may be deposited as hydroxide, $Mn(OH)_4$ or as the oxide, MnO_2, and be in the form of concretions, nodules, and slabs. Such precipitation is more apt to occur in deep water. These masses of manganese dioxide vary in size from a few millimeters to a foot or more (Fig. 5-16). Deposits of manganese dioxide nodules are fairly abundant over certain parts of present day ocean floors and warrant commercial exploitation. These nodules contain chiefly manganese oxides but also some iron and lesser amounts of copper, nickel, and cobalt. The manganese in these nodules very likely is derived from submarine volcanic eruptions by subaqueous weathering.

The chief manganese minerals of sedimentary origin are the hydroxide manganite, $MnO(OH)$, and the oxides pyrolusite, MnO_2, and psilomelane, $(Ba,H_2O)_2Mn_5O_{10}$. Manganese-bearing rocks usually take on a brown to black color. Rocks of these types are not as common as iron-bearing sediments.

Evaporates

The evaporates are composed of sedimentary rocks which are formed as the result of evaporation of restricted bodies of seawater as well as desert

Fig. 5-17: Pink to gray rock gypsum from Grand Rapids, Michigan, made of fine gypsum crystals.

waters. The most common evaporates are rock salt, rock gypsum, and anhydrite, and thus evaporates are saline deposits. Other less common evaporates are a few carbonates and the rare soda salts and nitrates.

Rock gypsum is composed predominantly of the mineral gypsum and varies from coarsely crystalline to finely granular in texture (Fig. 5-17). Often-times it displays bedding planes, but it also occurs as compact bodies of material with no bedding.

Anhydrite deposits, composed dominantly of the mineral anhydrite, along with *rock gypsum* are found in fairly thick and extensive accumulations often associated with shales or clays. Both are usually in fine granular form, but they occasionally take on a fibrous or coarsely granular appearance (Fig. 5-18). The colors of rock gypsum vary from white to pink to tan to gray to dark earthy reds, while anhydrite is commonly tan to gray. Anhydrite ($CaSO_4$) is often replaced by gypsum ($CaSO_4 \cdot 2H_2O$) in certain portions or a rock bed. In rock gypsum, crystals may be small and large, but in many cases the large ones have formed later than the smaller ones. Gypsum often forms from anhydrite simply by hydration, i.e., by adding $2H_2O$. Fossils are rare or lacking altogether, due to the high salinity of the waters from which the rocks are formed. Varying quantities of bituminous matter, iron oxides, and clay may be present.

Rock salt (halite-NaCl) is usually massive, coarsely crystalline rock, often

Fig. 5-18: Anhydrite rock showing bedding and curious surface appearance probably due to minor gypsum development.

Fig. 5-19: Rock salt from Stassfurt, Germany, showing cubic halite crystals deposited by sedimentary processes.

interbedded with shale (Fig. 5-19). Rock salt is most commonly white when pure, but very transparent varieties occur. It often takes on a red color of varying tones due to the presence of small amounts of hematite. When clay becomes incorporated with the silt, gray tones result. Rock salt beds are very commonly associated with gypsum and/or anhydrite beds or shales. Fossils are uncommon, but small amounts of carbonaceous material and iron oxides may be present.

Precipitates

Silica (SiO_2) precipitated by chemical processes from marine or fresh water may become mixed with other sediments and may be in the form of opal, chalcedony, quartz, or cristobalite to form *siliceous rocks*. It may be deposited as nodules, as minute quartz crystals, as small rounded spherulites of chalcedony, and as irregular masses in limestones, as well as in other rocks. Most of these types of silica are referred to as *chert* or *flint*. Chert is a dense cryptocrystalline rock composed of chalcedony and cryptocrystalline quartz (Fig. 5-20). (Chalcedony is microcrystalline fibrous crystalline silica

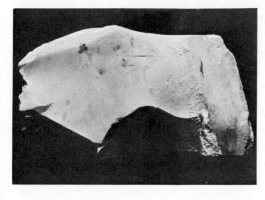

Fig. 5-20: Dense white chert (rock) consisting of crypto-crystalline chalcedony and quartz. Note the tough splintery to conchoidal fracture.

and microcrystalline fibrous amorphous silica or opal.) Chert possesses a tough splintery to conchoidal fracture. It varies in color from white, gray, green, blue, pink, red, yellow, brown, and black. *Flint* (dark chert) is the same as chert, but is that type found as artifacts. Other terms used synonymously with chert are *jasper* (red to black due to iron oxides), *jasperoid* (from the Tri-State region of Oklahoma, Missouri, and Kansas), and *novaculite* (from Arkansas). Nodules of chert are not concretionary because they lack radial or concentric structure. As a rule, chert nodules are irregular masses two or three feet in size. Shapes are very irregular, and the outside often looks knobby or warty.

Bedded cherts often are interbedded with marine shales, sandstones, and limestones. The beds may thicken and thin, or pinch out, or split into discontinuous beds.

Another type of bedded siliceous deposits are the *siliceous earths*. These include *radiolarites* and *diatomites*, both of which are porous and friable material and occur interbedded with shales and limestones. Radiolarites are composed predominantly of the remains of Radiolaria, but other siliceous organic remains also are present. Diatomites are predominantly made up of diatom organic remains composed of opaline silica. Both Radiolaria and diatoms are microscopic in size (Fig. 5-21).

Any rock composed predominantly of phosphate minerals is called *phosphorite, phosphate rock,* or *rock phosphate* (Fig. 5-22). Most phosphatic rock is dark brown or black and is composed of cryptocrystalline or amorphous phosphatic material. Minor amounts of quartz, pyrite, and marcasite may be present. The phosphatic material occurs as small granules, oolites, or pebbles, which are made up mostly of amorphous "collophane," a convenient term to include several very closely related phosphate substances. Most phosphate rock is formed by slow sedimentation under marine conditions, probably in partially closed basins.

Certain iron-bearing sediments are deposited under slightly reducing conditions, while others may be formed under highly reducing conditions. If seawater is only slightly reducing, then iron silicate in the form of glauconite, $K(Fe,Mg,Al)_2Si_4O_{10}(OH)_2$, greenalite, $Fe_6Si_4O_{10}(OH)_8$, or chamosite, $Fe_3Al_2Si_2O_{10} \cdot nH_2O$ may form. Chamosite is usually deposited in the form of oolites. The chamosite is usually very-fine-grained and often is mixed with a little siderite, $FeCO_3$. Some chamosite deposits contain finely disseminated limonite or magnetite. Such chamosite deposits are referred to as *clay ironstones*.

Glauconite is commonly deposited as green pellets which are relatively unstable. Glauconite sediments are closely related to chamosite iron-bearing sediments. The pellets are often transported, mixed with clastic material, and then redeposited in the form of *greensand* or *glauconitic sandstone* (Fig. 5-23).

Fig. 5-21: Diatomite, a siliceous rock nearly white in color, composed of microscopic-sized diatoms of opaline silica. (Lompoc, California.)

Fig. 5-22: Phosphate rock, called phosphorite, from southeastern Idaho. This dark brown rock consists of fine-grained apatite (collophane), a little calcite, and minor carbonaceous matter.

Fig. 5-23: Glauconitic sandstone from Monmouth, New Jersey. The rock is composed mainly of quartz, with substantial glauconite providing dark green color. Darker brown streaking is iron oxide.

Iron carbonate in the form of siderite, $FeCO_3$, also can be precipitated from surface waters under slightly reducing conditions. Although not common sediments, the most important iron carbonate rocks are ferruginous limestones. Sediments of this type are usually interstratified with other shallow marine types. Such beds are extensive but thin. They commonly take on granular, fossiliferous, oolitic, and concretionary aspects, but uniform beds are also known. Iron carbonate beds, called *bedded siderites*, quite often become intimately interbedded with chert or clay units. In some cases, the bedded siderite was deposited as oolites, which are often completely or partially replaced by silica.

Limestones contain more than 50 percent calcite, and *dolomites* contain more than 50 percent calcite and dolomite mineral. The remainder of the rock in each case is usually clastic quartz or clay. Other minerals in minor quantities in limestones and dolomites may include chert, feldspar, glauconite, collophanite, pyrite, and coaly material.

Pettijohn (1949, p. 293) discusses four major types of limestones: chemical, accretionary, clastic, and metasomatic (altered). Although a very rare type, the chemical limestones will be discussed first, because these are classed as precipitate deposits. However, clastic limestones are much more common and extensive than the others, and the accretionary types are also relatively important geologically.

Chemical limestones (precipitates) are those formed as a result of direct precipitation from water, but these are not particularly common. It appears that most limestones are formed as a result of organic action. Few places exist in oceanic areas where direct precipitation of calcite could occur.

Conditions for direct precipitation of calcium carbonate exist much more frequently in freshwater conditions such as in springs, seeps, rivers, lakes, and swamps, and without organic action. But such nonmarine limestones are rather rare. *Tufa* and *travertine* are limestones formed in freshwater environments by evaporation of spring and river water. Tufa (Fig. 5-24) is spongy

Fig. 5-24: Travertine (tufa) from Mumford, New York, showing porous nature and a few plant impressions.

Fig. 5-25: Travertine from Death Valley, California, showing coarse crystals of calcite in white and dark pink bands. Note vitreous luster of white calcite.

and porous, as well as friable, whereas travertine is dense and banded and is common as flowstone and dripstone in caverns (Fig. 5-25). Neither tufa nor travertine form extensive deposits, and have little geologic importance. *Caliche* is a lime-rich aggregate in soils in semiarid regions. It is formed as a result of capillary action drawing lime-bearing waters toward the surface, followed by evaporation and precipitation of calcium carbonate, often in concentrically banded masses. Other freshwater limestones may be deposited under lake conditions, as more or less continuous beds (Fig. 5-26). They may become dense or friable, but for the most part they are of small thickness and of limited areal extent. Marine fossils are absent, but fossils of certain molluscs and algae may be present.

Accretionates

The *accretionary limestones* (biochemical accretionates) form *in situ* as a result of slow accumulations of organic remains, mainly shells of calcareous composition. Animals that form such deposits take calcium into their tissues

Fig. 5-26: Fresh water limestone from west Texas showing dense nature of this nearly white rock.

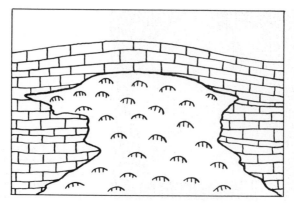

Fig. 5-27: Cross-section of a limestone reef.

in soluble form and then secrete $CaCO_3$ shell material during growth. There are three subtypes of such limestones. The first of these subtypes is the so-called *organic reefs* (or bioherms) which may be composed of algae, sponges, corals, stromatoporoids, bryozoans, tubiculous annelids, rudistid clams, and oysters (Fig. 5-27). A reef is a dome-like or mound-like mass made up of the sedentary (generally immobile) organisms and enclosed in normal rock material. Reefs vary in size, shape, and in types of organic remains. Bedding is generally lacking in a reef except in the peripheral area where very crude bedding grades out into normal bedding of flanking normal rock. *Algal limestones, coralline limestones*, and *crinoidal limestones* are abundant. Various fossiliferous limestones are shown in Figs. 5-28—5-30.

A second subtype of accretionary (biochemical) limestone is called the *coquinoid* (or biostromal) type. These are poorly bedded structures such as shell beds or crinoid beds made up of sedentary organisms (Fig. 5-31). These beds do not develop mound-like shapes. Such layers are usually crowded

Fig. 5-28: White fossiliferous limestone from Texas containing large mollusk shells and other organisms.

Fig. 5-29: Fossiliferous gray limestone from near Rochester, New York, containing numerous brachiopods. Note shiny pearly luster of shells.

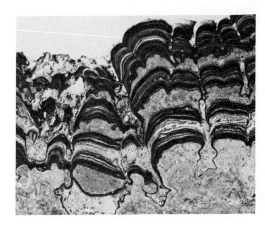

Fig. 5-30: Algal limestone composed of alternating algal and inorganic layers. Black and gray layers are due to presence of disseminated pyrite. (W. H. Bradley, U. S. Geological Survey, Prof. Paper 154, by permission.)

Fig. 5-31: Dark gray fossiliferous limestone composed of numerous broken shells of pelecypods.

Fig. 5-32: Very fossiliferous limestone with numerous bryozoans and other organisms. Matrix is light gray in color.

Fig. 5-33: Chalk, a white pelagic limestone from the Dover Cliffs, England, composed of shells of microscopic-sized organisms. Note dull earthy appearance.

Fig. 5-34: Coquina, a clastic limestone composed of fossil debris from accretionary limestone.

Fig. 5-35: Oolitic calcite limestone from near Bedford, Indiana, consisting of oolites about 1/4 to 1/2 mm in size, giving the nearly white rock a grainy texture.

with fossils and are rather porous. Beds of biostromal type often become mixed with clastic calcite or with shale (Fig. 5-32).

A third subtype of accretionary (biochemical) limestone is the *pelagic limestone*, formed by the accumulation of microscopic calcareous shells (tests) of floating organisms. The resulting limestones are thus very-fine-grained. As a rule, pelagic limestones are rather rare. The best-known example is the variety called *chalk*, such as the Niobrara chalk and Selma chalk.

Chalk nearly always consists of the tests of microorganisms, especially Foraminifera, cemented by finely crystalline calcite. The calcite content runs very high, usually well over 90 percent. Chalk is defined as being white or light gray in color, porous, fine-textured, and somewhat friable (Fig. 5-33). A few chalks are nearly devoid of microorganisms, and these probably are due to precipitation of calcite by chemical or biochemical processes.

The third major type of limestones, the clastic limestones (or bioclastic), probably the most widespread and most common type, consist of detrital carbonate material of subaqueous origin with or without calcite cement between clastic fragments. The percentage of carbonate detritus must be greater than 50. The carbonate detritus has been mechanically transported and deposited, and it develops textures and structures (including cross-bedding) of a clastic sediment. However, limestone composed of fragments of older limestone is rare. In the most common type, the clastic carbonate material partially or completely consists of fossil debris derived from the erosion of accretionary limestones; thus the rock is a *coquina* (Fig. 5-34). In other cases it may be made up partially of fossil debris with precipitated fragments of carbonate, or it may be made up of oolites to produce an *oolitic limestone*, as shown in Fig. 5-35. Clastic quartz may be mingled in varying proportions, and some clastic limestones also contain varying quantities of glauconite. Calcite cement may be present as a fringe on the detrital material or as a crystalline mosaic between grains. The most common of the coquina type are the crinoid coquinas. When the detrital carbonate material is excep-

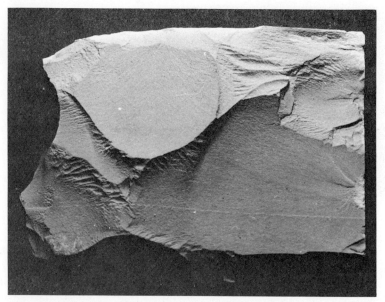

Fig. 5-36: Lithographic limestone from Solenhofen, Bavaria, composed of very fine-grained, dense, detrital carbonate material.

tionally fine-grained, dense, and homogeneous, the rock is called *lithographic limestone* (Fig. 5-36). The presence of significant amounts of clay may yield the name *argillaceous limestone*.

Altered Limestones

The fourth major type of limestone is called *metasomatic limestone,* i.e., those that have been altered in some way after they have initially formed. As a rule, limestones of all types are quite susceptible to alteration as well as to complete replacement. Calcite is often replaced by dolomite or silica and occasionally even by chamosite, siderite, limonite, hematite, or phosphate material. Depending upon the nature of the replacing mineral and the extent of replacement, a limestone may be altered to a dolomitic limestone or to a dolomite rock, to a silicified limestone, or to a phosphatic limestone, etc. During replacement, the original textures and structures of the initial limestone may or may not be preserved. The most common metasomatic limestone is *dolomite* (rock), which contains 50 percent carbonate, more than half of which is the mineral *dolomite*. Carbonate rocks containing both calcite and dolomite are fairly rare; most rock dolomites contain mostly the mineral dolomite with very subordinate calcite. Common, though, are dolomite beds interbedded with limestone beds. Grading of carbonate rocks from pure limestone (more than 95 percent calcite) through magnesian limestone through dolomitic limestone through calcitic dolomites through dolomites (less than 10 percent calcite) are known, but the intermediate hybrid carbonate rocks are not too common. Dolomites generally resemble limestones in gross features, but many detailed differences exist between them. The replacement process of dolomitization generally involves large-scale recrystallization, producing coarse or medium crystalline rock. Dolomite rock is thus a modified limestone, as shown by altered fossils, by dolomite in patches in limestone, by replacement of oolites by dolomite, and other features (Fig. 5-37).

Reduzates

Organic sediments are formed by the accumulation and preservation of organic compounds of animals and plants. The organic compounds in living matter undergo combustion, a process of converting organic matter to carbon dioxide, CO_2, and water, H_2O. The process continues even after death of the organisms, but bacterial decay begins along with combustion. If the decomposition processes go to completion, then no residue is left for accumulation. But if the decomposition is not complete, due to a covering of water which would prevent too rapid oxidation and destruction because very little free oxygen is available, then putrification and humification can occur in such an oxygen-deficient (reducing) environment. Some of the products which are not completely destroyed may become buried and preserved as sediment. Various types of organic material have different resistances to oxidation and decomposition. For example, proteins, sugars, and starches

Fig. 5-37: Dolomite rock composed of coarsely crystalline dolomite and minor calcite. Slightly pinkish in color.

Fig. 5-38: Dark chocolate-colored peat, made up of a loose, soft mass of twigs, leaves, bark, and other parts of vegetation. Note light colored fibrous portions.

are most easily destroyed; cellulose and fats are somewhat more resistant, but chitin, amber, resins, and waxes are most resistant.

These organic residues are referred to as humus, peat, and sapropel. *Humus* is organic residue made up of newly formed as well as partially oxidized material, together with other organic compounds in various stages of decomposition. *Peat* is partially decayed accumulations of grasses, mosses, and other woody plants (Fig. 5-38). Peat is brown in color, contains shreds of plant tissue, and is porous. Peat usually forms in lakes, swamps, lagoons, and deltas. *Sapropel* is organic-rich silt which develops on bottoms of water basins such as lakes, lagoons, and estuaries. It consists mainly of hydrocarbons, residual after the decomposition of phytoplankton and zooplankton types of organisms. Some ferrous oxide and ferrous sulfide may be present.

Coal is an accumulation of plant matter, in a swampy environment, which has been changed by both biochemical and physical processes to a bedded, consolidated rock. The biochemical and physical changes produce a series of products such as peat, lignite, bituminous, and anthracite, a gradational series which is nearly perfect from plant tissue to anthracite coal.

Peat and sometimes lignite are not always considered as coal. Peat forms by partial decay of grasses, mosses, etc. Similar material, more intensely altered after burial, is lignite, actually a type of brown coal. But true *coal* requires more intense alteration with the development of *black* color. Coal is opaque and noncrystalline, with dull to brilliant luster and a low specific gravity of 1.0 to 1.8. The hardness varies from 0.5–2.5. Coal is brittle and has a hackly to conchoidal fracture. Commonly associated with coal are small amounts of clay and varying quantities of pyrite, FeS_2, and marcasite, FeS_2.

Coal is classified according to *rank*, based on the degree of coalification. The ranks of coal which are commonly recognized are, in order of increasing rank, lignite (brown coal), subbituminous, bituminous, semibituminous, semianthracite, and anthracite. Coal is composed chiefly of carbon, hydrogen,

Fig. 5-39: Cannel coal from Cannel City, Kentucky, composed partly of spores and fragmental wood debris, showing dull, black color and conchoidal fracture.

and oxygen, with very minor amounts of nitrogen and sulfur. The proportions of these constituents vary with the rank. As the rank increases, the carbon content increases, but hydrogen, nitrogen, and oxygen decrease.

Lignite is a low-rank coal which retains wood structure of the original plant material. It is brown or brownish-black, and it contains high moisture. It gives off very little heat during burning and is smoky. Cannel coal, composed of spores and wood debris, is shown in Fig. 5-39. *Bituminous* coal (Fig. 5-40) contains a higher proportion of carbon and lesser amounts of water. And it is less smoky while burning and has a higher fuel ratio (heat value) than lignite. *Anthracite* coal, actually produced by metamorphism, has a bright, nearly metallic luster, and is harder than bituminous (Fig. 6-21). The carbon content is much higher than in bituminous coal, and consequently anthracite produces more heat and less smoke during the burn than bituminous.

Mixtures of organic and clastic or chemical sedimentary material are fairly common, producing such rocks as *black shale. Carbonaceous black shales* (see Fig. 5-14) contain humic material composed of plant fragments or carbonaceous flakes. Plant leaves and stems are usually incorporated. *Bituminous black shale*, on the other hand, consists of bituminous organic matter, mostly oily substances such as pollen, animal matter, and resins.

Fig. 5-40: Bituminous coal showing black color, compact somewhat shiny surface, and slight banding.

Fig. 5-41: Gray compact limestone consisting of very-fine-grained calcite and some dark organic material. Note subconchoidal fracture.

Fossils which are present in bituminous shale are bottom-dwelling types. Black shales are very dark gray to black in color. Finely divided FeS_2 (pyrite or marcasite) may be present.

A few limestones are dark and dense, and often contain abundant amounts of organic matter. Such a rock, called *bituminous limestone*, has a rotten odor when freshly broken. Bituminous limestones are associated with organic-rich black shale and seem to require stagnant conditions for preservation of the organic matter (Fig. 5-41).

If surface waters are highly reducing, then iron sulfide in the form of pyrite, FeS_2, or marcasite, FeS_2, is apt to be deposited if sulfur is abundant. These two minerals are frequently disseminated in black shales and in bituminous limestones. These iron sulfide minerals commonly are in the form of nodular masses, as crystals in geodes, or they may actually replace fossils. Coal beds often contain nodules of pyrite or marcasite. Soft, porous *bog iron ores* often contain iron sulfides along with iron hydroxide, iron silicate, and iron carbonate. Phosphorus is usually present as well as manganese in such sediments. The bog ores also contain varying amounts of peat and thus are spongy masses of differing shapes and textures. Bog iron ores of relatively small size are being formed today in the glaciated northern regions of North America.

Structures of Sedimentary Rocks

Sedimentary rocks usually contain various structures. These are the large features of the rocks rather than small features relative to the constituent particles. Many structures are formed at the time of deposition, while others are formed later. Sedimentary structures are usually referred to as *external* or *internal*. The external structures include size, shape, and nature of the boundaries of the rock unit itself, and also to types of folding within it. Internal structures include bedding, ripple marks, concretions, fossils, and the like. Structures formed at the time the rock is deposited are called "primary," while structures formed after deposition are called "secondary." Structures observed in clastic rocks are significantly more abundant and different from those of nonclastic rocks; consequently, most of the following structures are seen in clastic sediments.

External Structures

Sedimentary rocks are masses of material having length, breadth, thickness, and a finite size. Therefore, they have some type of geometric *shape*. The shape of a sedimentary mass may be described as a *sheet, lens, wedge, fan, delta,* or *shoestring.*

A sheet (or blanket) has a ratio of width to thickness greater than 1000: 1. If the sedimentary mass has a ratio of 5 to 50: 1, it is a wedge. A shoestring is a sedimentary mass which has a ratio of 5: 1. Many variations of these basic shapes exist and take on such names as fan, lens, prism, and delta.

Sedimentary bodies vary in size according to sedimentary conditions and supply of material. The following size terms are commonly used:

Total volume of more than 500 cu mi—large

Total volume between 1 and 500 cu mi—medium

Total volume of less than 1 cu mi—small

The contacts between sedimentary masses are often very sharp, both vertically and laterally, but gradational contacts or boundaries are also quite common.

Primary Internal Structures

Internal structures are present within the sedimentary mass itself. The larger internal structures are best seen at the outcrop of the rock, but certain of the smaller internal structures may be seen best in hand specimens.

Practically all sedimentary rocks are arranged in layers (called strata or beds). This structure, called *bedding* or *stratification,* is singularly the most universal feature of sedimentary rocks (Fig. 5-42). Bedding (or stratification)

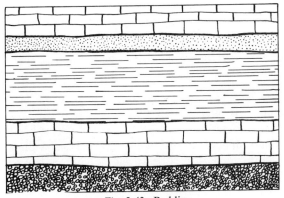

Fig. 5-42: Bedding.

is shown by units of rock material, each of which is generally tabular or lens-like in geometrical shape, and each of which has fairly uniform lithology. Such units of rock material are set off from each other by having some difference in physical makeup. These individual sedimentary rock units are of various sizes. The various units and their thicknesses are as follows: a *stratum*, an individual layer 1 cm or more in thickness, separated from strata above and below by a definite change in makeup or by a physical separation called a bedding plane; a *lamina*, an individual layer less than 1 cm in thickness; a *bed*, a rock unit composed of several laminae. Rock units may in some cases be of great lateral extent and of very uniform thickness. In other cases, rock units may change very rapidly and develop lens-like forms which thin out in short distances. Beds may range from a fraction of an inch to many feet in thickness. Sedimentary rocks formed in shallow seas tend to develop very uniform bedding with uniform composition. But rocks formed in rapidly subsiding sedimentary areas, such as geosynclines, tend to develop bedding which shows much local variation in thickness and areal extent. A single *bed* represents a single episode of sedimentation, and a *bedding plane* probably represents a period of cessation of sedimentation or a change of material being supplied. In the case of some thin laminae, there is a suggestion of seasonal changes during deposition. Thin laminae may also be related to the presence of platy minerals, such as mica, lying in parallel arrangements.

Graded bedding, as a structure in a sedimentary unit, is marked by gradual decrease in size of particles upward from the base to the top of the unit (Fig. 5-43). Individual graded beds may be fairly thin, of the order of 4 or 5 inches, but considerably thicker graded beds are also commonly encountered. Graded bedding may be caused by:

(1) variation in seasonal supply of sediment by progressively weaker streams;

(2) progressive settling of coarse to finer sizes in calm bodies of water;

(3) stirring up of bottom deposits by storms and then progressive settling of coarse to fine sizes; or

(4) stirring up of sediment by sliding and slumping due to earthquakes and then differential settling.

Cross-bedding is a fairly common structure in deposits formed under water or by wind. It is most commonly seen in sandy deposits, but it is not found exclusively confined to this type of material. Cross-bedding produced under water is very difficult to distinguish from cross-bedding produced by wind. This structure appears as concave (or arcuate), nearly parallel smaller beds inclined somewhat to the main larger bedding plane. The angles of

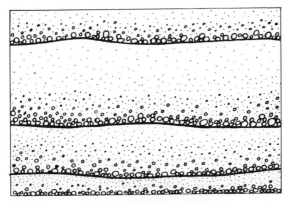

Fig. 5-43: Graded bedding.

inclination of the beds are usually less than 30°. Cross-bedding formed under water is apt to have the upper surface parallel to the lower surface, while wind-formed cross-bedding will display curved or inclined upper surfaces. Each bed slopes in the direction of current flow. The inclination of the beds is controlled by the rate of deposition and by the amount and coarseness of supply. The faster the material is deposited, the greater the volume supplied; and the coarser the particles, the steeper is the inclination of the bed. Long inclined beds are produced by fast currents, and short ones, by rapid deposition of coarse material (Fig. 5-44).

The tops and bottoms of beds of sedimentary rocks show considerable variation in smoothness and roughness. One of the most familiar structures seen on the upper surface of sediments is *ripple marks*. Ripple marks are subparallel ridges and hollows formed by waves or currents. Ripple marks may be seen along beaches and are formed by irregular transportation and disturbance of unconsolidated sediment over short distances, with final

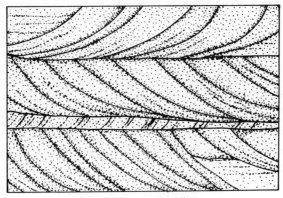

Fig. 5-44: Cross-bedding.

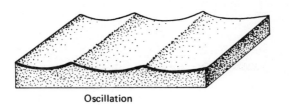

Oscillation

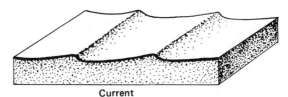

Current

Fig. 5-45: Ripple marks.

accumulation in ridges. Two types of ripple marks are common (Fig. 5-45). *Oscillation ripple marks* are caused by a to-and-fro motion of sediment along the bottom as a result of to-and-fro motion of water waves; hence these ripple marks have symmetrical shapes. *Current ripple marks*, on the other hand, are asymmetrical in shape being produced as fairly constant current tends to move grains up the gentle side of the ridge and down the other steeper side. With a fairly predominant current direction, this type of ripple mark moves slowly downstream.

Mud cracks may develop on the upper surface of fine-grained sediments, as drying may cause a shrinkage to occur. These shrinkage cracks, which taper downward, usually form polygonal patterns of various lengths and widths. The resulting pattern is partly determined by the presence of foreign material in the mud and partly by the degree of drying. Mud cracks develop in areas where drying out of water-laden clayey sediments may occur, such as along tidal flats, dried-up river courses, lake beds, and flood plains. However, mud cracks occur more frequently in continental than in marine sediments, because of the greater opportunity for drying out to take place.

Secondary Internal Structures

In many sedimentary rocks, there are various secondarily formed accretionary masses. These variously shaped and variously formed accretionary bodies fall into four groups:

(1) nodules;

(2) spherulites (and other regular crystal growths);

(3) concretions; and

(4) geodes, septaria, and other masses with partially or wholly filled interiors.

Nodules are irregular bodies of mineral matter different from the main mass of rock in which they are embedded. Nodules have knobby exteriors and commonly are slightly elongated along bedding. The most common type is chert (or flint) nodules in limestones. Nodules generally are structureless, dense, and homogeneous, but a few contain fossils. These nodules seem to form by irregular replacement of limestone by silica. Nodules made of phosphatic material also are fairly common.

Spherulites vary in size from microscopic to an inch or two in diameter. Their internal structure is described as radial, and hence these structures are more or less spherical in shape. However, they may also take on irregular shapes if two or more grow together. Several different minerals commonly occur in spherulites; such minerals as chalcedony, apatite, and marcasite have been observed. Growths of accretionary bodies which develop a floral-like arrangement are termed *rosettes*, and pyrite, gypsum, barite, and other minerals have been found as major constituents.

Concretions are concentric structures which accumulate about a center; hence they are usually spherical, subspherical, or disk-shaped. They usually have a nucleus, such as a fossil shell, leaf, or bone, which served as a center around which the structure accumulated. However, in many concretions there is no nucleus. Concretions vary in size from microscopic to as much as ten feet across. The size achieved is apparently determined by the permeability of the host rock. The minerals which have been most commonly observed in concretions are those usually making up the cement of the rock, such as silica, calcite, and iron oxides. Concretions are formed by waters percolating through the rock which collected mineral matter and redeposited it around centers of growth, such as a shell fragment or other object.

Geodes, which are hollow, globular masses, vary quite a lot in size. Most geodes range in size from two to three inches up to a foot or more across. They usually have a subspherical shape and a hollow interior (Fig. 5-46).

Fig. 5-46: Geode, about 6 inches across, showing hollow interior partly filled with quartz crystals. Outer layer is chalcedony, a compact microcrystalline variety of quartz.

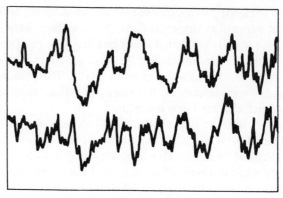

Fig. 5-47: Stylolites in limestone.

Geodes are commonly found in limestones but are occasionally found also in shales. Geodes are usually partly filled with crystals of quartz or calcite, but some contain aragonite, magnetite, hematite, pyrite, or other minerals. The outer shell is nearly always chalcedony. Geodes begin to grow in cavities, most of which are cavities in fossils. As growth of the geode continues, the surrounding limestone bursts. Silica of the geode finally dehydrates and crystallizes followed by shrinking and cracking. Mineral waters then enter the geode and deposit the final aggregate of projecting crystals over the chalcedony layer.

Others. A series of smaller-sized structures may develop in sedimentary rocks, and these will be defined only very briefly. *Conchilites* are bowl-shaped, oval, or circular objects, composed principally of limonite and goethite. *Septaria* are fairly large subspherical nodular masses with radiating and crisscross cracks. They are usually composed of clayey carbonate minerals. *Cone-in-cone* is merely a series of concentric cone-like masses, rare in occurrence, but seen in shales occasionally. *Stylolite seams* are surfaces of contact showing interlocking or mutual interpenetration of both sides of the stylolite seam, teeth-like projections on one side fitting into sockets on the other side (Fig. 5-47).

Chapter Six

Nature of Metamorphic Rocks

The term metamorphic (or metamorphism) is generally used to describe the sum total of changes in rocks brought about by physical and chemical changes of the environment. However, it does not include the weathering environments. In rocks undergoing change, whether they be igneous or sedimentary, the changes which take place predominantly involve recrystallization of minerals while the rocks remain solid; no melting takes place. The changes are mainly mineralogical and textural. Essentially no change in the overall composition of the rock occurs. New and different minerals may be produced, but no new elements are introduced into nor any elements removed from the rock; the elements which are present are just rearranged into new and different minerals. Factors which bring about such changes are rise of *temperature*, increase of *pressure*, and the *action of chemical agents*. These factors tend to upset the physical and chemical equilibrium in a rock, thereby causing minerals to become less stable. The constituent minerals are changed over to new and different ones which arrange themselves into structures and textures better suited (more stable) to the new conditions. This is metamorphism. It is mainly recrystallization of rock, partial or complete.

Metamorphic Processes and the Formation of Metamorphic Rocks

The three agents of metamorphism, heat, and pressure, and the chemical action of solutions and gases, generally cooperate in various degrees to produce metamorphism. The chemical factor is operative probably in all cases, and temperature and pressure effects are hard to separate. Thus, in many cases the effects of one are more readily observed than effects of the others, i.e., one is usually predominant.

Effects of Heat, Pressure, and Chemical Solutions

Rise of temperature causes thermal agitation of the atoms in the minerals, resulting in ultimate breakdown. However, the atoms become reconstituted into more stable minerals in this new heat environment. There are three possible sources of heat to produce these effects. At depth, heat may be supplied from the normal increase of temperature with increase of depth in the crust. The rate of increase varies considerably from about 1°C per foot near active volcanoes to 1°C in 400 or so feet of depth in other areas. A second source of heat may be supplied by mechanical energy produced during earth movements such as mountain-building processes. Finally, the greatest source of heat may be that emanating from large masses of magma which are present in various localities at depth.

Natural pressures in the Earth's crust are of two types: *hydrostatic* (uniform pressure in all directions) and *stress* (directed pressure operating chiefly in one direction). Hydrostatic pressure, as a function of depth, has a tendency to produce volume changes of minerals, with the result that more compact and higher density minerals will be formed. Stress or directed pressure is produced during great earth movements. Such stress produces changes in the form and orientation of the constituent minerals, and it is apt to produce mechanical breakdown (crushing and granulation), with little new mineral formation. However, with depth, heat becomes a more important operating factor in softening of the rock and in aiding chemical solutions. In local areas where frictional heat is high, some minor melting may occur.

When both directed pressure and increased heat are operative at considerable depth in the Earth's crust, metamorphism proceeds at a much greater rate than when only one of these factors is in effect. Under such a powerful combination of influences, the rocks involved tend to become completely recrystallized, and new structures and textures are apt to be developed in the rocks. Under conditions of this type, the melting points of minerals are often lowered, at least locally, and recrystallization can proceed more easily and rapidly.

Chemically active solutions in a region undergoing metamorphism consist of liquids and gases which occupy very tiny pores, fissures, and interstitial space in the rocks. The most important solution is water, often chemically charged locally by other substances such as carbon dioxide, boric acid, hydrofluoric acid, and hydrochloric acid. Such solutions, either as liquids or gases, act as solvents or catalysts and play the role of accelerating metamorphic reactions. These substances migrate through the minute pores and interstitial spaces of the rocks and produce changes as they slowly move. Any volatiles (gases and liquids) being emitted from hot magmas in the area may be distributed out into the surrounding rocks, carrying heat along. Water in the rocks may have come from seawater trapped with the sediments at time of deposition, from water molecules in minerals undergoing reaction, as well as the above-mentioned water from magmas.

Since metamorphic change takes place in the solid state, without melting or liquification, very small quantities of water or other solutions are necessary. These solutions act as a medium in which atoms (or ions) are transferred from one mineral to another during the recrystallization and may or may not enter into the formation of new minerals.

Any new minerals and new textures which develop in metamorphic rocks result from processes of recrystallization, of granulation, of foliation, and locally from additions or subtractions of ingredients. Pressures operating during regional metamorphism cause rock flowage to occur. Rock flowage is described as deformation of rock due to granulation and recrystallization acting in combination. The most intense metamorphic deformation is developed in deeper parts of the crust, in folded mountain areas, and in regions where large-scale intrusive pressures are aided by heat and vapors.

Recrystallization

Recrystallization simply means a change in texture or mineral makeup developed as material is lost by partial solution of some crystals and addition of material to the outside of others. The process usually involves the solution and fusion of some crystals while others are growing. Recrystallization can also occur without any liquid being produced. Metamorphism is presumably a solid rock reorganization, with only very small amounts of solutions of minerals undergoing change at any one time. The process of recrystallization apparently occurs a little at a time. Crystals of minerals are usually enlarged or new ones are formed; even crystals of new minerals often form. However, no change in bulk composition of the rock occurs, except perhaps hydration of some minerals or some small additions or deletions of components, mostly as a result of small amounts of water. Though water is not absolutely essential, practically all rocks contain water, and the water must certainly play a part. Small crystals of minerals commonly disappear, while large ones enlarge even more.

Granulation

Granulation refers to the crushing of rock under the conditions of directed pressure, with little or no increase in heat. This is sometimes referred to as cataclastic metamorphism. The dominant factor producing granulation effects is intense and rapidly applied stress, causing the deformation in a short time. Such action must necessarily occur only in upper levels of the crust where rocks are cold and brittle, because at lower levels in the crust heat becomes increasingly important, and rocks become softened and less subject to crushing. Granulation is thus associated with regions of intense mountain folding. Mineral grains become fractured, distorted, and crushed, but the confining pressure permits the rock to remain cohesive and the porosity to remain relatively low. Some recrystallization of rock into flinty, fine-grained types may occur, and rotation of fragments may occur on a limited scale.

Hard, brittle, insoluble minerals such as quartz and feldspar are more easily granulated than soft minerals like gypsum or talc. Calcite crystals of limestone are easily granulated, but they are also easily recrystallized, an action which may mask granulation effects. In addition grains of different colors may cause *banded rocks* to form as flattening of grains is produced.

Foliation

Foliation refers to the orientation of newly formed crystals, especially of platy minerals such as mica in planar directions or of elongated minerals such as amphiboles in planar directions or in linear patterns. Foliation is commonly produced by directed pressure of regional metamorphism, with the development or rocks such as slates and schists. The foliation may be plane, undulating, or lenticular. Foliation is characterized by the platy or elongated crystals having their long dimensions perpendicular to the direction of greatest stress. The foliation may develop as a result of slow movement of new minerals as they form. Movement may consist of pushing or rolling the crystals into orientations which offer the least resistance to movement. The orientations may be guided by planes of weakness in the rock. Not only do the flaky and elongated crystals have preferred orientations, but even equidimensional crystals of quartz, calcite, and feldspar become slightly elongated in the same orientations.

Apparently, foliation is developed by pressure in combination with heat, chiefly under regional metamorphic conditions. Such conditions would occur at lower levels of the crust in areas undergoing mountain construction. Foliation may be produced by very-fine-grained crystals and may be well developed, as in the case of slates. Coarse- or medium-grained crystals will produce folia which can be easily separated from one another, as in the case of schists. Coarse-grained crystals which lie in rough parallel arrangements or bands would produce gneisses.

Kinds of Metamorphism

Metamorphism may be of two broad types: *contact metamorphism* and *regional metamorphism*, according to the important factor, *heat* or *pressure*, respectively, which is most influential in producing the metamorphism.

Contact Metamorphism

In contact metamorphism, *heat* is the dominating factor. This type of metamorphism is produced locally around large igneous masses. Although heat is the most influential factor, pressure plays a significant role also. Pressure may be derived from the pushing aside of the country rock as the magma migrates, from expansion of rocks as they become heated, and by the weight of overlying rock material. Contact metamorphism around large masses of igneous rock is produced at relatively low temperatures. No change in the overall composition of the rock occurs. Mineralogical changes (formation of new minerals) are facilitated by gases and solutions which migrate from the magma. Very locally along the immediate contacts of magma with adjacent country rocks, high-temperature changes may occur with or without interchange of mineral material. This higher-temperature subtype of contact metamorphism is called *pyrometamorphism*. When fair amounts of magmatic substances are added to and alter the country rock, the term *additive metamorphism* is used. If very large amounts of material are added from the magma into the country rock, the term *injection metamorphism* may be used. Pyrometamorphism, additive metamorphism, and injection metamorphism all may be thought of as subtypes of *contact metamorphism*.

The effects of contact metamorphism usually extend a few tens or hundreds of feet from the magma, and rarely out to 2000 or 3000 feet (see Fig. 4-15). All kinds of magmas are capable of producing contact metamorphic effects in the surrounding country rock. The larger and hotter the igneous mass, the more extensive are the metamorphic effects. Large quantities of escaping vapors are apt to have greater effects than small quantities of vapors, but the rate of escape of vapors plays a part too. The intensity and extent of contact metamorphism also are controlled by the nature of the surrounding rocks because different rock types have different susceptibilities to metamorphic change. Porous and reactive type rocks are more susceptible to contact metamorphic effects than nonporous and nonreactive types.

Regional Metamorphism

Regional metamorphism is produced by great pressures and stresses acting over great distances, producing metamorphic effects in the rocks over a large region. Although pressures and stresses are important factors, most of the metamorphic action occurs in deep-seated levels of the Earth's crust, where temperatures are high and load pressures are high. Regional metamorphism occurs at temperatures somewhat lower than those producing contact metamorphism. Essentially no change in the bulk composition of the rocks takes place during regional metamorphism. Mineralogical changes (formation of new minerals) are facilitated by gases and solutions which occupy pore spaces in the rocks, just as in contact metamorphism. Rocks which have been regionally metamorphosed usually display a wide range of mineral transformations and rock structures, depending upon their original nature and on processes which affected them. The rocks are usually all crystalline and are tightly compacted as a result of tight interlocking of crystals resulting from the pressures. The rocks become laminated or foliated, regardless of their nature before metamorphism.

Regional metamorphism is usually associated with major mountain-making episodes. Pressures (and temperatures) of crustal movement bring about deformation with subsequent metamorphism. Crustal movements often result in cataclastic changes (breakage and granulation) in rocks at the surface, and they also aid in mineral transformations at depth where pressures are higher. Higher temperatures would ensue, recrystallization would proceed, and perhaps large bodies of magma would be generated and intruded into the region, with resultant contact metamorphic effects being produced.

The physical aspects of the rocks undergoing metamorphism have some influence on rate of metamorphic change. The more important factors are the resistance of the rocks to crushing and deformation, the grain size of the rocks, and the porosity and permeability of the rocks.

Resistance to deformation depends on physical aspects of the constituent minerals, i.e., strength of minerals, brittleness, and degree of cementation. Strong rocks offer more resistance to deformation than do weak rocks. Grain sizes of constituent minerals, along with resultant porosity and permeability, control the movement of gases and liquids through the rock. Sedimentary rocks generally are less resistant to regional metamorphism than igneous rocks.

The nature of a metamorphic rock produced during metamorphism is controlled by several factors. These are:

    (1) type of the rock being affected;
    (2) processes acting on the rock;
    (3) intensity of the processes; and
    (4) addition or subtraction of chemical material.

The importance of these factors in their role in metamorphism will vary from time to time and form place to place in particular circumstances.

Minerals of Metamorphic Rocks

Rocks which become metamorphosed by either *contact* or *regional metamorphism* may be both igneous and sedimentary. Consequently, minerals found in metamorphic rocks consist of two main types: those stable minerals which resist metamorphic effects, and new minerals formed by recrystallization and have thus become stable against metamorphism. In addition, a third type may be developed: new minerals formed by local chemical reaction involving addition or subtraction of ingredients.

The minerals of metamorphic rocks can be divided into three groups as shown in Table 6-1. Those listed in group A are generally of wide distribution. They are usually the chief (or essential) minerals and are commonly present in substantial amounts. Those listed in group B are of lesser importance which may occur as prominent accessory, but they are locally developed as chief components. Those in group C are present in only minor quantities, but they may be locally abundant. This grouping is true in a very broad sense, and its divisions may be arbitrary.

The listing of minerals in Table 6-1 occurring in metamorphic rocks does not mean that all are present in the same rock, but rather they are minerals occurring in different metamorphic rocks. However, many of them occur exclusively in metamorphic rocks.

Table 6-1: MINERALS OF METAMORPHIC ROCKS

Group A (Common)	Group B (Accessory)	Group C (Sporadic)
Quartz	Tremolite	Graphite
Orthoclase	Wollastonite	Tourmaline
Microcline	Grossularite	Idocrase
Plagioclase	Chondrodite	Hematite
Calcite	Epidote	Anthophyllite
Dolomite	Pyroxenes	Cordierite
Serpentine	Magnetite	Spinel
Chlorite	Kyanite	
Muscovite	Andalusite	
Biotite	Sillimanite	
Sericite	Actinolite	
Hornblende	Staurolite	
	Talc	

Even though all varieties of feldspar have been found in metamorphic rocks, the most common ones are orthoclase, microcline, albite, and the intergrowths, microperthite and perthite.

Many minerals of metamorphic rocks have elongated or flattened habits, such as mica, talc, or hornblende. The physical environment of metamorphism, particularly pressure, tends to favor the development of minerals having these shapes, although minerals with other shapes also are often present as shown in Fig. 6-1. These flattened or elongated minerals tend to lie with their longest direction at right angles to the direction of greatest pressure and thus produce directional types of structures. Other minerals in metamorphic rocks have shapes which are more equidimensional, such as quartz, feldspar, and garnets. Even so, these tend to become somewhat elongated due to continued flow or to recrystallization during regional metamorphism.

During contact metamorphism in which heat is more influential than pressure, elongated or flattened minerals are not so well developed. Because of these different environmental factors, equidimensional crystals are more apt to retain their shapes.

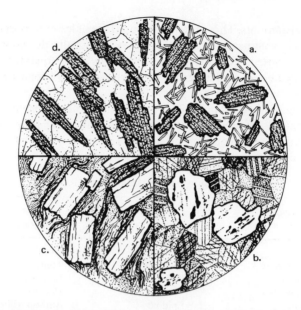

Fig. 6-1: Shapes of crystals in metamorphic rocks: (a) biotite flakes (flat) in hornfels; (b) large albite (flat) in marble; (c) elongate andalusite in schist; (d) elongate hornblende in schist.

Textures of Metamorphic Rocks

It is quite difficult to distinguish between texture and structure in metamorphic rocks. During metamorphism, different textures may become quite intimately intergrown. Textures of the original rocks may be preserved in varying degrees, and any new textures developed during metamorphism may become superimposed on original relic textures. Rock textures depend upon mineral shapes, modes of growth, and mutual arrangements. On the other hand, structures depend upon the interrelations of the textures in the rock.

A texture resulting mainly from recrystallization in which the new minerals develop fair crystal form is called *crystalloblastic.* However, well-shaped crystals usually do not develop; recrystallization most often results in poorly developed crystal faces, and the texture is referred to as *xenoblastic.* If the recrystallized minerals have taken on a fine granular habit, the term *granoblastic* is used to describe the texture. In some cases, a few minerals which have very strong crystallization powers may develop into large well-shaped crystals, embedded in a finer matrix. The large crystals are called *porphyroblasts.* Such a texture with some large crystals is called *idioblastic.* Lens-shaped porphyroblasts are called *augen* (eye-shaped). Regardless of the texture developed under different conditions of metamorphism, original textures of parent rock are often apt to be preserved in varying degrees. Metamorphic textures are illustrated in Fig. 6-2.

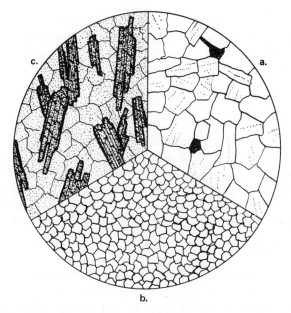

Fig. 6-2: Metamorphic textures: (a) crystalloblastic texture in quartzite; (b) granoblastic texture; (c) idioblastic texture with large hornblende crystals.

Structures of Metamorphic Rocks

During metamorphism, recrystallization of minerals and growth of new minerals take place in the solid condition. In a great many metamorphic rocks, recrystallization and crystal growth occur nearly simultaneously, and hence no particular sequence of development takes place. The minerals usually are associated with one another in indifferent manners and arrangements. Based on the arrangement of the constituent minerals, the metamorphic rocks are classified very broadly into two main groups: foliates (with directional structures) and nonfoliates (without directional structures).

Foliated Structure

Metamorphic rocks with a *foliated* structure display a thin layering arrangement of the constituent minerals. The minerals are either coarse-grained or very-fine-grained. The folia may vary from parallel, wavy, or lens-like, and are developed due to the presence of flattened or elongated minerals. These minerals take on a preferred orientation in which their longest dimensions are perpendicular to the direction of pressure. Three main types (and a fourth uncommon type) of foliation occur: slaty, schistose, gneissose, and migmatic.

Slaty Foliation

The presence of very-fine-grained or aphanitic minerals causes the rock to display a well-developed parallel foliation. The rocks have a tendency to split into thin parallel sheets. Rocks with slaty foliation and rock cleavage are referred to as *slates*.

Schistose Foliation

Coarse- or medium-grained minerals cause the rock to show a marked parallel to wavy foliation. The rocks have a fairly strong tendency for the folia to separate from each other rather easily. The presence of large numbers of mica crystals, lying with flat surfaces all parallel, are responsible for both ease of separation of folia as well as shimmering surfaces (as in phyllites) in varying degrees. Rocks with schistose foliation are referred to as *schists*.

Gneissose Foliation

Fairly coarse-grained minerals, lying in a series of parallel arrangements, cause the rock to display a roughly parallel-banded foliation. Bands are often made up of large numbers of a single mineral species such as quartz or feldspar. The bands are usually very cohesive. Enlarged crystals commonly may be present as individuals at various positions in the rock. Rocks having a *gneissose foliation* are referred to as *gneisses*.

Migmatic Structure

Near the contact of granite with the country rock, granite is injected as narrow dikes into laminated rock. An alternating light and dark series of mixed rock is produced, granite being light and country rock being darker. Thus, these mixed rocks are partly igneous and partly metamorphic. The resulting laminated structure allows them to be classed as metamorphic. Mixed rocks of this type are called *migmatites*.

Nonfoliated Structure

Metamorphic rocks with *nonfoliated* structures display mineral arrangements which are not directional, as shown in Fig. 6-3. The constituent minerals are either equidimensional and of generally uniform size or there may be both fine- and coarse-grained crystals. Crystals generally lie in the rock with no preferred orientation. Three main types of nonfoliated structures occur: maculose, hornfelsic, and granulose.

Maculose Structure

Generally fine-grained minerals serve as the main part of the rock, but other somewhat larger minerals are well developed too. If a significant

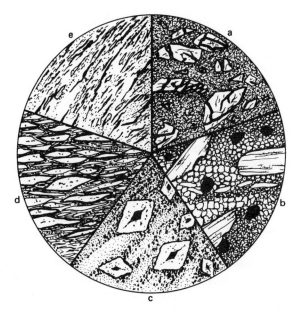

Fig. 6-3: Metamorphic structures: (a) cataclastic structure, (b) granulose structure; (c) maculose structure; (d) flaser structure; (e) mylonitic structure.

amount of these larger crystals are present, the rock may appear spotted. Such a structure is developed under contact or thermal metamorphism. It is sometimes referred to as *porphyroblastic structure*. Rocks having maculose (porphyroblastic) structure are generally referred to as *spotted rocks*.

Hornfelsic Structure

Clayey rocks, as well as many other types, under thermal metamorphic conditions, become extensively altered or changed. Clay minerals become converted to micas and chlorites, usually arranged in random orientations. In addition, these and other minerals recrystallize at about the same time, with no stress to produce any directional structures. Hornfelsic structure usually is sugary in appearance, since most of the resultant crystals are of sugar-grain size and even-grained. A few enlarged grains of less common minerals may be present. A rock with hornfelsic structure is referred to as *hornfels*.

Granulose Structure

Rocks possessing a predominance of equidimensional minerals have a granulose structure. Elongated, flattened, or lamellar minerals are either absent or only in very subordinate amounts. As a result, these rocks do not possess directional structure. Crystals of minerals are usually fairly coarse, and only on rare occasions will banding, streaking, or lensing of mineral masses occur.

Another type of structure in metamorphic rocks which is neither directional nor nondirectional, but is related to a very different type of physical aspect, is the *cataclastic structure* shown in Fig. 6-3. Cataclastic structures (or fabrics) consist of broken and fragmented rocks produced by cataclastic metamorphism. The production of this type of structure is called granulation (crushing of hard, brittle rocks). Hard rocks are usually shattered. Breakage may occur along grain boundaries, and intense mechanical deformation may produce pulverized or very-fine-sized rock debris. Fracturing and distortion of minerals may accompany crushing, producing flaser and mylonitic structure (see Fig. 6-3).

Classification of Metamorphic Rocks

The names of many metamorphic rocks actually are structure or fabric names; for example, a rock having a schistose structure is called schist. However, a number of the metamorphic rocks are not so named. The classification of metamorphic rocks is not an easy one; many difficulties are encountered. For example, under different kinds of metamorphism, the same material may be changed to rather different metamorphic rocks. Opposite to this, different kinds of original material may be metamorphosed to identical rocks.

Classification Scheme

The classification herein set forth is a simplified version of that of Travis. The metamorphic rocks are classified into two major groups on the basis of the structure which they possess: directional, and nondirectional, as presented in Table 6-2. The two fundamental changes which occur during metamorphism are mechanical structural changes (chiefly directional), brought about mainly by pressure, and recrystallization and reconstitution (chiefly nondirectional), brought about mainly by heat. Consequently, the structure of the rock more or less indicates the kind of metamorphism, and the mineralogy indicates the intensity (or grade) of metamorphism. It is not to be assumed by the reader, though, that classification along these lines is at all clear-cut or definite. Many kinds of transitions occur from one group to another, and it is often difficult to categorize a particular rock because its structure or mineralogy cannot be accurately determined.

Nondirectional Metamorphic Rocks

The rocks in this major group in the classification are produced chiefly by contact metamorphism but also are frequently formed by regional metamorphism. The most important factor in the development of rocks with non-

Table 6-2: CLASSIFICATION OF METAMORPHIC ROCKS

Nondirectional Structure				Directional Structure			
Chiefly Contact Metamorphism			*Mechanical Metamorphism*	*Chiefly Regional Metamorphism*			*Plutonic Metamorphism*
Low Grade	Medium Grade	High Grade	Very Low Grade	Low Grade	Medium Grade	High Grade	Extreme Grade
	quartzite marble hornfels serpentinite soapstone		crush breccia flaser rock mylonite augen gneiss	slate phyllite	schist amphibolite	gneiss granulite	migmatite

329

Fig. 6-4: Coarse-grained quartzite, light gray in color, from eastern Quebec, with quartz crystals badly shattered.

directional structures is heat associated with igneous intrusions; pressure is subordinate and not very influential. The rocks may be produced under varying intensities of heat effect depending upon distance from the igneous mass or amount of heat dissipated. If only relatively small amounts of recrystallization have taken place, the grade of metamorphism is called *low*. If recrystallization is more intense and hydrous minerals are formed, the grade is *medium*. When recrystallization is more advanced and anhydrous minerals are developed, the grade is *high*.

Quartzite

A metamorphosed sandstone is called *quartzite*. If original sedimentary structure of the sandstone is retained, the quartzite may be called a *metaquartzite*. Quartzites are formed due to sufficient quantities of heat to cause recrystallization of quartz and feldspar, predominantly. Other constituents of the original sandstone may also be recrystallized, but quartz and feldspars are the more common. Subsequent pressure on quartzites may cause them to be shattered, as in Fig. 6-4. They may develop a granoblastic texture, often with complete obliteration of clastic features, as shown in Fig. 6-5. The rock often develops a vitreous luster. Metamorphic quartzites are quite hard due to the presence of siliceous cement (usually crystalline quartz) deposited around larger quartz grains, producing a tight interlocking of constituent grains.

Fig. 6-5: Fine-grained quartzite from Wisconsin. Note gritty but tough appearance of this recrystallized sandstone. Minor hematite yields a purplish color.

Fig. 6-6: Quartz pebble quartzite containing quartz pebbles (1/2 to 1 in. across) in a recrystallized brown sandstone matrix.

The rock would break through the sand grains instead of around them. Other minerals sometimes recorded in quartzites in minor quantities are apatite, zircon, epidote, and hornblende, as well as others.

Quartzites may be formed by contact or regional metamorphism, by heat and pressure effects on various types of sandstones, cherts, quartz veins, and quartz-rich pegmatites. They may be banded, thin-bedded, thick-bedded, or laminated. Original bedding, laminations, or cross-bedding may be retained, and even fossils may be preserved. Colors of quartzites range from white to brown to nearly black. Most are fairly light colored though. The presence of hematite or chlorite in minor amounts may give the rock a pink or greenish color. A pebbly sandstone may be metamorphosed to *pebbly quartzite*, and a conglomerate may be changed to a *conglomerate quartzite* (Fig. 6-6). The presence of certain minerals in significant amounts may permit variety names to be used; for example, feldspar in fair amounts, but subordinate to quartz, may yield *feldspathic quartzite*. Other varieties are *micaceous quartzite* and *ferruginous quartzite*.

Marble

A *marble* is a metamorphosed limestone, either a plain calcite limestone or a dolomite limestone. Marbles are formed chiefly by recrystallization of the calcite (or dolomite), with the result that the rock marble becomes usually coarser grained than the original limestone (Fig. 6-7). Marbles produced

Fig. 6-7: Pink marble from Tate, Georgia, composed of coarse recrystallized calcite. Note dark streaking by mica.

from dolomitic limestones are usually called *dolomite marbles*. Bedding is often contorted or obliterated as a result of metamorphic pressures and recrystallization processes. Streaking by serpentine and other minerals may cause a type of banding.

Marbles and dolomite marbles may be formed by contact or regional metamorphism and may be found associated with other metamorphic rocks such as phyllites, slates, schists, and metaquartzites. Structures (fabrics) vary from very-fine-grained (sugary) to very-coarse-grained. In contact types, preferred orientation of crystals is lacking, but in regional types, slight amounts of preferred orientation among elongated crystals may occur as a result of directional pressure.

Accessory minerals in marbles vary considerably and may include tremolite, forsterite, periclase, diopside, wollastonite, brucite, spinel, feldspar, and garnet, depending upon the kind of material in the original rock. Some of the accessory minerals may become enlarged to produce a porphyroblastic texture. Marbles are usually white or light colored due to the presence of calcite or dolomite, but they may take on various tones of gray, green, red, brown, or mixed colors, depending upon the nature of accessories. The coloration may be quite uniform, but often it is spotted, blotched, or veined producing a "marbled" effect.

Brecciated marbles are known. Dark streaks of graphite yield *graphite marbles*, but serpentine usually produces *green marble*, the name serpentine not being used. The presence of any accessory mineral in sustantial quantities may permit names like *tremolite marble*, *talcose marble*, and *phlogopite marble*.

Hornfels

A clayey rock altered by contact metamorphism is called *hornfels*. Hornfelses are also formed less commonly by recrystallization of other minerals in other kinds of original rock. The resultant rock (hornfels) is very-fine-grained and usually gives a massive appearance (Fig. 6-8). The constituent minerals

Fig. 6-8: Diposide hornfels showing fine, sugary-textured rock composed of white to light green diopside and a substantial amount of plagioclase and hypersthene. Note absence of foliation.

Fig. 6-9: Yellowish-brown tactite from Sonora, Mexico, showing coarse granular garnet crystals in this nonfoliated rock.

do not achieve a preferred orientation since stress is not an important factor. The texture is usually granulitic (or finely granoblastic) and schistosity is absent. More appropriately, the texture is hornfelsic. Hornfelses are commonly light gray to dark gray in color, due to the presence of some dark minerals such as biotite and pyroxenes. But light minerals such as quartz, feldspars, and calcite are usually abundantly present. In addition, other contact metamorphic minerals such as andalusite, cordierite, and garnet may occur as large porphyroblasts. Since most hornfelses are derived from clays and shales, other aluminum-bearing minerals such as spinel, corundum, and aluminous silicates may be found. Hornfels derived from shales may retain bedding features, while hornfels derived from volcanics may retain porphyritic or amygdaloidal textures.

Several varieties of hornfels are common. The presence of andalusite or cordierite in fair amounts yields *pelitic hornfels* (derived from clayey or pelitic sediments). Hornfels derived from sandstone, arkose, and silicic volcanics may be called *arenaceous hornfels.* If an argillaceous limestone is metamorphosed by contact action, a *calc-silicate hornfels* may be developed, but the composition may be variable because some of the limestone beds may be converted to marble where clayey material is lacking. Basalt and andesite, upon contact action, may produce *mafic hornfelses.* A hornfels composed of pyroxene and garnet, along with carbonate minerals, is called *skarn.* Lime-silicate rock of this type is called *tactite* (Fig. 6-9). Hornfelses usually occur in contact zones close to igneous intrusive rock, especially granitic types.

Special Silicate Rocks

In many cases, limestones contain substantial quantities of silica. During metamorphism, such limestones are apt to be converted to *serpentinite, soapstone, garnet rock, garnetite,* and *jade.* However, some of these rocks are produced in other ways as well.

Serpentinites are composed predominantly of the mineral serpentine

Fig. 6-10: Serpentinite from Vermont showing massive green serpentine with white veins of magnesite and fibrous serpentine.

derived mainly from hydrothermal alteration of rocks containing large quantities of magnesium silicates such as olivine and pyroxene of basic rocks. The serpentinites have a variety of textures according to their mode of origin, but they are usually fine-grained or fibrous (Fig. 6-10). All are very soft. Colors vary from green to red, but most are green in tone. Other minerals are not common in serpentinites, but hornblende and magnetite are often reported.

Talc-carbonate rocks are often associated with serpentinites, and serpentinites often grade into talc rocks, such as soapstones, which have large quantities of the mineral talc mixed with lesser amounts of chlorite. Soapstones (also called steatites) are produced in the same manner as serpentinites. Quartz veins and quartz crystals often are found in soapstones. Soapstones are not particularly common, but they often have high commercial value due to the abundance of talc.

Directional Metamorphic Rocks

The rocks in this major group in the classification are produced chiefly by regional metamorphism. The most important factor in the development of rocks with directional structures is pressure in combination with heat. These are associated at lower levels of the crust or with nearness to igneous intrusions of large size in areas undergoing mountain construction. The rocks may be produced under varying intensities of pressure, heat, and water effects. If effects are slight, deformation is minor, and the grade is called *low*. Progressively more intense effects are produced under greater deforming forces, and the grade increases up to *medium* and *high*, indicated by the mineralogy and structural features. Heat effects sufficient to develop mixed rocks (called migmatites) by injection of igneous material into laminated metamorphic rocks indicate that metamorphism has been *extreme.*

Crush Breccia, Flaser Rock, Augen Gneiss,
Cataclasite, and Mylonite

Rocks formed by mechanical (or cataclastic) metamorphism display varying degrees of crushing and granulation. Recrystallization is not important in these rocks. Fairly strong stress is applied at shallow depths and at low temperatures to produce cataclastic metamorphism. At greater depths and higher temperatures, cataclastic metamorphism passes gradually to normal regional metamorphism.

When strong lateral pressures are applied in hard and brittle rocks, the rocks are deformed by fracturing and crushing. The broken mineral and rock fragments are rolled out and crushed to varying degrees in accordance with the pressures and the resistance offered by the rocks and minerals. If shearing forces are not dominant during cataclastic metamorphism, the rocks are simply shattered and pulverized. The resultant mass may be structureless aggregates of fragmental material of various sizes. This material is called *crush breccia*, as shown in Fig. 6-11. However, if significant dislocation and

Fig. 6-11: Fault breccia in the Caballon novaculite (chert) near Marathon, Texas, showing shattered and broken fragments of novaculite formed by rubbing and crushing during faulting. Coarse material is cemented by finer grade crush material and other mineral cement.

Fig. 6-12: Augen gneiss from Harrisville, New York, showing large feldspar crystals drawn out into eyes (augen) by metamorphic pressures acting on granite.

mass movement of material takes place during crushing, lenticular and parallel structures are developed. These are called *flaser rocks* and *mylonites*, respectively.

Crush breccias may be fissured and shattered in all directions. Angular fragments so produced may be separated from one another by pulverized material in the fissures. If stress is increased considerably, flattening, elongation, and some granulation of minerals and rock fragments may occur. If large porphyroblasts, usually feldspars, are deformed into lenticular shapes, an *augen gneiss* (eye-shaped crystals) is formed, shown in Fig. 6-12. Further cataclasis produces nearly completely pulverized rock powder, but preservation of a few porphyroblasts in the powdered rock produces *cataclasite*. Complete pulverization of the rock and rolling out (milling) into parallel structures by differential movement along major fault zones produces *mylonites*, which are usually fine-grained, hard, and compact, as well as laminated (Fig. 6-13).

Slate

Slate is a homogeneous fine-grained to aphanitic rock having a secondary planar schistosity (cleavage) independent of bedding. The metamorphically

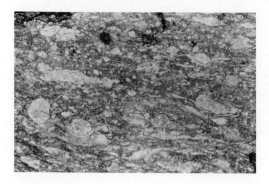

Fig. 6-13: Coarse-grained mylonite gneiss in the Towaliga fault zone near Auburn, Alabama. Large white feldspars have been crushed and rotated into parallel structures. (M. W. Higgins, U. S. Geological Survey, Prof. Paper 687, by permission.)

Fig. 6-14: Gray slate showing good slaty cleavage developed by very-fine-grained micaceous minerals.

produced schistosity usually is at some angle to the original bedding and allows the rock to be split into thin layers. Slates are developed by deep-seated, low-grade deformation of clays, shales, basalts, or other fine-grained rocks. The term slate is a structural one and is not related to composition (Fig. 6-14).

Original bedding of shales is almost never preserved, while slates produced from basalts may retain relict igneous features such as amygdaloids. The secondary (slaty) cleavage of slates is produced by micaceous minerals such as chlorite, illite, sericite, and micas, which are flattened and elongated (rotated) in the plane of cleavage. These are usually very-fine-grained and usually must be determined microscopically. A few coarse-grained minerals such as quartz, feldspar, chlorite, biotite, magnetite, hematite, calcite, and others may be present in the original shale and be retained. Slates are of various colors, ranging mostly from red to green to gray to black. Red colors are due to the presence of hematite, while green colors are due to chlorite. Gray and black colors are due to varying amounts of carbon, either in organic matter or as graphite (Fig. 6-15). Purple slates are known.

Recrystallization is of minor importance in slate development, though not altogether lacking. The clayey rocks are especially susceptible to chemical change during metamorphism, and some recrystallization may start early with development of muscovite. Slates derived from graywacke sandstones

Fig. 6-15: Black slate from Ontario, Canada, showing good slaty cleavage. Note original bedding (fine black lines).

Fig. 6-16: Tourmaline phyllite containing small shiny crystals of muscovite and abundant black tourmaline crystals lying parallel to mica flakes. Note silvery satiny surface of mica flakes.

are referred to as *graywacke slates.* High calcium content in slates may yield *calcareous slates.* Other common varieties are *graphitic slate* and *ferruginous slate* due to the presence of graphite and hematite, respectively. The presence of coarse crystals of pyrite may yield names such as *pyritic slate.*

Phyllite

With increased metamorphic activity by rise of temperature or greater degrees of recrystallization, slates grade into phyllites. Phyllites are composed predominantly of micaceous minerals such as visible mica, sericite, and chlorite flakes, resulting in a satiny sheen on the schistosity surfaces. Phyllites are coarser grained than slates (Fig. 6-16). No sharp boundaries separate phyllites and slates on the one hand, or phyllites and schists on the other, according to grain size and mineralogy. Muscovite, mica, sericite, and chlorite flakes are usually abundant and usually well-oriented. Quartz is usually present as elongated stringers. Minor accessories in phyllites usually are magnetite, hematite, graphite, and tourmaline, although still others have been reported. Most phyllites are called *sericite phyllite, chlorite phyllite,* or *sericite-chlorite phyllite.* Colors of phyllites range from silver-white to reddish or greenish tones. Phyllites are brittle and often have a slightly greasy feel.

Phyllites are produced by low-grade regional metamorphism chiefly from clays and shales but also from tuffs and tuffaceous sediments.

Schist

Schist is a crystalline, foliated (or laminated) metamorphic rock produced by regional metamorphism of a grade higher than for phyllites. Schists are generally coarser grained than both slates and phyllites, but they are usually finer-grained than gneisses. The foliation is developed by platy or long prismatic crystals (Fig. 6-17). Common platy minerals are chlorite, sericite, muscovite, biotite, and talc. Common prismatic minerals are actinolite, kyanite, hornblende, staurolite, and sillimanite. Occasionally, a schist will

Fig. 6-17: Muscovite schist showing coarse shiny muscovite flakes in a finer grained schistose material.

consist of just one mineral, for example, *talc schist*. However, in most cases, two or more minerals are present, such as *calcite-sericite-albite schist*. Schists often contain minerals which are neither flaky nor prismatic, but are equigranular, like garnet and feldspar. These equigranular minerals often present a porphyroblastic texture.

Schists are produced by regional metamorphism and may be derived from gabbros, basalts, ultrabasic rocks, tuffs, shales, and sandstones. The mineralogical changes apparently take place early in the metamorphic episode, perhaps with the onset of pressure. If any original textures are retained because pressure has not been strong enough to develop a true schist, rocks may be called metabasalt, metagabbro, etc. High-grade metamorphism produces normal schists because original textures are destroyed.

Schists are probably the most abundant regional metamorphic type of rock. Numerous varieties exist in accordance with the dominant ferromagnesian minerals present, such as hornblende, chlorite, biotite, talc, actinolite, and others (Fig. 6-18). These minerals tend to produce the structure of the rock, but they are not always the dominant constituents. In many schists, minerals such as quartz, calcite, or feldspar may be dominant, but these non-

Fig. 6-18: Chlorite schist from Vermont. This dark green rock is composed of chlorite and very minor white tremolite. Note good foliation on left side of specimen.

Fig. 6-19: Tourmaline mica schist from near Custer, South Dakota. The white rock is composed of orthoclase, quartz, mica, and abundant large black prismatic crystals of tourmaline.

platy and nonprismatic minerals do not produce the schistosity. The platy or prismatic minerals which produce the schistosity are chlorite, muscovite, biotite, hornblende, talc, or actinolite, and these often are interleaved with the nonfoliated minerals Common varieties are *biotite schist, muscovite schist* (Fig. 6-19), *chlorite schist, actinolite schist, talc schist, graphite schist* (Fig. 6-20), *sericite schist, hornblende schist,* etc. If quartz is fairly abundant, the rock may be called *quartzose schist.* The presence of abundant calcite or dolomite yields *calcareous schist.* Any green mineral such as chlorite or epidote may permit the name *green schist.* A *hornblende schist* in which the foliation gives place to a granoblastic texture is called amphibolite, discussed in next section.

Anthracite coal is produced by metamorphic pressures acting on bituminous coal. It is hard coal and usually shows bright and dull bands of vegetative masses flattened by pressure. Anthracite is compact, fairly dense, and usually jet black in color. It commonly displays a conchoidal fracture (Fig. 6-21).

Fig. 6-20: Graphite schist from central Texas showing rather poor foliation formed by graphite and minor layers of quartz.

Fig. 6-21: Anthracite coal revealing the very compact nature, black color, shiny luster, slight banded structure, and subconchoidal fracture.

Amphibolite

Like schists, *amphibolites* are quite common rocks of regional metamorphism. Amphibolites (Fig. 6-22) require moderate to high-grade metamorphism. This type of rock is composed predominantly of hornblende (amphibole mineral). Plagioclase feldspar and biotite may or may not be abundant. Textures are not especially variable in the amphibolites, consisting mostly of granoblastic types; but in a few, some foliation resulting from the alignment of hornblende and biotite is obvious. In those lacking mica, foliation may be less obvious. Grain size in amphibolites varies from coarse to fine, and in many cases a spotted or streaked appearance may develop by segregation of dark minerals, and they may become banded to limited degrees. Colors of amphibolites are commonly green to black (Fig. 6-23).

Amphibolites are derived from mafic to ultramafic igneous rocks, from limestones, or from tuffs and tuffaceous sediments. During metamorphism, the introduction of magmatic materials such as silica, magnesia, and iron seems to be required in most cases.

Fig. 6-22: Amphibolite from Minnesota showing banding produced by black hornblende and light feldspars and quartz.

Fig. 6-23: Amphibolite containing abundant long prismatic crystals of bronze-brown anthophyllite in a matrix of fine-grained anthopyllite and mica.

Gneiss

A *gneiss* is a coarsely crystalline metamorphic rock, usually banded by segregation of different minerals, thus forming a secondary crude foliation. Gneisses are usually formed by deep-seated, high-grade, regional metamorphism associated with fold mountain structures. Gneisses are derived principally from siliceous igneous rocks such as granites, quartz monzonites, syenites, and granodiorites, but they may also be produced from rhyolites, tuffs, arkoses, and feldspathic sandstones. The chief minerals in gneisses are quartz and feldspars, but other minerals commonly found are hornblende, biotite, and augite. The colors of gneisses are varied and depend upon the colors of the dominant minerals present.

Banding is produced by alternation of light and dark layers of minerals, and some banding may be produced by slight variations in grain size. Banding may be thick and coarse or extremely thin. Grain size may be uniform in some gneisses or may vary from coarse to fine in others, as shown in Fig. 6-24.

Fig. 6-24: Coarsely banded pink and black granite gneiss showing curved bands.

Fig. 6-25: Sillimanite-garnet gneiss from Warren County, New York, showing large garnet crystals in compact granular matrix of abundant feldspars and lesser sillimanite. Note mottled appearance and poor gneissose structure.

Gneisses sometimes contain other metamorphic minerals such as garnet, epidote, tourmaline, graphite, and sillimanite (Fig. 6-25). If igneous rocks are the parent rock types, certain mafic minerals may be developed such as olivine, serpentine, augite, hornblende, and biotite. If the igneous parent rock types can be definitely recognized, the names may be modified to *gabbro gneiss*, *syenite gneiss*, or *granite gneiss*, depending upon the mineralogy. By far the most common type is granite gneiss (Fig. 6-26). Gneisses derived from recognizable sedimentary rocks may be called *quartzite gneiss*, *conglomerate gneiss*, *pelitic gneiss* (from clayey sediments), or *calc-gneiss* (from siliceous limestones and dolomites). If sufficient hornblende is present, this factor may yield *hornblende gneiss* (Fig. 6-27).

A gneiss developed by injection of igneous rock material into the foliation to produce a mixed rock with resultant alternating bands of mafic minerals and thin dikes of quartz-feldspathic igneous material is called an *injection gneiss*. It is formed by a lit-par-lit injection process. Injection gneisses are quite widespread and probably form a large proportion of gneisses. Even though they are partly metamorphic and partly igneous, they are classed as

Fig. 6-26: Granite gneiss showing reddish-pink orthoclase-quartz bands alternating with dark bands of biotite and hornblende.

Fig. 6-27: Granite gneiss (metamorphosed granite) showing coarse banding developed by pink orthoclase plus white quartz and black hornblende plus some biotite.

metamorphic because the metamorphic foliation guided the igneous injection. These may be more properly discussed under *migmatites*, following later in this chapter.

Granulite

A *granulite* usually is a light-colored rock containing large deformed lens-shaped masses of quartz alternating with layers of feldspar, garnet, or finer-grained quartz. In many granulites, pyroxene and garnet may be more abundant than the quartz and feldspar. Other minerals often seen in granulites are sillimanite, kyanite, spinel, and hornblende. Granulites are developed at high temperature and high pressure under deep-seated conditions during regional metamorphism from a variety of rock types. At great depth, directed pressure is less pronounced and less influential, but heat may be more important at the lower levels.

A parallel banded (or somewhat schistose) structure is commonly developed in granulite by elongated quartz masses and by alternating layers of other minerals. Many granulites have a quartz-feldspar composition and commonly the aluminum-bearing minerals sillimanite, kyanite, and spinel are developed. Such granulites are derived from silica and alumina-rich sedimentary rocks such as sandy or silty shales, or from arkoses, as well as from igneous rocks such as granites, granodiorites, and quartz monzonites. Other granulites with pyroxene and garnet have a gabbroic composition.

Several varieties of granulite are distinguished, according to the composition or according to the general fabric, e.g., *granitic granulite* has a composition similar to a granite. Textural types are numerous, such as *foliated granulite* or *massive granulite*.

Migmatite

In a preceding section of this chapter, discussion of *injection gneiss* was made. Closely spaced injection of magmatic material into foliation of

344

Fig. 6-28: Migmatite, partly metamorphic and partly igneous, showing injected igneous mass into foliated gneissose rock.

metamorphic material results in bands of igneous material alternating with metamorphic minerals (Fig. 6-28). If igneous injection is somewhat more intense, small dikes and veins may be developed, running in many directions through the metamorphic material.

If a relatively large mass of magma is brought into play and the metamorphic rocks are more or less thoroughly assimilated and soaked, much exchange of material occurs, and a fairly complete mixing takes place. Schistose or gneissose structure may even develop (Fig. 6-29). Such extreme metamorphism of this type produces *migmatites* (mixed rocks).

Contacts between the different types of material are usually gradational in migmatites. Streaks, irregularly elongated masses and veins of igneous material, will become localized parallel to the schistosity of foliated rocks or will develop irregular and contorted folding in nonfoliated rocks.

Fig. 6-29: Possible migmatite. Large feldspar crystals appear to have been enlarged by material added from adjacent magma into gneissoid rock.

Chapter Seven

Utility of
Earth Materials

Our natural resources include two major types of materials: renewables, including such materials as food, natural fibers, and forest products; and nonrenewables, including minerals, metals, rocks, and fuels.

The renewable resources come primarily from plants and can be regenerated seasonally. As long as the soil and natural waters are properly cared for, the supply of renewable resources can be extended endlessly. However, the increase in world population may eventually regulate the rate of production of renewables.

The nonrenewable resources present a different type of situation. The nonrenewable minerals, metals, rocks, and fuels cannot be regenerated after being removed from the earth. These materials are present in the earth in fixed quantities, and most can be used only once. Any metals, such as lead or zinc at best may be only partially recovered in scrap form and reused. Limestone used in portland cement is completely consumed and not recoverable again. Very few of the nonrenewables are ever recycled for a second use. One of these, water, can be used more than once, but it eventually returns to its natural evaporation-precipitation system.

In general, the nonrenewables are being consumed at ever increasing rates, especially in the United States. The constantly increasing population requires more products for its existence. In addition, the expanding technology which supports our industrial and social welfare requires more nonrenewable raw materials. Consequently the United States must import many raw materials. Rates of consumption vary considerably among the different nonrenewables, and they vary annually too. In terms of world geography, consumption rates of nonrenewables also vary considerably, since each nation's industrial potentials and capacities are so uneven.

The United States has been very fortunate in possessing a vast quantity and variety of mineral resources. No nation of the world has comparable rock and mineral resources. The ability to use vast resources has made possible the leading industrial position of the United States among all nations of the world. Mineral resources are highly valuable to any country, and except for agricultural resources, they lead any country in the promotion of its national welfare. The future of the American mineral industry, basic to the national economy and welfare, actually rests upon a continued supply of rock and mineral materials. Nearly every nation in the world uses progressively more rock and mineral resources as industry demands more and more supply. Such rock and mineral resources are nonrenewable because the Earth does not continually regenerate these materials. As a consequence, sources of supply which formerly appeared adequate now appear small, and the sources are fewer. Search for further reserves must increase.

The nonmetallic (industrial) minerals and industrial rocks have a much lower unit value than metallic resources, and hence the industrial rocks and minerals must be utilized not far from their source of supply. The metallic minerals (ores) are generally more valuable and may therefore be processed and sold at considerable distances from their point of discovery. The high price paid for these metallics usually is sufficient to cover costs of shipment as well as processing.

Mineral resources (both nonmetallic and metallic) are accumulations or concentrations of one or more of Earth's materials which have some commercial utility important to mankind. But valuable mineral deposits are generally very sparsely distributed in the Earth's crust; they are not just everywhere. On the other hand, rock and rock materials are everywhere in the crust, but most of them have no commercial value. Although rocks are actually composed of minerals, they generally are not called mineral deposits.

Historical Development of Uses

Mineral resources are very valuable to any nation. These resources include minerals and rocks of many types, whether they be of inorganic or organic origin, as long as they can be used for the benefit of man. In fact, the historical development of man, the discovery of gold, and the utilization of mineral resources are intimately interwoven through the ages. In addition to pure water, primitive man, between 300,000 and 50,000 years ago, used several nonmetallic materials such as flint, quartz, soapstone, and limestone most likely for weapons of war, for tools, for cracking nuts, for carving, for hunting instruments, for preparing skins, and even for agricultural purposes.

Copper was discovered about 20,000 years ago, but gold was used even before that. Apparently clay was used rather extensively for figurines as early as 30,000 B.C. in Moravia, and later for pottery and brick. Probably since about 10,000 years ago, various ornamentations came into use as native copper, silver, and gold were found in stream placers. The use of pigments grew as hematite, limonite, and cinnabar were discovered. Many metals became important soon after man learned to smelt ore and make metallic tools. Slowly, implements and weapons of war made of copper or iron replaced stone ones, and it can be said that those peoples with metal weapons usually defeated those without.

Construction of the pyramids of Egypt took place in the 2600s B.C. and required large volumes of stone, as well as incredible engineering skills. Many different types of construction materials were quarried in various parts of Egypt at that time and later.

Copper mining took place in Cyprus as early as 2500 B.C. Iron objects were known in the Hittite empire of Asia Minor and Syria about 1500 B.C., and iron became a very important metal by 1200 B.C. in Egypt as well. Gold was being mined as early as 2500 B.C. in China, and silver was well known during the Yin Dynasty in 1760–1100 B.C. Tin was used in the eastern Mediterranean area before 1500 B.C., but where it came from is not known with certainty. Iran is a likely source. The Phoenicians (1300–300 B.C.) carried out a great deal of metal trading, particularly tin, gold, silver, copper, iron, and lead, metals which were obtained in Africa, Greece, Spain, and England. Both gold and silver became prominent as currency several hundred years before Christ. Herodotus reported gold in Greece as early as 425 B.C.; even earlier than this, in 480 B.C., silver from Laurium, Greece, was used by the Athenians.

Gemstones became desirable and gem mining became an important art among Egyptians, Assyrians, Babylonians, and Indians as early as 3500 B.C. The colors of the gems were generally more desirable than the actual substances, such colorful minerals as turquoise (blue), malachite (green), amethyst (purple), lapis lazuli (blue), and carnelian (red) being those chiefly sought. Other gemstones of interest included agate, beryl, garnet, and chalcedony, which were made into faceted or round beads. Both turquoise and copper mining had taken place in the Sinai Peninsula as early as 5300 B.C., and emerald mining has been recorded in 1250 B.C. along the Red Sea coast of Egypt. Diamonds were known in India many years before Christ, but none were taken to Greece until about 480 B.C. Many additional gems such as jasper, opal, ruby, sapphire, and topaz were mined up to the time of Christ in the eastern Mediterranean region.

Asphalt is reported as being used as a cement for art works by the Sumerians in the pre-Babylon Euphrates Valley in 3000–2500 B.C. This same material was used by Persians in 2800–2500 B.C. and by the Egyptians in 2500 B.C. to preserve mummies, and later as construction material. The Egyptians are also reported as artisans of glass blowing as early as 3500 B.C. Sulfur was used by the peoples of Mesopotamia (Iraq) in 2000 B.C. as a bleaching agent for textiles. Apparently, salt was an important commodity around the Persian Gulf in the days of Herodotus in the fifth century B.C., as well as in other areas at later times, particularly during the days of the Roman Empire.

The expansion of the Roman Empire depended partly on the availability of gold and silver, as well as other metals and rock materials. Various rocks such as limestone, marble, granite, travertine, and basalt were used extensively during the days of the Roman Empire. These were used primarily as dimension stone and were often moved long distances by water. Such dimension stone was used for roads, stadiums, temples, theaters, viaducts, aquaducts, and even for public buildings. The possible presence of gold and silver in Britain resulted in invasion by the Romans, but lead became more important as a commercial metal instead.

Iron was also extensively used early by the Romans. Mercury was used as a curiosity metal and did not gain importance until about the 1500s. But the red mercury mineral, cinnabar, had been used as a pigment for many years. Coal was used extensively by the Romans occupying Great Britain, and it had been used rather abundantly in China even prior to Christ. Gold in China was known to be rather scarce for many years after Christ. Iron coins were used in the Far East around 525 A.D., and copper became one of the important metals during the ninth century A.D.

After the collapse of the Roman Empire in the fifth century A.D., production of many building stones and minerals suffered very greatly. Mines closed, trade declined, and except for salt and gold mining, the rock and mineral industry ceased in Italy and parts of Europe, but not in Great Britain and eastern Germany. Although many Spanish mines were reopened in the eighth century, the stone and mineral industry generally remained poor for three or four centuries throughout the remaining Dark Ages. However, the Dark Ages saw continued salt mining and trading by various Germanic peoples, with the eventual discovery of important silver deposits at Freiberg in 1170. Other metals were soon discovered nearby. But in the fourteenth and fifteenth centuries, a new renaissance in mining, arts, commerce, and philosophy gripped Europe. Production of minerals and metals increased substantially, beginning even as early as 1200 A.D. from Central Europe. Important mining and export of silver, copper, and lead took place in southeastern Germany between 1480 and 1570.

Early Theories of Ore Genesis

Great strides in mining and metallurgy were made in the sixteenth century, partly as a result of Agricola's ideas and writings on those subjects. In addition to writing about mining and metallurgy, he also wrote about ores and rocks and even classified them into various types. In fact, the theory of ore genesis was born when Agricola described mineral deposits, particularly lodes, as being generated by heated underground waters in his famous book, *De Re Metallica*, in 1556. His writings on ore genesis were quite advanced and influenced the thoughts of later writers to a marked degree. Descartes, in *Principia Philosophae*, published in 1644, proposed that ores were derived from a hot interior and were deposited as veins or lodes in fissures. Many modifications of this theory were subsequently forthcoming.

However, Werner, at Freiberg Mining Academy in 1755, objected strenuously to the interior origin of ores. He believed that minerals were deposited by descending ocean water, and thus Werner became a "Neptunist." Werner's thoughts prevailed for a number of years, and he had a great following. Vigorous objection to Werner's views was raised by Hutton in 1788, who believed that ores and igneous rocks were of magmatic origin, much as Descartes had believed a hundred years earlier. The ideas of Werner and Hutton were debated for a number of years, and detailed field and laboratory studies eventually proved that Werner was wrong and Hutton was generally right in terms of present-day knowledge. De Beaumont further suggested that mineral deposits were simply just one phase of igneous activity.

Minor modifications of de Beaumont's theory soon followed, and Posepny in 1893 further explained details of the association of ores and igneous rocks. About the turn of the century, Kemp, Lindgren, and Weed added considerable strength to Posephy's ideas, and various genetic classifications of ore deposits were forthcoming. And now, modern theorists pretty much agree that most ore deposits are formed by hot mineralizing solutions derived from magmatic sources.

Classification of Ore Deposits

Ore is generally considered as an accumulation of ore minerals and gangue minerals sufficiently large enough and of high enough grade that one or more metals may be metallurgically extracted at a profit on a continuing basis. The item of profit depends on many factors such as the amount and price of the metal, the cost of mining, treating, transporting, and preparing the metal for market. An *ore mineral* is one which is valuable for the metals it contains. Ore minerals are present, along with gangue minerals (valueless), in sufficient quantity for the total mass to be called ore.

Ore deposits are of many varied types, and it is important that they be classified in one manner or another. Various classifications of ore deposits have been devised in the past, but some of them have not been sufficiently usable. The most usable ones have been the genetic classifications which group ore deposits according to their manner of formation. With very few exceptions, the world's important ore deposits can be placed in one of the groups of a genetic classification such as that presented in Table 7-1.

Table 7-1: GENETIC CLASSIFICATION OF ORE DEPOSITS*

1. Magmatic Deposits
   a. Gravitational segregation deposits
   b. Magmatic injection deposits
   c. Disseminated deposits
   d. Late magmatic deposits
2. Pegmatite Deposits
3. Contact Metamorphic Deposits
4. Hydrothermal Deposits
5. Sedimentary Deposits
   a. Chemical precipitate deposits
   b. Placer deposits
6. Secondary Enrichment Deposits
   a. Residual deposits
   b. Supergene enrichment deposits

*Modified after Bateman, 1950

Relation of Ore Deposits to Magma

Most ore deposits are formed as a result of igneous activity involving the crystallization of magma. Included in ore formation are the various magmatic processes, pegmatitic processes, contact metamorphic processes, and hydrothermal processes. Magmas are considered as being masses of hot molten rock material in the Earth's crust, and the composition is best described as being a mutual solution of silicates, silica, metal oxides, and variable amounts of dissolved volatile substances. The composition varies from magma to magma, but magmas nevertheless play an important role in supplying the liquids, the metals, and the minerals which make up the ore deposits. The temperatures of magmas vary from about 600°C up to perhaps 1250°C, depending on composition, depth, and pressure. Some magmas are very silicic (acidic) and some are very basic. Common volatiles which may be dissolved in the magmatic solutions are water, carbon dioxide, sulfur, chlorine, fluorine, boron, oxygen, nitrogen, hydrochloric acid, and a few others. In most magmas the volatile fractions are extremely important in dissolving common metals such as copper, tin, gold, lead, zinc, iron, nickel, barium, manganese, and cobalt. Consequently, the volatiles play a very important role in the formation of certain metallic ore deposits. Since magmas are generated below the surface where pressures are great, the volatiles act as very mobile gases and liquids, while the major part of the magma apparently behaves as a thick viscous mass. When the magma cools and crystallizes, the volatiles remain as mobile portions even until the last major igneous rocks form. They are responsible for the development of pegmatites, certain contact metamorphic deposits, and hydrothermal deposits, thus acting as mineralizing solutions at these varying stages in their existence. But such mineralizing solutions do not play this role in all mineral deposits of igneous origin.

Magmas form in pockets and are considered as local temporary bodies, because they are usually forced upward, and then crystallize and consolidate. Magma pockets form where great superincumbent rock pressure is relieved, such as in weak zones of the Earth's crust where crustal disturbances such as faulting and folding are active. In areas where no crustal forces are active, no magma pockets occur because the pressure of overlying rock prevents melting even if the temperatures at depth are otherwise high enough.

Magmas generally move upward to regions of lesser pressure produced by crustal disturbance, or they move by being squeezed to upper levels by these movements. Magma is not generated all at once; it develops over some time as more and more heat becomes available to cause continual melting of the heated rocks. The newly melted material thus becomes assimilated into the magma. Magma may also encompass large blocks of country rock, allowing them to sink into the magma in a "stoping" process. In many cases, magma becomes squirted under great pressure into fractures, bedding planes, or other openings to form laccoliths, lopoliths, sills, dikes, and other igneous bodies.

In most cases, magma is not homogeneous in composition. Very likely, some parts may be rich in sodium or potassium, in silica, in volatile content, in ferromagnesians, or in still other substances. It seems probable too that a magma's composition undergoes changes as various chemical reactions occur. Consequently, magmas are not considered as being static or in a constant state of equilibrium.

When the temperature falls and the magma begins to cool, it begins to crystallize into different fractions by various processes of differentiation generally producing igneous rocks in accordance with Bowen's theory. Important metallic elements often become concentrated in certain of these fractions and locally may form ore deposits. Differentiation may occur in one or more phases, and it may even be deep-seated in the magma chamber. These processes may occur after the magma has been transferred to upper levels of the crust or even out on the surface. Some fractions which have already separated from the parent mass may be injected into other levels and crystallize into uniform masses. During differentiation, certain metals such as chromium, nickel, and platinum become associated with and enriched in fractions high in magnesium and iron (mafic fractions). However, other metals such as tin, zirconium, and thorium become enriched in silicic fractions (high in silica). Metals such as iron and titanium, however, usually do not preferentially associate with mafic or silicic fractions; they become localized as parts of many different igneous rock types.

Magmatic Deposits

Gravitational Segregation Deposits

During crystallization of magma, certain of the early formed minerals such as magnetite, olivine, and chromite, are heavy and will sink in the magma, a process known as *crystal settling*. They will continue to settle toward lower portions of the magma, through some time, and accumulate into sizable ore bodies. Any lightweight minerals which form early would tend to rise toward the top of the magma and essentially cause a similar segregation. Large deposits of magnetite and chromite have been formed by this gravitational crystallization process, as in the Bushveld Complex in the Union of South Africa. Nickel-bearing sulfides, such as pentlandite and pyrrhotite, along with chalcopyrite, also may form fairly extensive ore deposits in much the same manner. These sulfides may separate out as heavy droplets, which may move to lower levels and crystallize after much of the magma.

Ore deposits developed by gravitational segregation form compositional layers and usually are of relatively small size. In some cases, they take the form of layers and bands, enclosed by the host rocks, and extending for several miles in length. They may take the form of disconnected pod-shaped lenses, stringers, and irregular masses. Ore deposits formed by gravitational segregation may occur fairly early or rather late in the crystallization of magma.

Magmatic Injection Deposits

Ore minerals which become concentrated by the various differentiation processes may be injected into openings in the surrounding country rocks to form a magmatic injection deposit. Since the injected material has moved from its original place of formation, the deposits usually cut across or transect the structures of the country rock, taking the form of dikes or irregular bodies in accordance with the dominant openings available. The mineral-rich solutions are squirted toward regions of less pressure and become localized. The injected material often gathers up broken rock fragments in its course before coming to rest. Very hot liquid-ore masses injected into the adjacent rock may even thermally metamorphose the rock. The large deposit of magnetite at Kiruna, Sweden, apparently was formed as an injection of iron-rich solution.

Disseminated Deposits

In a massive igneous rock, ore minerals may become concentrated in large or very small amounts and thus may or may not be ore, depending upon the

economic situation. When valuable minerals become crystallized as disseminations (scattered crystals), they simply are part of the igneous rock and are
present as isolated crystals. For example, magnetite may be disseminated
through an igneous rock as an accessory mineral, but it may be economically
feasible to mine it as ore. In this situation, the disseminated minerals were not
concentrated by geologic processes, and the whole igneous rock mass would
need be mined as ore, so long as the magnetite was recovered at a profit.

Years ago at Tahawas, New York, iron ore was mined at very high cost
because of admixed ilmenite (with titanium). The smelting of titanium was
very expensive. But now that titanium is a very valuable industrial metal, the
ilmenite is mined for its titanium content, and iron is obtained as a by-
product.

Late Magmatic Deposits

Masses of minerals which form late during the magmatic crystallization
stages are termed *late magmatic deposits*. Magmatic solutions producing
these deposits result from the various processes of crystallization differentiation, and they generally are residual solutions left over after consolidation of
the igneous rocks. These deposits commonly form after the bulk of the igneous-rock-making silicate minerals are crystallized, and so they cut across,
embay, or even react with the preceding silicates. Some geologists believe that
granitic pegmatites should be included in this type of deposit.

Late magmatic deposits are usually associated with dark ultrabasic
igneous rocks, and in many cases the ore solutions are enriched in iron and
titanium. The ore solutions may accumulate in some portion of the magma
chamber, and the resulting deposits will generally take the shape allowable
by available spaces and density controls. They might even be injected upward
into areas of lower pressure. Ore minerals commonly found in these deposits
are magnetite and ilmenite.

In certain basic magmas, iron-nickel-copper sulfides may separate out as
drops of metal sulfides. These heavy liquid sulfide drops will move downward
and accumulate at lower levels. They usually remain liquid until most of the
silicate minerals crystallize, and then crystallize themselves a little later. Such
heavy sulfides produce the nickel-copper ores containing pyrrhotite, pentlandite, and chalcopyrite. And in a few cases, both platinum and gold are
formed. These ore bodies occur as streaks, layers, or pockets along the lower
portions of the magma chambers, especially where depressions are present.
In some cases, these sulfides may be injected into sheared or fractured areas
in the adjacent country rock to form injection deposits, taking the shape of
dikes and irregular masses.

Pegmatite Deposits

Igneous pegmatites form from the residual liquid fraction of magmas, generally granitic magmas. Pegmatite solutions are usually very silicic and are made up of low-melting silicates, water, various other compounds, volatiles, and metals not deposited during the magmatic stage. Temperatures of pegmatite solutions are of the order of 300–550°C. The volatiles are usually quite abundant and consist of compounds of fluorine, chlorine, sulfur, phosphorus, and still others. These volatiles tend to keep the solutions very fluid and also aid in crystallization processes.

Pegmatites commonly are formed in deep-seated high-pressure regions around the margins of magma chambers or just outside the margins in available openings. Most pegmatites are relatively small, achieving sizes up to only tens of feet. Rarely do they approach a thousand feet in length. Pegmatites may take any shape, but apparently most are dike-like or lens-like. They are coarsely textured, light-colored rock bodies, often containing unusual minerals.

These rock masses are classed as simple or complex, according to the mineralogy and structure. *Simple* pegmatites have simple mineralogy, being made up chiefly of quartz, feldspar, and minor amounts of mica, all producing a rather uniform body overall. They usually do not contain minerals of economic value.

Complex pegmatites have more complex mineralogy which often makes them economically important. These pegmatites are usually zoned, the minerals being arranged in somewhat concentric zones within the pegmatite body. The zones are referred to as border, wall, intermediate, and core zones. However, not all zones are always present, and in many cases additional subzones are present. Giant crystals are frequently present in the inner zones where minerals such as quartz, feldspar, micas, apatite, spodumene, beryl, tourmaline, and others are up to tens of feet in size. Such large crystal development is probably due to the volatile content which permitted rapid growth as ions moved rapidly in the fluid.

Border zones of complex pegmatites are usually thin, fine-grained, and often devoid of valuable minerals. Wall zones are coarser, thicker, and contain feldspars, quartz, mica, and minor amounts of tourmaline, apatite, beryl, and garnet. The intermediate zone usually contains most of the valuable minerals and metals, along with large quartz, feldspar, and mica crystals. Core zones usually contain quartz, feldspar, and minor tourmaline and spodumene, but metallic minerals are usually absent.

Complex pegmatites are frequently mined for various metallic or rare-earth minerals. In the case of certain minerals, pegmatites are the only source for them. Common minerals and their metals which are found in pegmatites are uraninite (uranium), lepidolite (lithium), spodumene (lithium), tourmaline (boron), beryl (beryllium), tantalite (tantalum), and monazite (rare-earth metals). Even gem minerals such as topaz and garnet are sometimes present.

Contact Metamorphic Deposits

Heat emitted from a magma usually has a strong effect on the adjacent country rock. In some cases, heat alone may cause textural changes by recrystallization processes. In other cases, minerals are broken down and new ones regenerated. Textural changes and development of new minerals take place without the addition of outside materials, the effects being solely thermal effects. These changes are referred to as *contact metamorphism*.

However, high-temperature gases emanating from a consolidating magma may have a more profound effect on the surrounding country rock. The hot gases are usually silicic in nature, and they cause strong chemical reactions to occur in certain types of country rock, often producing ore deposits. Chemical reactions produced in this way are called *contact metasomatic*, and ore deposits so formed are *contact metasomatic deposits*. Nearly all metamorphic zones around a magma show both contact metamorphic and contact metasomatic effects.

Contact metasomatic effects are best produced in limestones and limey shales, but more strongly in limestones. Limestones are very reactive rocks against the silicic gaseous emanations. Entire beds of limestone may become changed to silicate rocks called tactites or to skarns when iron oxides and garnets are developed. Temperatures in the range of 500–100°C seem to be required. In addition to the frequent development of metallic and nonmetallic mineral deposits, associated effects may include recrystallization of limestone to marble, of carbonaceous beds to graphite, or sandstone to quartzite, and of shale to hornfels.

Contact metasomatic effects depend on the composition as well as on depth and size of the magma. Silicic gases issuing from quartz monzonite, monzonite, and quartz diorite magmas are more apt to produce ore deposits, while gases from gabbro or diorite magmas rarely produce them. Contact metasomatic effects are partly controlled by structures in the affected rock, such as bedding and faults which act as channelways for the solutions. Ore deposits frequently developed by contact metasomatic effects include magnetite, ilmenite, corundum, spinels, graphite, gold, scheelite, and wolframite. These ore deposits become localized close to the contact and become irregularly scattered as disconnected bodies along the contact. They commonly take irregular shapes, ranging in size from 100 to 800 feet in length.

Hydrothermal Deposits

Residual magmatic solutions left over after consolidation of magma usually are laden with hot water and dissolved mineral material, especially metals. Temperatures of such hydrothermal (hot water) solutions range from about 500 to 50°C, depending upon their depths, pressures, and other factors. They are generally cooler than pegmatitic solutions. As these hydrothermal solutions migrate through the rocks, they react with the adjacent rocks and sometimes deposit ore minerals. Or they may move away from the magma source, thereby becoming cooled and thus deposit ore minerals by a temperature decrease. The solutions often become more diluted as they enter groundwater zones.

Hydrothermal solutions usually carry significant amounts of dissolved metals such as gold, silver, tungsten, iron, nickel, arsenic, copper, lead, zinc, tin, antimony, barium, and mercury, along with large amounts of sulfur. These generally rising solutions may become localized in different types of openings in the adjacent country rock, such as fissures and fractures. If the solutions are not too hot, they will simply fill up the fractures and deposit ore minerals by additional cooling. But if the solutions are still quite hot and under some pressure, they will enter the open spaces and dissolve away (replace) adjacent country rock and deposit ore minerals by a replacement process. Such deposits are referred to as fissure fillings and replacement bodies, respectively. However, both are referred to as "veins." Other types of hydrothermal deposits are also known.

Ore minerals frequently found in hydrothermal deposits are fairly numerous and may include pentlandite, pyrrhotite, scheelite, wolframite, gold, silver, arsenopyrite, pyrite, chalcopyrite, galena, sphalerite, cassiterite, enargite, bornite, tetrahedrite, chalcocite, stibnite, and cinnabar. In addition to ore minerals, other gangue (valueless) minerals such as quartz, calcite, dolomite, fluorite, micas, topaz, apatite, hornblende, and feldspar are common.

The so-called "porphyry copper" deposits are hydrothermal in nature. They are associated usually with monzonites and monzonite porphyries, which, after being intruded and consolidated in an area, become shattered or crackled, and hydrothermal solutions enter these small fractures and deposit various minerals. Minerals such as chalcopyrite, bornite, sphalerite, molybdenite, chalcocite, and other sulfides become disseminated throughout a large volume of rock. Hydrothermal deposits of this type are usually low grade, but they are operated on a large scale at low cost.

Sedimentary Deposits

As rocks in continental areas become weathered, they give up a great variety of ions in solution as well as varying amounts of clastic material. Both types of materials are gathered by rivers and are transported to sedimentary basins and deposited. These materials may be deposited mechanically, chemically, or biochemically, but the manner of deposition depends on a variety of factors. Some of these materials are of economic importance. Most ore deposits of sedimentary origin are formed in shallow marine seas, but occasionally in open sea areas. Small sedimentary deposits may also be formed in swampy areas, in bogs, or in lakes.

Chemical Precipitate Deposits

The important *sedimentary ores* are precipitated from surface waters by chemical and biochemical processes. Many metallic elements can be precipitated from oceanic waters, but only iron and manganese seem to be important. Once these ions reach the ocean, they may be precipitated in accordance with the conditions in the area, such as acidity or alkalinity, amount of oxygen, depth of water, amount of salts, and types and amounts of bacteria in the water. Biochemical (bacterial) processes play a role in the precipitation of sedimentary ores in various ways. They sometimes act as catalysts for oxidation reactions, or they may produce hydrogen sulfide (H_2S) which may bring about precipitation of ores.

The dissolved metals are brought to the ocean by streams; so the deposits are usually thickest near shore and develop best along shore lines. The deposits commonly take the form of reef-like masses, but they frequently are deposited in broad sheets.

By far the most important ores of chemical precipitates are sedimentary iron ores. Iron may be precipitated as siderite, goethite, hematite, glauconite, chamosite, or greenalite. The famous Lake Superior iron ores were originally deposited as iron-bearing sediments consisting of siderite, greenalite, and hematite, along with chert and jasper. These deposits have been weathered somewhat and much of the silica has been removed, leaving enriched iron oxides. The oolitic iron ores in the Clinton formation from Alabama to New York were deposited as iron oxides (hematite) in shallow marine seas. The iron was carried into shallow marine waters, was slowly oxidized, and was precipitated along with other sediments.

Fairly important deposits of manganese, phosphorus, sulfur, and other elements are also formed by sedimentary processes.

Placer Deposits

Since continental rocks are continually undergoing weathering, they often yield heavy stable minerals as lightweight and unstable minerals are washed away. The heavy stable minerals stay behind, but later get moved around by streams and accumulate downstream as *stream placers*. Such concentration is a mechanical one because the surface waters merely free the stable minerals and then concentrate them. A similar type of action occurs at the seashore where heavy minerals are separated from light ones by waves and shore currents, leaving behind *beach placers*. Both types of placers often form important ore deposits.

In order to accumulate placers, the water velocity must be favorable. If the velocity is too low, the lightweight minerals will not be removed, and if too great, the heavy minerals will also be washed away. Once the heavy minerals are being moved, they will be deposited where water velocities are reduced. Stream velocity is apt to be reduced at meanders, at places where gradients are lessened, or where riffles obstruct the flow, and it is at these types of places where alluvial placers accumulate.

Alluvial (stream) placers are the most important type and have yielded gold, cassiterite, platinum, ilmenite, rutile, and precious stones such as diamonds, rubies, and sapphires. Beach placers have also yielded gold, ilmenite, rutile, magnetite, and diamonds.

Secondary Enrichment Deposits

Rocks at the surface continually undergo breakdown as chemical weathering processes operate on them. Unstable minerals are broken down easily and soluble constituents are carried away by surface waters; the more insoluble portions accumulate. Surface waters are charged with carbon dioxide and oxygen and are potent agents of decomposition due to their acidity. Some minerals are very stable and suffer little or no chemical change. However, these more resistant minerals become loosened and may be moved to another location by water or wind.

Chemical weathering proceeds most rapidly in areas of warm, humid climate where moisture is abundant and organic acids derived from plants are available to assist in chemical reactions. Long-continued and deep weathering may have very profound effects on exposed rocks and ore deposits. If the exposed rocks contain minor amounts of valuable metals, the metals may become concentrated into ore deposits by these enrichment processes. If ore deposits become exposed to weathering, they too will undergo leaching and enrichment processes, which may change certain minerals significantly. In these ways, weathering may produce an ore deposit from otherwise barren or near barren rock, or it may destroy a previous ore deposit, or it may simply produce new minerals and thus enrich an already existing deposit.

The depth to which weathering proceeds varies in accordance with climate, structure of the rocks, permeability, and depth to the water table. If the water table is high, oxidation processes are shallow. If the water table is deep, oxidation processes reach to greater depths. The ease with which metal ions are removed in the weathering area depends partly on the composition of these circulating waters as they move down from the surface and partly on the country rocks which may supply certain ions to these waters.

Residual Deposits

Residual deposits are best developed in tropical climates where leaching proceeds more rapidly and more intensely. It is in these regions where iron and aluminum oxides and hydroxides remain as stable constituents to form laterites and remain at the outcrop. Such iron-rich laterites form by the weathering of iron-bearing rocks where rainfall is heavy. The area must be relatively flat so as to retain the waters to promote leaching processes. Iron-rich laterites often serve as ore deposits.

In areas where aluminum-rich rocks such as nepheline syenite are undergoing deep weathering, the residual laterite deposit becomes enriched in aluminum oxides and hydroxides. The deposits are called *bauxites* and are the chief ore deposits for aluminum, such as those near Little Rock, Arkansas. Similar residual concentrations of manganese, clays, nickel, and tin have also served as ore deposits.

Supergene Enrichment Deposits

When an ore deposit becomes exposed by erosion, it becomes weathered just like any rock mass. Surface waters which percolate down through the ore deposit may cause supergene processes to work on the ore minerals. Supergene processes are of two general types: oxidation and supergene sulfide. Oxidation processes take place above the water table, while sulfide enrichment takes place below the water table. The downward-moving water may oxidize some of the ore minerals and dissolve others. The deposit becomes oxidized as well as leached of certain minerals, but generally only down to the water table. As the dilute solutions percolate downward, their metallic contents may be redeposited as oxide types of minerals, thereby enriching this part of the ore body. If the percolating solutions continue down to and beyond the water table, the dissolved metals from above may be redeposited as secondary sulfides in the supergene sulfide zone. The newly formed sulfides may become localized in any open spaces in the primary ore, thus enriching this part of the deposit.

Although many types of ore deposits undergo supergene enrichment of both types, supergene processes are most important in copper and silver deposits. Supergene enrichment is frequently effective in concentrating dispersed metals, and it has played a part in the deposition of secondary minerals of the "porphyry copper" deposits, upgrading the upper portions of the disseminated ore considerably.

Many important ore deposits are of the hydrothermal vein type, occurring as tabular or lenticular bodies. They commonly contain various sulfide minerals such as arsenopyrite, pyrite, galena, chalcopyrite, molybdenite, sphalerite, cinnabar, stibnite, chalcocite, bornite, and enargite, as well as cassiterite, wolframite, scheelite, gold, and magnetite. Associated with these ore minerals are a host of gangue minerals such as fluorite, tourmaline, topaz, quartz, calcite, siderite, rhodochrosite, aragonite, and barite.

To show the effects of supergene enrichment, let us take a typical vein of primary sulfide minerals such as sphalerite, chalcopyrite, galena, and pyrite. As oxidizing waters percolate through the vein, leaching of metallics occurs. The upper surface of the vein becomes leached of its sulfide ores, with the result that only a stable residual porous mass of goethite remains, a mass

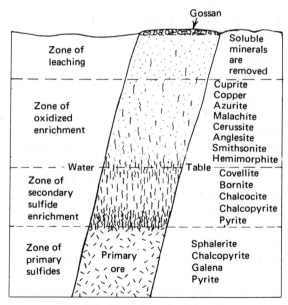

Fig. 7-1: Alteration of sulfide vein.

called a gossan. Metalliferous material which is removed is carried downward varying distances and becomes redeposited above the water table in the oxidation zone. Sphalerite alters to hemimorphite and smithsonite, while galena changes to anglesite and cerussite. Copper sulfides alter to native copper, azurite, malachite, and cuprite.

Percolating copper-bearing waters which reach into the zone of groundwater are apt to react with preexisting sulfides, such as pyrite and chalcopyrite, and deposit other sulfides. Many of the secondary sulfides such as covellite, bornite, and chalcocite are richer in copper, forming sulfide-enriched ore. Such enriched zones grade into primary ore (Fig. 7-1).

Classification of Commercial Earth Materials

Earth materials, rocks and minerals of many types, which man removes from the earth for the purposes of industrial or personal utility, may be classified in several ways. Some classifications group the materials according to mode of formation, others according to the most common industrial use of the material; still other classifications call on the value of the materials as a basis for grouping. The materials may be most readily grouped into *types* as follows:

(1) metals and metallic ores;
(2) nonmetallic (industrial) minerals and industrial rocks;
(3) mineral fuels, and
(4) ground and surface water.

The *metals* and *metallic ores* such as gold, copper, nickel, and iron deposits are valuable for the metals they contain. As we shall see later, these are further subdivided into various groups. Minerals not classed as metallic ore minerals usually are referred to as *industrial minerals*; they differ from ore minerals in that they are used in their raw state, after suitable processing, without requiring a metal to be extracted. The *industrial rocks* include materials such as granite, limestone, and sandstone, which are used in a great variety of industrial applications, mostly in the raw state. Both the industrial minerals and industrial rocks are valuable because of their physical rather than their chemical properties. *Mineral fuels* include coal, petroleum, natural gas, and uranium deposits, all of which owe their value to their content of utilizable energy. *Groundwater*, as well as surface water, ranks high as a mineral resource because civilization and indeed mankind itself are dependent on it. A classification based on the above types is presented in Table 7-2.

Table 7-2: CLASSIFICATION OF COMMERCIAL EARTH MATERIALS*

1. Metallic Ores (Metals)
   a. Iron and Ferroalloy Metals

(1)	Iron	(5)	Molybdenum
(2)	Manganese	(6)	Tungsten
(3)	Chromium	(7)	Cobalt
(4)	Nickel	(8)	Vanadium

   b. Nonferrous Metals

(1)	Copper	(3)	Zinc
(2)	Lead	(4)	Tin
		(5)	Aluminum

   c. Precious Metals

(1)	Gold	(2)	Silver
		(3)	Platinum group

   d. Minor Metals

(1)	Beryllium	(3)	Mercury
(2)	Magnesium	(4)	Titanium

2. Nonmetallic (Industrial) Minerals and Rocks
   a. Industrial Minerals

(1)	Nepheline syenite	(11)	Asbestos
(2)	Feldspar	(12)	Talc
(3)	Mica	(13)	Vermiculite
(4)	Lithium minerals	(14)	Diamond
(5)	Beryl	(15)	Diatomite
(6)	Quartz crystal	(16)	Potash
(7)	Fluorite	(17)	Soda
(8)	Barite	(18)	Borates
(9)	Magnesite	(19)	Nitrates
(10)	Graphite	(20)	Sulfur
		(21)	Phosphates

   b. Industrial Rocks

(1)	Granite	(7)	Sand and gravel
(2)	Basalt	(8)	Sandstone
(3)	Pumice (Pumicite)	(9)	Clay
(4)	Perlite	(10)	Limestone and dolomite
(5)	Slate	(11)	Rock gypsum
(6)	Marble	(12)	Rock salt

3. Mineral Fuels
   a. Coal
   b. Petroleum
   c. Uranium
4. Water

*Modified after Flawn, 1966, and Bates, 1960

Exploration and Discovery
of Minerals and Rocks

The exploration for valuable mineral and rock deposits today involves high capital investment, followed by long, careful investigation. A detailed study of probabilities of discovery and the associated risks requires scientific thinking if discoveries are to be made. Mineral deposits (and not so much rock deposits) are not found by accident or luck. Perhaps in the old days of the West when prospectors occasionally discovered gold, silver, or copper deposits, seemingly with very little scientific approach, a great deal of luck came into play only after the prospector had spent many a long day tramping the hills. Even today, exploration is physically difficult.

The finding of mineral deposits today, as well as in the future, requires the combined effort of scientists and engineers, including geologists, mining engineers, geophysicists, land evaluation appraisors, process engineers, and chemical engineers. Included in the efforts will be a detailed plan for appraisal, reconnaissance, land acquisition, geological and geophysical study, drilling, locating, extending, evaluation of discovery, detailed exploration, mining, and processing for market. Important new discoveries will not be made without detailed planning of this type if we are to have a continuing supply of mineral resources. Long-term plans of this type may require several years before actual testing of potential targets is made. And the cost will be high. Furthermore, the probability of discovery is low.

Mining and Quarrying

Once a mineral deposit (or valuable rock) is discovered and assessed, the next task is to remove the material from the ground. Naturally, it is desirable to get the material out as cheaply as possible. It is also desirable to separate the valuable material from the waste material as easily and cheaply as possible. Surface or very shallow mineral deposits of very large size are mined by open-pit methods called quarrying. The open pit proceeds downward as the valuable material and waste rock are removed by truck or rail. In the case of mineral deposits occurring as gravel accumulations (placers), mining is conducted by a dredge which loosens the wet material and pumps it to a separatory plant. Draglines, on the other hand, can reach over steep embankments and haul the material up to be carried away by truck or conveyor. In the case of mineral material that has very high mobility, such as gas or liquid or easily dissolved minerals, it can be brought up to the surface through a borehole. Many mineral deposits occurring at some depth below the surface require underground mining for removal of the valuable minerals. Underground mines usually consist of workings called adits, drifts, inclines, shafts, crosscuts, galleries, raises, stopes, and winzes. These are merely names given to various types of passageways and working areas, in and surrounding the mineral deposit, for the purpose of reaching the deposit, removing the valuable material, and conveying it to the surface. Just which technique is used depends on many factors such as size and shape of the ore body, the distribution of high-grade portions, and the general nature of the surrounding host rock.

For removal of rock and certain types of other valuable deposits, quarrying is the usual method. The ground is usually blasted down in benches which are then made to retreat on two open sides, or the quarry may just extend on into the nearby hills.

Processing, Beneficiation, and Refining

Once removed from the ground, minerals and rocks are considered to be in the crude state. They must be processed or beneficiated in some way before they become salable or ready for a market. Even the highest-grade ores usually must be concentrated before smelting and refining. Any processing operation is designed to separate the valuable material from waste material, or to clean the valuable material, or to convert it to more usable forms. However, any process on the material to make it more desirable, including necessary handling, transportation, and equipment costs, means a higher price, which is finally absorbed by the consumer.

Any processing to be done may be completed by the miner or by a processor who buys material from several miners. Preliminary processes may prepare the material for a second type of processing operation, such as in a smelter or in a refinery. In the case of large mining companies, mining, processing, smelting, refining, and preparation for market of finished mineral product are accomplished by one large integrated operation.

Metallic ores generally require a whole series of complex beneficiation processes. After crushing to required sizes, separation of valuable mineral from waste may be achieved by various properties such as specific gravity, magnetism, conductance, surface chemistry, and solubility. To effect separation, the plants (mills) commonly use jigs, shaking tables, heavy media tanks, flotation tanks, magnetic separators, electrical separators, and leaching tanks, depending upon the nature of the ore.

Most other raw materials generally require much less complex upgrading processes than the metallic ores to make them salable. Large-sized pegmatite minerals can be upgraded by simple hand cobbing (sorting). Sand and gravel requires only washing and screening. In crushed stone operations, crushing and grinding may be sufficient. Certain others like lime, cement, and lightweight aggregate require simple heat treatment in a kiln, usually sintering.

Utilization of Earth Materials

Metallic Ores (Metals)

Metals have high importance in technological applications because they have distinctive properties such as malleability, ductility, electrical conductivity, thermal conductivity, and luster. The metallic ores are those which are mined chiefly for the valuable metal or metals they contain. The metals are extracted from their ores by metallurgical processes in smelters and refineries. The smelting process involves the melting of metallic ores in blast furnaces or reverberatory furnaces to obtain the valuable metals. The raw ore is heated with coke or natural gas, after a flux is added, and melting takes place, producing a slag. Then the heavy metals settle to the bottom where they are tapped off. After the smelting operation, most metals must be further processed in a metals refinery. The refining serves two main purposes: all deleterious ingredients are removed, and, having been thus purified, the valuable metals are recovered. Two types of refining processes are commonly used: electrolytic refining and fire refining.

Iron and Ferroalloy Metals

Iron has been the most important metal of modern civilization because it is the basis for the steel industry; we are extremely dependent upon it. Not only is iron ore important in the steel industry, but a series of other metals are utilized in making various metal alloys with iron in various aspects of the steel industry. These are known as the *ferroalloy metals.*

The ferroalloy metals are those which are used principally for alloying *with* iron to make special steels having certain desired characteristics. However, they are not used exclusively in steel making, for they have various other uses as well. When they are added in small quantities to steel, the properties of hardness, toughness, durability, temper, and resistance to corrosion are generally improved. The ferroalloy metals are manganese, chromium, nickel, molybdenum, tungsten, cobalt, and vanadium. Still others are aluminum, boron, calcium, cerium, copper, niobium, phosphorus, selenium, tantalum, titanium, and zirconium.

Iron. Iron is a silvery-white metal which is highly malleable and ductile. It can be easily magnetized, and it is also a very reactive metal, uniting with oxygen to form rust. This metal melts at 1540°C (2804°F). Iron has been the most important metal of the industrial age because it is the metal which supports the steel industry. More than 20 times more tons of steel (chiefly iron) is consumed annually in the United States than copper, lead, zinc, and aluminum metal combined. Worldwide, iron accounts for 95 percent of all metals consumed.

Iron is easily removed from its oxide ores in a blast furnace by a simple chemical process. The iron oxide ore, which usually contains some impurities, is mixed with coke (carbon) and limestone (calcium carbonate). This mixture is then heated to high temperatures, as a blast of heated air is introduced, permitting the carbon to react with the iron oxide to form a gas (carbon dioxide), leaving metallic iron. The limestone acts as a flux which combines with the impurities, forming a slag which is easily removed from the top part of the molten iron. The metallic iron so obtained is called pig iron from which various steels are made. Each ton of pig iron requires about 600 pounds of limestone and 2440 pounds of coke.

Steel is made by removing the impurities in pig iron and then adding the proper amounts of carbon, manganese, silicon, nickel, chromium, tungsten, vanadium, and molybdenum. These metals are added accordingly for the manufacture of special steels, that is, steels with special properties. Steel itself is made in a Bessemer converter or in an open-hearth furnace.

In the Bessemer method, molten pig iron is poured into a Bessemer converter, and a blast of air is blown through it. Carbon, manganese, and silicon are thereby oxidized, raising the temperature considerably, burning out the impurities. Then the required amounts of carbon and manganese are added. The furnace is then tipped and the molten steel poured out.

In the open-hearth method, pig iron is mixed with some scrap iron and iron ore and placed on the hearth of a gasfired furnace. Oxygen in the iron oxide ore causes the carbon, silicon, and other impurities to be oxidized. These impurities are then absorbed by the furnace lining, which is made of lime and magnesia brick. After the impurities are removed from the pig iron, carbon and manganese are added in proper amounts to make the steel.

Steel industries often develop in areas away from the source of iron ore because the other required raw materials, such as coal and limestone, are too inexpensive and bulky to ship long distances. The higher-priced iron ore usually can more easily absorb the shipping costs. World consumption of iron ore has risen from 1900 to 1973 and no doubt will continue to increase as more nations become more industrialized and populations increase. No substitute for steel is anticipated, except for certain "special" metals with "special" properties. Fairly large deposits of iron ore are located in various parts of the world and will become available in the future. In addition to a continuing program of exploration and discovery of new deposits, another important development has taken place in recent years which essentially has enlarged our iron ore resources. In order to use extensive sedimentary low-grade siliceous iron ore, called iron formation or taconite, of the Lake Superior iron region (and other areas as well), the material is concentrated and pelletized (agglomerated). This pelletized material thus becomes the high-grade feed to the blast furnace, and it is very easily handled. In fact, it even performs better in the blast furnace than the high-grade raw iron ore.

In addition to the all important use of the metal iron in steel manufacture, this silvery-white metal is used in minor other ways. These include *cast iron*, *wrought iron*, and various *iron alloys*. Iron is also used in medicine in the treatment of iron-deficiency anemias.

In the United States, iron ore comes from four regions. The Lake Superior region produces the largest quantities annually, from several districts in Minnesota, Michigan, and Wisconsin. Second in importance is the Northeastern region, including New York, New Jersey, and Pennsylvania. The Southeastern region, consisting of Alabama and Georgia, ranks third. The fourth region, called the Western States, includes Arkansas, Colorado, Idaho, Missouri, Montana, Nevada, New Mexico, South Dakota, Utah, and Wyoming. The chief economic iron ore minerals mined (or processed) in these various areas are as follows:

Mineral	*Composition*	*Percent Fe*	*Commercial Name*
Magnetite	Fe_3O_4	72.4	Magnetic black ores
Hematite	Fe_2O_3	70.0	Red ores
Limonite	$HFeO_2$	59.0–63.0	Brown ores
Siderite	$FeCO_3$	48.2	Black band or clay ironstone
Iron formation (raw)	Siliceous iron ore	20.0–30.0	Taconite
Agglomerates (finished)	High-grade iron oxide	61.3	Pellets

Approximately 94 percent of the iron ore produced in the United States annually is hematite; magnetite runs about 5 percent, and brown ore runs about 1 percent. Although magnetite is the richest in Fe, it is not very abundant, and the iron industry thrives on hematite. The recently devised concentrating and pelletizing process of taconite appears to allow a continuing increase in tonnages, which is now rapidly approaching crude ore production in the Lake Superior region.

Manganese. Silver-gray manganese metal is relatively soft. It is fairly unstable and, hence, reacts easily with other chemicals to form compounds; it also oxidizes easily in air. Manganese is brittle and harder than glass. It has a melting point of 1245°C (2273°F), but the pure metal can not easily be fabricated.

Manganese is the most important of the ferroalloy metals. The bulk of manganese metal is used in making steel; indeed, it is essential for all carbon steel. Its function is to remove oxygen and sulfur and also to harden the steel. Some 14 pounds of manganese metal are required for every ton of carbon steel manufactured. Approximately 95 percent of the metal is used metallurgically because it alloys well with iron and with the nonferrous metals copper, aluminum, magnesium, and nickel. Hard and tough steels containing manganese are used for bridge steel, crushers, car wheels, projectiles, and armor plating. The remainder of the metal, chiefly as compounds, is utilized in other applications such as in dry-cell batteries, in the glass industry, paints, pigments, dyes, fertilizers, and in the chemical industry. Certain manganese compounds are used in the photographic industry and in insecticides.

The Soviet Union produces about 40 percent of the world's manganese. Other large producing nations are India, Ghana, Republic of South Africa, Brazil, Morocco, and China. United States' production is small (2 percent of world production) coming mainly from Montana, New Mexico, Nevada, and Arizona. The most important manganese deposit in the United States is at Butte, Montana, where rhodochrosite ($MnCO_3$) occurs in association with important hydrothermal copper ores.

The chief manganese ore minerals in the various deposits are as follows:

Mineral	Composition	Percent Mn
Pyrolusite	MnO_2	63.0
Manganite	$MnO(OH)$	62.4
Psilomelane	$(Ba,H_2O)_2Mn_5O_{10}$	45.0—60.0
Hausmannite	Mn_3O_4	72.5
Rhodochrosite	$MnCO_3$	47.6
Rhodonite	$MnSiO_3$	41.9

Sedimentary and residual types of deposits yield the largest quantities of manganese, but some manganese is obtained from hydrothermal and metamorphic deposits. Large potential resources of manganese, as black manganese oxide nodules, are present on the ocean floors.

Chromium. Chromium is a hard and brittle metal. It melts at 1550°C (2822°F), resists high temperatures, and is not affected by salt spray.

The most important use for the gray metal chromium is in the metallurgical industry. However, some chromium is used in the refractories industry, and lesser amounts in the chemical industry. In the metallurgical industry, chromium is used to make various alloys with iron, nickel, and cobalt. These are referred to as chrome iron, chrome steels, stainless steels, high-speed tool steels, and chrome alloys. Chromium imparts strength, hardness, toughness, heat resistance, corrosion and oxidation resistance, and other properties to alloys; thus these alloys are used in airplane engines, safes, and armor plate. Chromium can be plated on other metals such as iron and steel, and it finds extensive use in the auto industry for plating of radiators, pistons, valves, and bearings. Chromium-plated steel engravings are used for printing of United States paper money and postage stamps. Chrome refractory bricks are used to line furnaces in the ceramic industry because they are stable and chemically neutral. In addition, chromium chemicals are used for dyeing, bleaching, textiles, leather tanning, pigmenting compounds, and as a photographic film hardener.

Chromium comes chiefly from South Africa, Russia, Turkey, Southern Rhodesia, and the Philippines. Only one chromium-bearing mineral serves as the ore mineral. It is chromite, $(Cr,Fe,Al)_2O_4$, carrying approximately 68 percent Cr_2O_3, which is usually associated with ultrabasic igneous rocks. The ore is melted in an electric furnace to ferrochrome, which in turn is marketed.

Nickel. Nearly 85 percent of the world's nickel comes from magmatic deposits near Sudbury, Ontario, Canada. The remainder comes from Russia, New Caledonia, and Cuba. A very-low-grade residual nickel deposit is located in Oregon. This nearly white metal, nickel, is magnetic, takes a high polish, and is tarnish and corrosion resistant. It is also malleable, hard, and tough like iron. It has a melting point of 1455°C (2651°F) and is nontoxic. Nickel is used chiefly in making nickel steels and nickel cast-irons, but it is also used in the chemical industry, especially in vegetable oil preparation. It is used extensively as an alloying metal for a very great variety of wearing parts for heavy mining and steel-making machinery and transportation vehicles, and also in agricultural machinery. The use of nickel in coins, in electroplating, and in sheet metal is fairly extensive. Nickel-cadmium alloy is used in storage batteries. Nickel stainless steels are corrosion and stain resistant and hence are used for kitchenware and food-processing ware. Nickel steels are also used in textile and paper mills and in oil refineries. The glass and ceramics industries also use nickel compounds.

The chief ore mineral of nickel, pentlandite, $(Fe,Ni)_9S_8$, a sulfide mineral carrying 22 percent nickel, is always associated with pyrrhotite and chalcopyrite in basic igneous rocks. In Oregon, the nickel ore is garnierite, a nickel-bearing serpentine whose composition is $(Mg,Ni)_6Si_4O_{10}(OH)_8$. A little nickel is obtained from millerite, NiS, and from niccolite, $NiAs$, carrying 64.7 and 43.9 percent nickel, respectively.

Although annual increased demand for nickel is about 2 percent, the United States will continue to obtain Canadian nickel for some time to come.

Molybdenum. Molybdenum is a silver-white metal with a melting point of 2625°C (4760°F). This metal has high strength and high electrical and thermal conductivity. It is resistant to many, but not all, common acids and salts. Molybdenum metal alloys well with steels and other metals, especially iron. Nearly all of the molybdenum produced is used as an alloying element in steel, acting as a hardening and strengthening agent. About 4 pounds are used in a ton of steel. Such hard and strong molybdenum steels are used in aircraft engines, automobile engines, cutting tools, drills, and a great variety of other metallic parts of machinery. Small amounts of molybdenum are used in electrical equipment, X-ray tubes, heating elements, grids, and anodes. It also is used in inks, dyes, catalytic agents in the chemical industry, and in the glass industry. Molybdenum compounds are used as plant nutrients, pigments, and lubricants.

The United States produces about 85 percent of the world's molybdenum. Most of the domestic molybdenum comes from hydrothermal veinlets in igneous rocks at Climax, Colorado, and minor amounts come from hydrothermal veins at Questa, New Mexico. There are other minor producing areas in Arizona and New Mexico, mainly associated with porphyry copper deposits. The chief ore mineral of molybdenum is molybdenite, MoS_2, which accounts for most of the world's production. Molybdenite carries about 60 percent molybdenum metal. Two other minerals, wulfenite, $PbMoO_4$, and ferrimolybdite, $Fe_2O_3 \cdot 3MoO_3 \cdot 8H_2O$, are minor in importance. Wulfenite contains about 39.3 percent molybdenum trioxide.

Tungsten. Although tungsten deposits are located in many parts of the world, the bulk of it is found in China, Burma, Thailand, Korea, and even in Malaya and Indonesia. In the United States, tungsten deposits are located in North Carolina and Virginia in the East, and in California, Montana, and Colorado (as well as other states) in the West. Two common tungsten minerals, scheelite, $CaWO_4$, and wolframite, $(Fe,Mn)WO_4$, are found in high-temperature hydrothermal veins and in contact metasomatic deposits associated with granitic rocks. Scheelite contains about 80.6 percent tungsten trioxide, and wolframite carries about 74 percent tungsten trioxide. Two other relatively common tungsten minerals are ferberite, $FeWO_4$, and huebnerite, $MnWO_4$.

Tungsten has very high strength but is ductile. It also is relatively inert to most acids and alkalies. The metal alloys well with the ferrous metals such as iron, molybdenum, and chromium. Tungsten, often called wolfram, is used principally for making tungsten carbide for use in high-speed cutting tools. Such steels also contain several percent of chromium and vanadium. Tungsten, usually present from 2 to 12 percent, acts as a superior hardener. Tungsten carbide is a substance nearly as hard as diamond and thus can be used as an abrasive. Because this silver-white metal has the highest melting point of all metals, 3337°C (6035°F), and is very hard, tungsten filaments are used in light bulbs, in electrical apparatus, in X-ray equipment, and in a variety of other applications. Other uses of tungsten include automobile parts, fireproof clothes, pen points, dental and surgical instruments, furnace windings, rocket fuel nozzles, space vehicle reentry surfaces, drill bits, and cutting saws.

Cobalt. This silver-white metal, cobalt, is stronger than iron and does not rust or tarnish, but it is attacked by several common acids. It melts at 1495°C (2723°F). Cobalt is about as hard as nickel and iron, and it is ferromagnetic.

It is used chiefly in the manufacture of special steels, particularly in magnet steels, and for metal cutting. Cobalt alloys are used in jet engines and gas turbines because they remain hard up to 1800°F. Cobalt is also used in a variety of alloys for use in magnets for radios and television sets. It is used as a bonding agent for tungsten carbide abrasives, and it also is used in electroplating. Cobalt gives a blue color to glass and enamels, while other cobalt compounds are used for decorator colors. Another minor use is as a catalyst in the chemical industry. Cobalt-60, a radioactive isotope of cobalt, has largely replaced radium in the treatment of cancer because it is cheaper and easier to use.

Cobalt is produced mainly in Africa, from the Congo, Zambia, and Morocco. A little comes from Canada and the United States. The United States' production is from Cornwall, Pennsylvania, as a by-product from magnetite iron ores. In fact, most of the cobalt obtained anywhere is a by-product of the mining of copper, nickel, and iron. Production of cobalt varies with the production of these other metals. At present the demand for copper, nickel, and iron is good, so cobalt is in good supply.

Ore minerals of cobalt are uncommon: linnaite, Co_3S_4, cobaltite, CoAsS, smaltite, $(Co,Ni)As_3$, and black oxide. Cobalt minerals are usually found in hydrothermal veins, often associated with nickel, arsenic, and copper minerals.

Vanadium. Vanadium is a very hard silver-white metal with a melting point of 1900°C (3452°F). It is ductile and does not readily tarnish. Vanadium is relatively rare and is obtained chiefly as a by-product from uranium ore; but it is found in ores of other metals as well. The United States produces about 70 percent of the world's supply. Smaller amounts come from South West Africa, Finland, and the Republic of South Africa. The United States' production is chiefly from the uranium areas of the Colorado Plateau (Colorado, Utah, Arizona, New Mexico), Wyoming, and South Dakota, but large reserves of vanadium are present in certain titaniferous magnetite deposits and phosphate rock deposits in other parts of the United States.

The chief ore mineral of vanadium is carnotite, $K_2(UO_2)_2(VO_4)_2 \cdot 3H_2O$, carrying about 20 percent V_2O_5. Other vanadium minerals are vanadinite, $Pb_5(VO_4)_3Cl$, and descloizite, $PbZn(OH)(VO_4)$. These minerals commonly form by weathering and are often associated with sedimentary rocks.

Vanadium alloys well with other common metals such as aluminum, cobalt, copper, iron, and manganese. Vanadium is a steel toughener and also offers great resistance to strain or fatigue. These steels require about 1 percent vanadium. Vanadium is used in a variety of high-temperature services. It finds extensive use in tool steels and high-strength structural steels. Vanadium steel is used for frames, gears, axles, and springs of automobiles. Its compounds are also used in the chemical industry, in the electrical trade, in ceramics, in glass, in dyes, and in the printing industry.

Nonferrous Metals

The nonferrous metals are used in large quantities and are next in importance to iron as metals of industry. However, the nonferrous metals are not used for alloying with iron. The nonferrous group includes copper, lead, zinc, tin, and aluminum, Demand for these metals is high, and discovery of new deposits and production generally lag well behind utility. Consequently, exploration is always active.

Copper. The world's largest producer of copper is the United States. In addition, the United States imports substantial quantities of copper from Chile, Canada, Peru, and the Republic of South Africa. Other large production is from Russia, Zambia, and the Congo. Production of copper in the United States is mainly (98 percent) from Arizona, Utah, Montana, New Mexico, Nevada, and Michigan, in that order.

Copper is localized into important mineable hydrothermal sulfide deposits in porphyry copper deposits associated with acidic to intermediate igneous rocks. The porphyry copper deposits are large and mineable at low cost. About 50 percent of the world's production comes from the porphyry copper deposits. Hydrothermal veins also yield large quantities of copper sulfides. Copper also occurs less abundantly in massive sulfide deposits in shear zones in metamorphic rocks, and also as native copper in lavas. Large deposits also occur as copper-rich sediments such as at White Pine, Michigan. The chief ore minerals of this reddish-orange metal occurring in these various types of deposits are as follows:

Mineral	Composition	Percent Cu
Native copper	Cu	100.0
Chalcopyrite	$CuFeS_2$	34.5
Bornite	Cu_5FeS_4	63.3
Chalcocite	Cu_2S	79.8
Covellite	CuS	66.4
Enargite	Cu_3AsS_4	48.3
Tetrahedrite	$Cu_{12}Sb_4S_{13}$	52.1
Tennantite	$Cu_{12}As_4S_{13}$	57.0
Cuprite	Cu_2O	88.8
Tenorite	CuO	79.8
Malachite	$Cu_2(CO_3)(OH)_2$	57.3
Azurite	$Cu_3(CO_3)_2(OH)_2$	55.1
Chrysocolla	$CuSiO_3 \cdot 2H_2O$	36.0
Antlerite	$Cu_3SO_4(OH)_4$	54.0
Brochantite	$Cu_4SO_4(OH)_6$	56.2
Atacamite	$Cu_2(OH)_3Cl$	59.4

Those in which copper is combined with sulfur are the most valuable.

Copper is one of the most important metals of industry, especially in the electrical industry. This metal has a melting point of 1083°C (1981°F). Copper is a good conductor of heat and thus can be used in cooking utensils, radiators, and refrigerators. Copper is very malleable, and it may be hammered, stamped, or spun into various shapes. It can be rolled into thin sheets. The metal can be drawn into wires thinner than a hair. Copper is also corrosion resistant, and it may be welded, brazed, or soldered. It is used in a great many products, for example, as pure copper in wire and electrical equipment because of its excellent conductivity. It is used in generators, motors, electric locomotives, in switchboards, in light bulbs, telegraphs, air conditioners, and in ships. Copper alloys well with zinc in brass, and as other alloys with nickel, aluminum, and other metals. Areas of utilization include the electrical, automobile, construction, electronic, ammunition, and coinage industries. Copper is important as a munitions metal, and it is constantly being developed in various military applications.

Lead and zinc. Lead and zinc commonly occur together under the same geological conditions and hence are discussed together here. The chief ore minerals of each, galena and sphalerite, respectively, are found in the same deposits nearly everywhere in the world. Copper and other important metals are often associated. The United States is one of the world's leading producers of lead and zinc, and during some periods it is the world leader, alternating with Australia and Russia. Most of the United States' lead production comes from Missouri, Idaho, Utah, and Colorado. However, zinc production comes mainly from Tennessee, Idaho, New York, Colorado, and scattered other states. In these areas, lead-bearing and zinc-bearing minerals commonly occur as cavity fillings and replacement bodies formed by low-temperature hydrothermal solutions, and the deposits are found usually as veins and masses in limestones and dolomites. A number of lead and zinc deposits are possibly sedimentary. The chief ore minerals for lead and zinc are as follows:

LEAD

Mineral	Composition	Percent Pb
Galena	PbS	67.0
Cerussite	$PbCO_3$	52.0
Anglesite	$PbSO_4$	54.2

ZINC

Mineral	Composition	Percent Zn
Sphalerite	ZnS	67.0
Smithsonite	$ZnCO_3$	52.0
Hemimorphite	$Zn_4Si_2O_7(OH)_2 \cdot H_2O$	54.2
Zincite	ZnO	80.3
Willemite	Zn_2SiO_4	58.5
Franklinite	$(Fe,Zn,Mn)(Fe,Mn)_2O_4$	15—20

Lead is corrosion resistant and extrudable. However, it flows under sustained load and hence has restricted used. Lead is a blue-gray, soft, heavy metal which melts at 327°C (621°F). Compounds of lead are poisonous. Lead is used for storage batteries, as an additive in gasoline, as red lead and litharge, for caulking compound, for alloys, for sheaths and coverings of cables, for ammunition, water pipes, in the building trade, for X-ray shielding, in the glass and ceramics industries, in pewter, in solder, in bullets, and in dyes.

Zinc, a bluish-white metal, is malleable and ductile when heated to 200–300°F. This metal melts at 419°C (787°F). It is relatively corrosion resistant. If pure, zinc is not attacked by common acids, but impure zinc is readily attacked. Zinc alloys well with iron; it is also used with copper for alloys such as brass. Zinc is fine for galvanizing, for die castings (making special shapes for automobile door handles, fuel pumps, etc.), and rolled zinc (sheet metal), finding application in roofs and gutters. Zinc is also used in the chemical industry, in medicines, in paints, in rubber goods, in cosmetics, as dental cements, and in wood preservatives.

Tin. Tin deposits are formed under hydrothermal or pegmatitic activity. The chief tin mineral, cassiterite, SnO_2, often called tinstone, contains 78.6 percent tin. Another ore mineral, stannite, Cu_2FeSnS_4, is much less common. Cassiterite often is released from primary deposits by weathering processes and becomes accumulated in various types of placer deposits because it is hard and resists chemical breakdown.

The largest deposits of tin are in Asia, principally in Burma, Indonesia, Malaysia, and Thailand. Additional deposits are in the Congo and Nigeria in Africa. North America is very deficient in tin; however, small reserves are located in Alaska. Sporadic production has come from Alaska, California, Colorado, and a few other states. The United States will probably continue to import its needed supply.

Tin is a bluish-white metal which melts at 232°C (449°F). This metal is not ductile, but it is extremely malleable. It is also corrosion resistant, nontoxic, and tarnish resistant.

The chief use for tin is for tin plating for steel food containers (so-called "tin cans"). Although making up about 1 percent of the entire can, the tin serves as a thin anticorrosion coating, is not affected by foods, and is cheap. It is also used as an alloying metal in brass and bronze. Minor uses include babbitt metal, type metal, die castings, collapsible tubes, tinfoil, solder, wire, paper decorations, chemicals, friction bearings, pewter, tracer bullets, shells, and torpedoes. Low-priced mirrors are often coated with tin-mercury mixture or with tinfoil. Tin fluoride is used in toothpaste as a decay preventative. Other tin compounds are used in the manufacture of agricultural compounds.

Aluminum. Aluminum is present in many rock materials such as soils, clays, aluminum-bearing silicates, and bauxite. However, this silvery-white, soft metal is obtained almost exclusively from bauxite because it is hard to remove from most of the others. But it is also obtained in minor amounts from high-alumina clays. Other future potential sources for aluminum are nepheline syenite, alunite, and andalusite. Refining of aluminum requires large amounts of power; so the largest producing nations are those with higher power-generating capacity.

The United States leads the world in aluminum production, bauxite being produced chiefly near Little Rock, Arkansas, and minor amounts in Alabama and Georgia. The largest reserves in the world are in Surinam, Brazil, Venezuela, France, Greece, and Hungary. Additional large deposits are in Costa Rica, the Dominican Republic, Haiti, and a few other localities. The United States imports bauxite from Jamaica and Surinam.

Bauxite, the chief ore, is formed by tropical or semitropical weathering of aluminum-bearing rocks. The weathering processes allow aluminum minerals to accumulate because they are very resistant to decomposition. Bauxite usually consists of varying quantities of the aluminum minerals, diaspore, $HAlO_2$, gibbsite, $Al(OH)_3$, and boehmite, $AlO(OH)$, and it actually is a rock rather than a simple mineral. It takes about four pounds of bauxite to make two pounds of alumina, (Al_2O_3), from which one pound of aluminum metal is obtained. The chief ore minerals of aluminum are:

Mineral	Composition	Percent Al
Boehmite	$AlO(OH)$	45.0
Diaspore	$HAlO_2$	45.0
Gibbsite	$Al(OH)_3$	34.6
Kaolinite	$Al_2Si_2O_5(OH)_4$	20.9
Kyanite	Al_2SiO_5	33.5
Anorthite	$CaAl_2Si_2O_8$	19.4
Nepheline	$NaAlSiO_4$	18.4

Aluminum metal is light, has high strength, resists corrosion, and has a melting point of 660°C (1220°F). This metal can be welded, soldered, brazed, bolted, or riveted. It can be rolled and stretched into nearly any shape. Aluminum's chief use is in casting and wrought alloys, and it alloys well with copper, magnesium, manganese, zinc, nickel, tin, cadmium, bismuth, cobalt, vanadium, zirconium, boron, beryllium, and sodium. It is used in aircraft, buses, trains, ships, furniture, and in cars. The automobile industry uses aluminum in pistons, cylinders, and crankcases. It is even used in washing machines. The metal's rust resistance allows it to be used in storage tanks and pipelines, while its corrosion resistance properties allow aluminum to be used in chemical tanks, pipes, and vessels. Aluminum's high electrical conductivity permits its use as wires for transformers and generators, and its good heat conductivity means it can be used in cookware. This metal is used for storm doors, exterior panels on buildings, roofs for chicken houses, and for insulation. When flaked, aluminum can be used in paint. Because it is nonpoisonous, greaseproof, and moisture proof, aluminum is used as foil in food packages, in toothpaste tubes, in milk tanks, in beer barrels, and brewers kegs. The military uses aluminum mess kits.

Precious Metals

Metals which are considered most valuable in the jewelry trade are called *precious metals*. These metals are gold, silver, and platinum. Because of special properties, these metals have also found use in coins and in dentistry. Gold and silver are still alloyed with copper for coinage and for brazing alloys. Even so, these metals still remain of great value in the jewelry trade. Gold remains the chief metal in jewelry because it has an attractive color and it can easily be shaped and worked. Gold, of course, is still internationally recognized as a standard of value. It is closely tied to politics and is under a very rigid price control. Most of the gold produced is for such monetary purposes.

Gold. The larger part of the world's known gold reserves are in the Witwatersrand deposits of the Republic of South Africa, with much lesser amounts in Russia, Canada, Australia, and the United States. The largest producer (60 percent of world production) is the Republic of South Africa. The United States' production of gold comes mainly from a large deposit (Homestake deposit) in the Black Hills of South Dakota, and subordinate amounts come from Utah, Alaska, Arizona, California, Nevada, and Washington. Much gold is produced as a by-product of basemetal mining, such as copper. Gold mining is carefully supervised by the governments of all gold-producing countries, and 60 percent of all gold is held by banks and governments.

Gold deposits are formed chiefly by hydrothermal processes associated with igneous activity, and a little gold is formed by contact metasomatism. This primary gold often is secondarily concentrated (as nuggets, grains, and flakes) as valuable placer gold. Gold minerals which serve as ore minerals are *native gold*, and small amounts of gold tellurides, namely calavarite, $AuTe_2$, sylvanite, $(Au,Ag)Te_2$, and also as electrum, naturally combined with silver. Ores rarely contain more than 1–2 ounces of gold per ton of rock.

Gold is used primarily for monetary purposes, i.e., as a standard of value for money. Gold has been desired by man since prehistoric times. Value was early assigned to it, and it was exchangeable for services and goods. It is believed that about one-fourth of all the gold and silver ever mined lies in rotted hulks of Spanish galleons and pirate ships sunk in the world's waterways. Gold bullion is kept in reserve at Fort Knox for notes issued, as well as for coinage. Because of nontarnishing properties, industrial uses of gold include jewelry and arts, dentistry, gold plating, lettering and gilding, ornamentation, and plate glass. This metal finds use as a barrier to solar radiation in space equipment. Gold has a lovely yellow color, is ductile, is malleable, and is resistant to rust. Gold melts at 1063°C (1945°F). It is relatively inert and corrosion resistant.

Gold jewelry is usually marked with a karat mark indicating the amount of gold present. For example, 24K is pure gold, and 12K is half gold, the rest being another metal added to give a desired color or hardness. Gold rings usually are 10K, 14K, or 18K. Gold coinage previously used in the United States was 21.6K.

Natural gold usually is mixed with a little silver. Purity of natural gold is called "fineness," expressed as "parts per thousand." For example, 900 fine means 900 parts gold, the remaining 100 parts commonly being silver, possibly with a little copper or iron alloyed with the gold. The 21.6K gold coins were equivalent to 900 fine.

Silver. Most of the world's reserves of silver are in Canada, the United States, Mexico, and Peru. In fact, the Western Hemisphere and Australia have produced about 80 percent of the world's silver, with Mexico being the leading producer. Silver occurs frequently with gold, but about 70 percent is commonly associated in basemetal sulfide deposits of lead, copper, zinc, and is obtained as by-product; consequently, the production rate of silver is controlled by the production of the associated metals. Such deposits are chiefly hydrothermal in origin. The chief producing states are Idaho, Utah, Arizona, Montana, and Colorado. The chief ore minerals of silver are as follows:

Minerals	*Composition*	*Percent Ag*
Native silver	Ag	100.0
Argentite	Ag_2S	87.1
Cerargyrite	$AgCl$	75.3
Polybasite	$Ag_{16}Sb_2S_{11}$	75.6
Proustite	Ag_3AsS_3	65.4
Pyrargyrite	Ag_3SbS_3	59.9

Silver is both malleable and ductile, has very high heat conductivity, and good electrical conductivity. It is harder than gold but softer than copper. This lustrous metal melts at 960°C (1761°F). Silver is not altered by moisture, dryness, or alkalies, but it tarnishes very rapidly in air containing sulfur. Silver has high heat conductivity and high electrical conductivity.

The chief use for silver for a long time was coinage; but now the largest amounts are used for coating photographic film (silver bromide and silver iodide). Other uses include silverware, electrical and electronic equipment (switches, points, in computers, and tabulators), various alloys, in jewelry and arts, as well as in the chemical industry. Its use in jewelry and arts is based on the fact that silver is the brightest of all metals. The United States' silver coins are 90 parts silver and 10 parts copper. In colloidal form, silver is used in certain medicines. It finds use in food preservation, in beverages, and in dentistry. Silver alloys generally are hard and tough. Sterling silver is 92.5 percent silver and 7.5 percent other metal, usually copper. Yellow gold is 53 percent gold, 25 percent silver, and 22 percent copper.

Platinum group. The platinum group of metals includes platinum, palladium, rhodium, ruthenium, osmium, and iridium. They are commonly associated together geologically and often alloyed together.

Platinum most commonly occurs in mineral deposits as the native metal, but it is present sometimes as sperrylite, $PtAs_2$, carrying 56 percent Pt, cooperite, PtAsS, with 85 percent Pt, and braggite, PtPdNiS (palladium-bearing). Palladium is also found as stibiopalladinite, Pd_3Sb. The metals of this group usually are naturally alloyed with one another. Native platinum is never pure; it is always naturally alloyed with others in the group. However, most of the platinum produced is obtained as a by-product from the refining of nickel and not from specific platinum minerals.

Platinum deposits are chiefly of magmatic origin and are associated with ultrabasic rocks, in association with nickel. World production is chiefly from Russia, the Republic of South Africa, Canada, and Columbia. United States' production and reserves are small, being mostly in placers in Alaska. Platinum, slightly heavier than gold, is about 2.5 times more valuable than gold.

Platinum can be shaped and worked in much the same way that gold and silver can, i.e., it is malleable and ductile. Platinum does not tarnish, it is nonreactive to most strong acids, and it has a melting temperature of 1773°C (3224°F). It has a high electrical resistance and alloys well with other metals such as copper, gold, and silver. Because platinum resists most chemicals, except caustic alkalies, and is heat resistant, it is used as chemical dishes and containers. However, it reacts with carbon, phosphorus, silicon, lead, arsenic, and antimony. This metal's strength, hardness, color, and tarnish resistance warrant its use as gem settings. Some surgical instruments are made of platinum. Pen points are frequently made of platinum, and certain salts of platinum may be useful in photography. The chief uses for this silver-white metal are in jewelry, dental, electrical (as contact points), and chemical industries, with the latter two applications progressively more in demand.

Minor Metals

A large number of metals which are used in small quantities or have limited application in industry are called minor metals. However, some are very important in certain industries. The four most important metals of this group are beryllium, magnesium, mercury, and titanium, but the remainder are vital in selected applications. Others in this group are antimony, bismuth, cadmium, cesium, cerium, columbium, tantalum, germanium, lithium, rare-earth metals, rhenium, selenium, strontium, tellurium, thorium, zirconium, and hafnium.

Beryllium. Beryllium, a gray-white metal, is obtained from the mineral beryl, $Be_3Al_2Si_6O_{18}$, commonly found in pegmatites. Beryl contains about 14 percent beryllium oxide. Commercial sources are in the Black Hills of South Dakota, Colorado, New Mexico, Maine, New Hampshire, and other states. The metal's light weight permits it to be used in aircraft, rockets, and spacecraft. Beryllium has a high melting point of 1284°C (2345°F), has high electrical conductivity, resists oxidation well, and has high strength. Beryllium alloys well with copper, nickel, and iron, imparting high strength and fatigue resistance. This metal is used in gyroscopes, in accelerometers, in computer parts, in X-ray tubes, in fluorescent lamps, in cyclotrons, as reflectors in nuclear reactors, and in heavy-duty brake drums.

Magnesium. Magnesium is a grayish-white metal with a melting point of 650°C (1202°F). Being one of the lightest of metals, magnesium finds its chief use as a corrosion-resistant alloying metal for aircraft engines, wheels, floors, etc., and in automobiles. Magnesium is stable in air and is resistant to most alkalies and acids, but is attacked by some. This metal has a low structural strength. Magnesium is also used in catcher's masks, skis, snowshoes, boots, and in wheels of racing cars. It finds use in missile and space vehicles, electronics, and industrial machinery. Magnesium is used to protect hulls of ships, underground tanks, and certain pipelines. It also provides great strength in metals usable on microscopes, cameras, luggage, surveying instruments, cookware, bobbins, typewriters, musical instruments, and in television equipment. It is used also in medicines. Magnesium also burns at low temperatures and is used in flashlights, photography, fireworks, and flares. Magnesium compounds are used in refractory brick, ceramics, fertilizers, in rubber, plastics, medicines, and cosmetics. Supplies are very large. The metal is obtained from seawater (chiefly in Texas) and also from natural brines, from magnesite, $MgCO_3$, from dolomite, $CaMg(CO_3)_2$, and from brucite, $Mg(OH)_2$ in other localities. Magnesite contains 47.6 percent MgO, dolomite contains 21.7 percent, and brucite contains 69.0 percent respectively.

Mercury. The main ore mineral of mercury, cinnabar, HgS, contains 86.2 percent mercury. It is formed under shallow, low-temperature hydrothermal conditions. The world's leading producers are Spain and Italy. United States' production comes mostly from California and Nevada, but these are nearly exhausted. Reserves are small and by-production from other metals is limited. Mercury is the only metal that is liquid under ordinary temperatures. It freezes at $-39°C$ ($-38°F$). Mercury is dense, has a high electrical conductivity, and is toxic if fumes are inhaled or absorbed into the skin. The uses for this silver-white metal are many, including the chief ones of electrical apparatus (switches and relays), utility in producing electrolytic chlorine and caustic soda, explosives, special paints, and for various industrial and agricultural chemicals. Mercury, sometimes called quicksilver, finds minor uses in thermometers, dental preparations, pharmaceuticals, and batteries.

Titanium. Titanium is a silver-gray metal, with a melting point of 1670°C (3038°F). It is also ductile. Although titanium is a relatively abundant metal, the cost of smelting and refining it is high. The metal has very desirable properties similar to steel and aluminum, i.e., high strength, corrosion resistance, and light weight. It has a high luster, but not as high as chromium or stainless steel. It is particularly resistant to corrosion by marine air or sea spray. Titanium is used extensively as an alloying metal for high-speed and chrome steels, and other alloys with most other metals. An important use is in aircraft and jet engines, and in space capsules. It makes the whitest of all paints, which is in high demand as pigment for toilet articles, inks, glazes, etc. Titanium dioxide finds use in linoleum, rubber, paper, and textiles. Titania crystals are used for gems. The ore minerals are ilmenite, $FeTiO_3$ (31.6 percent TiO_2), and subordinate sphene, $CaTiSiO_5$ (40.8 percent TiO_2), and rutile, TiO_2 (60 percent Ti), minerals often secondarily accumulated in placer deposits. However, some ilmenite is found in basic igneous rocks, such as at Allard Lake, Quebec, the largest-known titanium deposit in the world. Other producers are New York State, Norway, India, Brazil, and Japan.

Nonmetallic (Industrial) Minerals and Rocks

The industrial minerals and rocks are those which are mined and quarried chiefly for their value as industrial materials. This group, called nonmetals, has two divisions: *industrial minerals* and *industrial rocks*. They differ from the metallic ores quite distinctly. The metallic ores contain valuable metals which must be metallurgically removed before their utility is achieved. The nonmetallics, even though they may contain certain amounts of metals, are utilized because of their physical properties which remain unchanged after the material has been used. Many of the industrial rocks and minerals are bulk commodities, bringing a low price per ton. Consequently, mechanized mining and quarrying equipment are usually used. A few of these industrial rocks and minerals, obtained in small quantities, are mined only by small-scale operations employing hand labor.

The nonmetallics, both rocks and minerals, are more commonplace and familiar to us than the metallic ores. They are quite abundant but also less expensive than the metallic ores; hence they must be used near to their source. They generally cannot support long-distance transportation to areas of utilization. Specifications for utilization vary quite widely and actually are determined by the uses of the material. The industrial rocks and minerals are used essentially in the form in which they were removed from the ground. The annual gross value of all the industrial rocks and minerals exceeds that of the metallic ores.

Industrial Minerals

Nepheline syenite. In geologic circles, nepheline syenite is an igneous rock, but actually it is used as an industrial mineral. Nepheline syenite contains little or no quartz (silica), but it does contain the minerals nepheline and feldspars. Nepheline, $NaAlSiO_4$, and the feldspars, $KAlSi_3O_8$ or (Na,Ca) $AlSi_3O_8$, have similar valuable properties. Since the rock, nepheline syenite, is a mixture of the two, it is used as a bulk raw material. The high alumina content and the high alkali content of both nepheline and feldspar permit these minerals to be used in ceramics, whiteware, and also as a vitrifying agent in glass. Nepheline syenite must have a low iron content in order to be commercially valuable.

Nepheline syenite is a relatively rare rock. One important commercial deposit is the Blue Mountain deposit in Ontario, Canada. Several noncommercial deposits in the United States are located at Magnet Cove, Arkansas, Beemerville, New Jersey, and the Bearpaw Mountains, Montana.

Feldspar. The feldspars are a group of aluminosilicates of potash, soda, and lime. The common ones are orthoclase, $KAlSi_3O_8$, microcline, $KAlSi_3O_8$, and plagioclase, $(Na,Ca)AlSi_3O_8$. The feldspars are usually white, gray, or pink in color. The feldspars are fairly hard (6.0—6.5) and have two good cleavage directions. They occur very abundantly in igneous rocks of nearly all classes. Feldspar is the most abundant mineral in the Earth's crust, but most igneous rocks are not commercial deposits because of the presence of other minerals which increase the cost of obtaining the pure feldspar. Commercial deposits are the granitic pegmatites, commonly only a few hundred feet in size, with abundant feldspar. Fortunately, in the pegmatites, feldspar and other minerals occur in large-sized crystals from several inches to several feet across, and removal of waste minerals is thus somewhat cheaper than from common igneous rocks which are much finer-grained.

United States' production of feldspar comes mainly from North Carolina (Spruce Pine district) followed by Colorado, South Dakota, and the New England states. Powdered feldspar is used chiefly in ceramics (pottery, china, tile, porcelain, glaze, and enamels) because of its alkali content. The high alumina content allows feldspar to be used in the glass industry, imparting resistance to impact, bending, and thermal shock to the glass. Feldspar is also used in toothpaste.

Mica. Mica is a group name referring to eight minerals, all of which are silicates of aluminum and alkalies. All contain hydroxyl (OH) in addition to either iron, magnesium, lithium, or fluorine. Only two are known as commercial micas, muscovite, $KAl_2(AlSi_3O_{10})(OH)_2$ and phlogopite, KMg_3 $(AlSi_3O_{10})(OH)_2$, of which muscovite is considerably more important. The remaining micas are used commercially only in very small quantities or not at all.

Muscovite, usually clear, also may be colored with red or green tints. The perfect basal cleavage of mica allows it to be split into sheets, or films, down to $\frac{1}{1000}$ inch in thickness. Low heat conductivity, high dielectric strength, and infusibility permit thin sheets of mica to be used as sheet mica. *Sheet mica* is obtained chiefly from granitic pegmatites. Crude book mica is cleaned up, split into smaller sheets called block mica, film mica, and splittings. Any of these sizings can be used for punch mica, from which special shapes may be punched. *Ground mica*, the second main type, includes scrap from the sheet mica preparation and also low-grade sheet mica. This mica is ground to a powder. Sheet mica has high flexibility and high strength. It is used for electrical and electronic application (in capacitors, spark plugs, condensers, in tubes, as insulation), as windows in metallurgical furnaces, and in computers. *Ground mica*, less valuable than sheet mica, is used in wallpaper designs, paint, filler in rubber, roofing materials, stucco, glitter, and others.

United States' production of *sheet mica* is small, our requirements coming mainly from India. The United States produces mostly *ground* (scrap) *mica*, and North Carolina leads in production of both sheet and scrap mica, mainly from the pegmatites of the Spruce Pine district. Other ground mica-producing states are New Hampshire, Maine, Georgia, South Dakota, and several western states.

Lithium minerals. Lithium is present in many minerals, but only three are of commercial importance: spodumene, $LiAlSi_2O_6$, lepidolite, $KLi_2Al(Si_4O_{10})$ $(OH)_2$, and amblygonite, $(Li,Na)Al(PO_4)(F,OH)$. Spodumene is the most important and most extensively used. All three of these igneous minerals occur chiefly in pegmatite dikes, notably in the Kings Mountain district, North Carolina, and in the Black Hills of South Dakota. Other producing states are Colorado, New Mexico, and Maine. Lithium compounds also are obtained from brines of Searles Lake, California.

Lithium is a silver-white metal; it is soft and is the lightest of all metals. It melts at 186°C (367°F). Lithium metal is alloyed with chromium, copper, and other metals to make special alloys, and it promises to be very important in nuclear energy fields as atomic fuel. It will undoubtedly have importance in rocket propellant, too. Lithium minerals are added directly in glassmaking and ceramics to provide special properties. Lithium carbonate is used in glass and ceramics, while lithium hydroxide monohydrate is used to produce lithium greases. Other minor uses for lithium compounds include air conditioning, industrial drying, dry-cell batteries, and in welding, medicines, paint, fungicides, bleaching agents, and in pharmaceuticals.

Beryl. Besides being the sole commercial source of beryllium metal (see previous section on metals), beryl is an important gemstone. Emerald, which is the most valuable of the precious stones, is a clear variety of beryl with emerald green color. Beryl, $Be_3Al_2Si_6O_{18}$, usually is colored various shades of blue and green. The clear green variety is emerald; the blue or sea green variety is called aquamarine. Yellow-colored stones are called yellow beryl or heliodor, and pink beryl is called morganite.

Beryl for gemstones is found almost exclusively in commercial quantities in pegmatites. Value is established on the basis of size, needing to be 1 inch or so across. Pegmatites usually are worked for other minerals such as mica or feldspar, and the beryl is obtained as a by-product because pegmatites do not contain large quantities of beryl. Fine emeralds are obtained in Columbia, Transvaal, Russia, Brazil, India, and Madagascar. Commercial emeralds are now synthesized. But commercial beryl is obtained in many places, including pegmatites of South Dakota, New Mexico, Colorado, and New Hampshire. An important use for beryllium oxide, BeO, is in the refractory industry, chiefly spark plugs and insulators.

Quartz crystal. Clear quartz (SiO_2) is an important mineral for use in radio circuits, radar, and in ultrasonic equipment. Quartz can be used in this application because it possesses the property of *piezoelectric effect*, the development of electric charge on the surface when the crystal is compressed. However, only high-quality quartz crystal, which is clear and free from impurities, warrants high demand for electronic application. Crystals must contain a minimum of one cubic inch of flawless material. Only a minute fraction of the quartz crystals found can qualify in electronics. Quartz crystal which does not meet the highest specifications for electronics can be cut into prisms, wedges, and lenses for microscopes and other optical instruments.

Several varieties of quartz are used as gemstones. Most in demand are clear rock crystal, amethyst, citrine, smoky quartz, and rose quartz; these are found in a number of places. Other varieties are in lesser demand because of special colors or appearances. Some are used as lenses in microscopes. The largest producer (and practically the only one) of radio-grade quartz crystal in the world is Brazil. Here, the quartz deposits are in the form of veins, pipes, pockets, and lodes of hydrothermal origin. Many placer deposits with quartz have also been developed in the area. Very minor deposits in Arkansas have been worked in the past.

Fluorite. The major use for fluorite (called fluorspar in commerce) is as a flux in the making of steel. About 6.5 pounds of fluorite are required to make one ton of steel. Its function is to lower fusion temperatures and also to enable sulfur and phosphorus to be slagged off more easily. This mineral is also used in the smelting of lead, copper, gold, and silver. Fluorite of metallurgical grade must contain a minimum of 85 percent CaF_2. It must be very low in sulfur and silica. Fluorite is also used extensively in the manufacture of hydrofluoric acid and synthetic cryolite (Na_3AlF_6), both of which have extensive application in other industries, particularly by the aluminum refining industry. It is used fairly extensively in the production of high-octane gasolines. From fluorite, CaF_2, the chemical industry extracts fluorine, which is then used to make hydrofluoric acid. For the manufacture of acids, a minimum of 98 percent CaF_2 is required. Acid-grade fluorite is now consumed in larger amounts than metallurgical-grade fluorite. Fluorite is used in various types of glass and ceramics. Fluorine is also used in refrigerants, insecticides, and plastics.

Fluorite varies in color from clear through blues, greens, purples, and yellows. It varies from translucent to transparent in its ability to transmit light. The mineral commonly is of hydrothermal origin, occurring in veins and as replacements in limestones. The United States is the world's leading producer and consumer, and fluorite is in very high demand. Much fluorite is imported from Mexico, Italy, Spain, Germany, and Newfoundland. United States' production is chiefly from southern Illinois and western Kentucky, centering around Rosiclare, Illinois. Here, fluorite occurs as replacement veins and also as horizontal blanket replacements.

Barite. Barite, $BaSO_4$ is an inert and stable mineral. In commerce it is called heavy spar and barytes. It is also heavy, a property which does not permit it to be moved great distances economically. However, its weight (S.G. = 4.6) has the advantage of allowing pulverized barite to be used with a clay-water medium as a heavy circulating fluid to control reservoir pressures and to prevent oil-well blowouts. About 90 percent of the United States' production is for that purpose. Much of the remainder is used to make barium chemicals which are used in many other industries. Barite is ground for use as a filler in rubber and in certain paints. Other uses include barium sulfate in paper and in linoleum. Barium chloride is used in leather and cloth. Other barium compounds are used in ceramics, glass, metallurgy, and sugar manufacture.

Barite is a white to light-gray mineral; it is opaque and has a vitreous luster. In mineral deposits, barite is granular or in crystalline masses. It is called hard or soft barite, depending on its hardness which ranges from 2.5 to 3.5.

Barite occurs commonly with fluorite in hydrothermal deposits or replacement deposits with or without copper, lead, and zinc minerals. Barite occurs also interbedded with sedimentary rocks. Major domestic barite production comes from Magnet Cove, Arkansas, where bedded deposits in shale are present. The original source of the barite was probably magmatic. Another important source is Washington County, Missouri, where barite occurs in residual clays from underlying dolomites.

Magnesite. Magnesite usually occurs as crystalline masses or as cryptocrystalline masses. It is usually white to gray in color and has a hardness varying from 3.5 to 4.5.

Magnesite, a magnesium carbonate mineral, $MgCO_3$, is formed in two main ways: as replacement of limestone or as vein deposits resulting from the breakdown of serpentine. Rarely, it forms as sedimentary beds.

When magnesite is heated, it loses carbon dioxide. Ninety percent loss of CO_2 produces "caustic magnesite" and 100 percent loss of CO_2 produces "deadburned magnesite." Nearly all of the magnesite output is deadburned to produce refractory magnesia, MgO, which is then used to make special refractory brick for lining open-hearth furnaces in the steel industry. Such fire-resistant and heat-resistant brick are also used in smelters, cement kilns, and other high-temperature furnaces. Caustic magnesite is used for special cements and special floorings. Magnesite also finds application in rubber and in ceramics.

The principal magnesite-producing countries are Russia, Austria, United States, and Greece. United States' magnesite comes mainly from Stevens County, Washington, and from the Gabbs district, Nye County, Nevada. California also produces significant amounts.

Graphite. Commercial graphite is classified into two main types: crystalline and amorphous. The crystalline type is coarser grained than the amorphous type, which is not actually amorphous but is very fine-grained or cryptocrystalline. The chief use for graphite is for foundry facings; the graphite provides a smooth surface of molds and allows metal castings of many shapes to be easily removed. This black mineral, graphite, is also used in the steel industry, and in metallurgical crucibles for melting the nonferrous metals brass and aluminum. The high melting temperature of 3700°C (6700°F) of graphite allows high-temperature products such as crucibles to be made. Graphite's softness allows it to be used as a lubricant. It is also used in dry cells, batteries, and pencils. Graphite is inert to most acids and reagents, and it is a good electrical conductor. Graphite usually has a dull or bright luster, but sometimes it is metallic and opaque. Usually it occurs as flakes or as bladed masses.

Most graphite (pure carbon) originates through metamorphism and is found in schists and marbles. But some graphite is found in hydrothermal deposits. The leading producer is Korea, followed by Austria, Mexico, and Russia. United States' production is fairly small, coming from New York, California, Alabama, Georgia, Tennessee, Texas, and Nevada. There is also a manufactured graphite made from coal or coke, which is used for electrodes. However, the manufactured graphite is no threat to natural graphite for demand.

Asbestos. Asbestos is a term applied to fibrous minerals belonging to two groups: serpentine and amphibole. The serpentines include chrysotile and picrolite; the amphiboles include anthophyllite, crocidolite, amosite, tremolite, and actinolite. Nearly 95 percent of the commercial asbestos is chrysotile, and it alone will be discussed. The most important property of asbestos is its noncombustibility. Asbestos also resists acids, alkalies, and other chemicals. It does not conduct electricity.

Chrysotile is a hydrous magnesium silicate, $Mg_6Si_4O_{10}(OH)_8$, having green or yellowish-green color when fibers are massively aggregated, or appearing white when disaggregated into fine fibers and filaments. The high flexibility of the long, strong, noninflammable "spinning" fibers permits their spinning into yarn and weaving into cloth. Shorter grades of asbestos are "nonspinning." Spinning fiber, after processing into yarn, tape, or cloth, is used in brake linings, fireproof curtains, safety clothing for firemen and foundry workers, blankets, conveyor belts for hot materials, and packing. Nonspinning fiber must be mixed with a binding material and molded into sheets which find a variety of applications in shingles, corrugated panels, floor tile, and insulation for cable covers, boiler coatings, and furnace coatings.

Chrysotile forms by metamorphic processes by the alteration of ultrabasic igneous rocks, such as peridotite or dunite, or by alteration of magnesium limestone. Nearly all of the asbestos supply is of the former type. The Thetford Belt in Quebec is the world's leading producer. Russia, Rhodesia, and Union of South Africa also are important producers. United States' production comes from Vermont and Arizona.

Talc. Talc, the softest mineral, is a hydrous magnesium silicate, $Mg_3Si_4O_{10}(OH)_2$, and feels greasy or soapy. It is commonly white, greenish-white, greenish-gray, or dark gray in color. It forms as an alteration product of magnesian minerals in rocks by metamorphism of limestones, ultrabasic igneous rocks, or in contact metamorphic zones near basic igneous rocks. The United States is the world leader, producing about 50 percent of the world supply. Korea, France, Italy, and Canada are also important producers. The United States' production is mainly in New York, from the Gouverneur district, with substantial amounts from Inyo and San Bernardino Counties, California. Other producing states are Vermont, Nevada, Georgia, Texas, and Washington.

In industry, talc is classed as hard or soft, and fibrous and flaky. Talc, including the massive variety called steatite and impure material called soapstone, is produced almost exclusively in the ground form. Ground talc has very good covering or hiding characteristics and is used as white pigment in paint and as whiteners in wall tile, porcelain, and dinnerware. It is also smooth and soft, two features which give it lubricating properties. It is used also in electrical apparatus. The softness and smoothness of talc permit its use in toilet powders, lotions, face creams, crayons, soaps, and as paper filler. Slabs of pressed talc are used for acid-proof board, chemistry table tops, sinks, and linings for furnaces and stoves, because of its chemical inertness and high fusion point.

Vermiculite. Vermiculite is a micaceous mineral similar to mica. Thin flakes with micaceous cleavage are not as flexible as mica however. The color of vermiculite is usually yellowish-brown, but black vermiculite is also known. Vermiculite has high heat resistance and is chemically inert. Vermiculite is a hydrous magnesium silicate, $Mg_3Si_4O_{10}(OH)_2 \cdot H_2O$, but it often has varying amounts of iron and aluminum. The most valuable commercial property of vermiculite is its capacity to expand considerably (6 to 20 times or more) when heated for a few seconds at near 1000°C, as steam is generated between layers. Once expanded, it has a low density. These two properties permit vermiculite to be used as lightweight aggregate suitable for monolithic floors, walls, precast panels, and many "special-shape" blocks. These two properties also allow vermiculite to be used as loose fill insulation in walls, ceilings, and roofs, as well as in refrigerators, incubators, and water heaters. It can also be used by nurserymen for special soil treatment.

The United States produces about 70 percent of the world's vermiculite, with South Africa producing the rest. Nearly a dozen states produce vermiculite, led by Montana and South Carolina. At Libby, Montana, vermiculite is located in alteration products associated with pyroxenite. The South Carolina vermiculite was developed by alteration of pyroxenite masses which lie within gneisses and schists.

Diamond. Diamond, consisting of elemental carbon in the isometric system, is the hardest substance in the world, much harder than silicon carbide and tunsgten carbide. However, synthetic boron nitride (borazon) is nearly as hard as diamond. Diamond is very durable and has a high index of refraction and dispersion ("fire"), properties which make it highly desirable as a gem. Diamond has been the most desired stone consistently for nearly 2000 years. Gem diamonds are clear or very faintly tinted blue, brown, or yellow.

Industrial (nongem quality) diamonds are darker, usually yellow-brown, dark brown, or black. The very great hardness of diamond is its greatest asset. However, diamonds can be broken with a hard blow because they lack toughness in certain directions. Diamonds are also resistant to acids, and they have a very high melting point.

The durability, "fire," and high hardness permit diamond to be of very great importance in the jewelry trade. To produce the greatest amount of brilliance in a diamond, 58 tiny facets must be cut and polished. Each facet must be the right size and must be in the right position. Sizes of these valuable crystals vary from microscopic to masses weighing several hundred carats. Only about 20 percent of the diamonds produced are of gem quality; the remaining 80 percent are of industrial grade.

The largest diamond ever found, the Cullinan, weighed 3106 carats (1 $\frac{1}{3}$ lbs), and measured 10 × 6.5 × 5.0 cm. It was found in 1905. The second largest diamond was found in 1893. It weighed 995.2 carats and was called the Excelsior. In 1972, the third largest diamond was found. It was called the Star of Sierra Leone, weighed 969.8 carats, and measured about 6.4 × 4.0 cm.

Industrial diamonds are used chiefly for many types of abrasives, for engraving, shaping metal and hard steels, gem cutting, mineral cutting for special purposes, diamond saws, diamond rock drills, and wire-drawing dies. The United States is the largest consumer of industrial diamonds.

Most diamonds are found in placer deposits, derived from primary kimberlite pipes. The Belgian Congo produces about 75 percent of world's production. Much of the remainder comes from other African localities, as well as Russia. A small amount comes from Brazil. Very small deposits of industrial-grade diamonds in Arkansas are occasionally mined.

Diatomite. Diatomite, also known as *diatomaceous earth*, is made up of the siliceous shells of microscopic aquatic plants called diatoms, most of which are floating types of algal organisms. Different species prefer fresh, brackish, and marine waters. After death, the diatoms sink to the bottom to become mixed with other sediments (clay, silt, or volcanic ash) of shallow seas, lakes, or swamps. Shells may accumulate in enormous numbers. One cubic centimeter of diatomite may contain more than 10,000,000 shells. The usual result is bedded diatomite with porosity of 75 percent or more. Most commercial diatomite is light in color due to the high content of silica (SiO_2), and tans and browns occur frequently. Silica content commonly runs 85—90 percent.

Its high porosity, absorptive power, and light weight permit diatomite to be used chiefly as a filter for various materials such as sugar, mineral oils, molasses, fruit juice, beer and wine, vegetable oils, liquid soap, water, sewage, gases, varnish, vinegar, medicines, and many others. Pulverized, it is also used as paper filler, as an abrasive, in paint, and in insulation. United States' production comes chiefly from the Lompoc district, California. Some is produced in Nevada, Oregon, Washington, and Arizona. The United States leads in world production, followed by Denmark.

Potash. Potash (potassium oxide, K_2O) is a compound which does not occur naturally nor is it manufactured. The term *potash* merely refers to the oxide equivalent of potassium in various potassium-bearing salts used commercially. These are sylvite, KCl, langbeinite, $K_2Mg_2(SO_4)_3$, carnallite, $KMgCl_3 \cdot 6H_2O$, and kainite, $KCl \cdot MgSO_4 \cdot 3H_2O$, the most important being sylvite. Many other potassium-bearing minerals, occurring more abundantly than these comparatively rare salts, are not mined because of low K content or great insolubility. These commercial salts are usually formed as sedimentary minerals by evaporation of natural waters and often are interbedded with rock salt and anhydrite.

The high solubility of the commercial potash minerals permits them to be used primarily in fertilizer. About 95 percent of the United States' production is so used. The remainder goes to the chemical industry to be used in soaps, glass, ceramics, matches, textiles, dyes, explosives, wallboard, drugs, and many others. Major producing countries are United States, West Germany, East Germany, Russia, Canada, France, Spain, and Italy. United States' potash comes principally from the Carlsbad district in New Mexico. Important supplies also come from Searles Lake, California, and reserves in North America are good.

Soda. Three sodium-bearing minerals are exploited commercially: trona, $Na_3H(CO_3)_2 \cdot 2H_2O$, mirabilite, $Na_2SO_4 \cdot 10H_2O$, and thenardite, Na_2SO_4. From these three natural minerals, the chemical industry obtains its two mainstay compounds, sodium carbonate, Na_2CO_3 (called soda ash) and sodium sulfate, Na_2SO_4 (called salt cake). Of these two alkalies, *soda ash* is more widely and abundantly used because of its greater versatility. Strange as it may seem, most soda ash is manufactured from salt, limestone, and coal. Much salt cake is manufactured, too. However, both natural salts are increasingly important for commercial sources.

Most soda minerals are formed as sedimentary materials in arid regions where soda-bearing waters are evaporated, causing precipitation of the very soluble compounds. From these alkali and bitter lakes, the salts are removed as brines or mined as solid salts, according to conditions. Soda salts are used in a great many ways: in soap, glass, dyes, cleansing agents, soil refining, insecticides, chemicals, detergents, baking soda, fire extinguishers, paper, textiles, metallurgy, and others. Large producers are United States, Germany, Canada, Peru, Chile, and Russia. Most of the United States' production is from Searles Lake, California, with Owens Lake, California, contributing substantial quantities also. New large deposits near Green River, Wyoming, continue to keep the United States self-sufficient in soda ash.

Borates. More than 95 percent of the world's production of boron-bearing minerals comes from the United States, chiefly from California. Other producing nations are Argentina, Chile, Peru, Bolivia, Turkey, Russia, and Italy. There are two large California deposits, at Kramer in Kern County and at Searles Lake in San Bernardino County. Borate minerals used commercially are borax, $Na_2B_4O_5(OH)_4 \cdot 8H_2O$, kernite, $Na_2B_4O_6(OH)_2 \cdot 3H_2O$, colemanite, $CaB_3O_4(OH)_3 \cdot H_2O$, and ulexite, $NaCaB_5O_9 \cdot 8H_2O$. Borax and kernite are the most abundant and most used. The white borate minerals commonly are obtained from brines of salt lakes, from deposits around playas and lakes, and from beds beneath old playas. Thus, the borates are of sedimentary origin. At Kramer in California, the borates occur in clay and shale. At Searles Lake, borates are present in the same brines which yield sodium salts.

These borates are readily soluble in water and are thus used in washing powders, water softeners, and soaps. Borax is easily fusible and has good fluxing capability. It can therefore be used in the manufacture of heat-resistant glass and vitreous porcelain enamel for bathtubs, stoves, refrigerators, and metal signs. Half of the borate production is so used. Other uses include disinfectants, preservatives, cleansing agents, coating of paper, plywood, paint, fertilizer, as a flux for welding, ointments and eye washes, in textiles, and in tanning leather. Elemental metal boron is used in metallurgical applications and as an antiknock agent in gasoline. Other boron compounds have potential as jet fuel and in shielding for atomic reactors; synthesized boron nitride (borazon) is nearly as hard as diamond and finds use as general abrasives.

Nitrates. The nitrates are various compounds of nitrogen, chief among which is anhydrous ammonia, NH_3. Others are ammonium sulfate, $(NH_4)_2SO_4$, ammonium nitrate, NH_4NO_3, nitric acid, HNO_3, and sodium nitrate, $NaNO_3$. Nitrogen compounds are used very extensively for fertilizer; nearly 85 percent of the production is used in this way. They are sometimes applied directly to the soil, but usually are mixed with phosphate or potash. In addition to several types of nitrate applications as fertilizers, nitrates have some utility in the chemical industry, in explosives, in chemical salts, in refrigeration, and in others. Ammonia itself is the most important nitrogen compound in the chemical industry because it is cheap and can be easily converted to other compounds. Ammonia is used in the manufacture of wood pulp and plastics. Nitric acid is employed as a solvent.

Nitrogen compounds are obtained from the atmosphere, from coal, and from nitrate deposits, the dominant source being the synthetic methods. The largest natural deposits of nitrates in the world are in Chile, where nitrate beds are actually gravels cemented by the valuable nitrate compounds. The beds lie on gentle slopes of the coast ranges. Very little nitrate occurs anywhere else. Only two small deposits occur in the United States, both in southwestern states. The largest consumer of nitrates is the United States.

Sulfur. Sulfur is pale yellow in color, but it may be greenish, brownish, or reddish. It is brittle and melts at 110°C (230°F). Sulfur burns at low temperatures with a pale blue flame. It is almost tasteless. Sulfur is insoluble in water and in many acids. This element conducts neither heat nor electricity.

Sulfur is one of the most important chemical minerals. The element occurs as native sulfur, in sulfide minerals (pyrite), and in sulfate minerals. However, the chief commercial sources are native sulfur, S, and pyrite, FeS_2, with native sulfur being first. Sulfur is obtained from upper portions of salt domes in Louisiana, Texas, and Mexico. Superheated water is forced down a well, melting the sulfur, which is then pumped up to the surface. This method is known as the Frasch process. Native sulfur is obtained also from bedded deposits and around volcanic areas, but not in the United States. Sulfur is also obtained from the mineral pyrite, FeS_2, found as massive sulfide deposits associated with ore minerals.

The United States leads the world in sulfur production, some 90 percent (raw sulfur) coming from Louisiana and Texas, and minor amounts from the Ducktown, Tennessee, pyrite deposits. Mexico, Russia, Japan, and Italy produce most of the remaining 10 percent. Most of the sulfur produced is used to make sulfuric acid and other sulfur compounds for industrial purposes. Raw sulfur is used in heavy chemicals, fertilizers, pulp and paper, paint, explosives, rubber, and dyes. Sulfuric acid is used as fertilizer, in oil refining, to make chemicals, in textiles, and also has general metallurgical applications. Sulfur is used in gunpowder, in matches, in insecticides, in certain medicines, and in photographic film processing.

Phosphate. Phosphate rock is any rock with sufficient phosphate content to allow the material to be used commercially for phosphate compounds and elemental phosphorus. Such rocks as phosphatic pebbles derived from underlying phosphatic limestone, marine phosphate beds, apatite deposits of igneous origin, and phosphatic marls are the chief commercial sources. Phosphate rock consists either of crystalline igneous apatite, $Ca_5(PO_4)_3(F, Cl, OH)$ in substantial quantities, or of noncrystalline sedimentary phosphorite with the mineral collophane (cryptocrystalline apatite or carbonate apatite). Over 90 percent of the world's production is phosphorite, occurring interbedded with shales or limestones, or nodular and pebbly phosphatic material mixed with sandy material.

The leading producing nation is the United States, followed by Russia and Morocco. United States' production comes mainly from Florida where pebble-phosphate sediments mixed with fossil teeth and bones of land animals yield large tonnages. The area of Idaho, Montana, Utah, and Wyoming yields fairly good phosphate rock, as does central Tennessee. Approximately 90 percent of the phosphate rock produced is used for fertilizer and plant food. Other phosphate chemicals are used for water softeners, in scouring powders, in insecticides, in beverages, and in ceramics. Oil refining and photography require some also. A fairly significant amount of phosphorus is used in steel, in fireworks, and in explosives.

Industrial Rocks

Granite. Commercially, the term *granite* refers to coarse-grained true granite, granite gneiss, granodiorite, and other coarse-grained members of the granite-gabbro series, including the porphyries. Even the dark rocks gabbro, diabase, and pyroxenite are referred to as *black granite* if they are suitable for polishing and cutting into dimension stone. However, most commercial granite is true igneous granite or granodiorite.

Granite has two major uses: as crushed granite and as dimension granite. These two industries have little similarity with respect to quarrying, milling, preparation for market, utility, and value. Granite is tough, hard, and sound (resists disintegration). After it is crushed to required sizes, granite may be used for concrete aggregate, road metal, railroad ballast, in filter beds, riprap (large chunks) in piers and breakwaters, etc. Pleasing and uniform colors (generally pink, gray, salmon, red, white) are desirable for dimension granite. High hardness, uniform texture, and other physical properties make dimension granite suitable for monuments, memorials, foundation blocks, steps, columns, etc. Large production of crushed granite comes mainly from California, with substantial amounts from Georgia, North Carolina, South Carolina, and Virginia. Dimension granite is produced in the Maine-Massachusetts-Vermont area, the industry centered chiefly at Barre, Vt. The North Carolina-Georgia area, chiefly at Mt. Airy, North Carolina, is another important center. The Minnesota-Wisconsin-South Dakota area, centering chiefly at St. Cloud, Minnesota, is the third major producing area.

Basalt. Basalts are fine-grained igneous rocks with dark color due to the presence of substantial amounts of ferromagnesian minerals such as augite, olivine, and magnetite, in addition to the more dominant (usually gray) plagioclase feldspar. Quartz and orthoclase are usually absent. The basalts are extrusive and often are amygdaloidal. Basalts are referred to as trap or traprock in commerce. A special type of basalt is called diabase. Both basalt and diabase are used almost entirely as crushed stone, of which about 80 percent is used as concrete aggregate and road metal. The remaining 20 percent goes for railroad ballast, roofing granules, and riprap. A small amount is used in concrete shielding around nuclear reactors. In addition, good-quality diabase occasionally is used as dimension stone.

Basalt is a tough, sound rock, and it resists weathering very well. Consequently, it is a dependable crushed stone. But it is a low-cost product and must be used not far from the quarry. Application in the general construction business and for highway construction supports the basalt industry. The United States' production of basalt comes from the states of Connecticut, Massachusetts, New York, New Jersey, and Pennsylvania in the East; and from Washington, Oregon, Idaho, and California in the West.

Pumice and pumicite. Silicic volcanic glassy pyroclastic rocks formed by explosive volcanism are referred to as pumice or pumicite. The difference in name arises from a difference in particle size: pumice is coarse (larger than 2 or 3 mm), and pumicite is fine (smaller than 2 or 3 mm, down to powder). Pumice, being coarser, is vesicular in texture, while pumicite is more compact because it is made up of much finer and more angular fragments. Both pumice and pumicite often become mixed during volcanic explosions.

Finely ground pumicite and unground pumicite are used in general scouring powders and for fine-polishing agents. The high hardness and sharp-ground fragments allow this application. The high porosity of pumice, giving it light weight, permits it to be used in lightweight concrete for construction purposes in place of more expensive materials. It is also used in pozzolan-portland cement, which is more resistant to salt waters than other cements. Other uses for unground pumicite and ground pumicite are as an ingredient in accoustical plaster, in insulation, as filter aids, as poultry litter, as soil conditioner, as insecticide carrier, and in highway blacktopping.

Pumice and pumicite are found around extinct, recent, or near recent volcanoes. The United States' production comes principally from Washington, Oregon, California, and New Mexico.

Perlite. Geologically, perlite is a volcanic glass with around 3 or 4 percent water and containing numerous curving shrinkage cracks. The term perlite is a commercial one referring to any glassy rock which expands considerably when heated. It includes true perlite as well as other glassy rocks which contain 3 to 4 percent molecular water. When heated, the molecular water is converted to steam, forming a light, fluffy, cellular structure appearing like pumice. In many cases, volume increase is at least ten times. The expanded material is commercially called *expanded perlite.*

About three-fourths of the expanded perlite is used as aggregate in plasters, particularly with gypsum plaster and wallboard. Small quantities are used in lightweight concrete. Minor uses include insulation, filtration, soil treatment, paint filler, in oil well drilling muds, and in packing material. The light weight, the insulating ability, the fire-resistant property, and the sound absorbing capability of perlite each play a role in its utility.

Expanded perlite is a relatively new commercial product, coming onto the market about 1946. It shows promise in many applications, and United States' reserves are large. Perlite occurs in regions of fairly young volcanoes. Nearly two-thirds of the United States' production comes from near Socorro, New Mexico. Other states producing perlite are California, Colorado, Nevada, and Arizona.

Slate. An important metamorphic rock in industry is slate. Slate is a micro-crystalline rock having a well-developed cleavage, resulting from mild dynamic metamorphic processes in shale. Common constituents are chiefly quartz and illite, with subordinate sericite (mica), calcite, and minor quantities of others. Slates are commonly black, gray, red, purple, or green, colors produced by organic material (carbon), hematite, chlorite, or ferrous oxide. Slate has high durability, high compressive strength, and low porosity. In addition to these commercial properties, uniformity and permanence of color are also very important. Any fading of color lowers desirability. The single property that makes slate a commercially important rock, though, is the slaty cleavage, developed by parallel alignment of flaky (micaceous) minerals as dynamic earth pressures were active. Such planar structure permits slate to be cut and fabricated into dimension stone.

Dimension slate is used for roofing slate, flagstones, sheets for switch-boards, mantels, steps, sills, blackboards, and billiard table tops. Crushed slate, as granules, is occasionally used for composition roofing. Pulverized slate (slate flour) has found utility as filler in linoleum and paint. Current research indicates promise as bloating slate for concrete aggregate and also for rock wool or mineral wool. Most (about three-fourths) of the United States' production comes from eastern Pennsylvania and northeastern New York. A little slate is produced in Maine, Maryland, Virginia, Arkansas, California, and North Carolina.

Marble. The geologic term *marble* refers to a metamorphic rock, composed chiefly of recrystallized calcite by thermal or dynamothermal metamorphism and having a fairly coarse texture and sparkling white appearance. The commercial term marble, however, includes any stone, except commercial granite, which has a pleasing appearance and can take a polish. In the marble trade, most of the rock utilized is true metamorphic marble, but fairly substantial amounts of crystalline sedimentary limestone are also used. Other very different types of rock, such as cave onyx, travertine, and verde antique (serpentinite) are used as marble in significant amounts as well.

Colors of marbles range from white to gray to black, as a result of varying amounts of carbonaceous matter; green tints are due to chlorite and other green silicates; pinks and reds are due to hematite or rhodochrosite; yellows and tans are due to limonite. Uniformity in color and color banding is desirable, and veining is acceptable for "book-cut" slabs (opposite sides of a saw cut are opened like a book to produce mirror images). Both coarse or fine textures are satisfactory.

Marble is used in two chief ways: dimension stone (cut and polished) and crushed stone. Dimension marble is used predominantly as architectural stone in such application as exteriors and interiors of buildings, columns, trim, window sills, wainscoating, floors, or steps. Memorial or statuary stone uses minor quantities of dimension marble.

The source of crushed marble is the waste material from the dimension-marble processing. Crushed marble is used similarly to crushed limestone, such as concrete aggregate, railroad ballast, road metal, filter beds, poultry grit, stucco, coal-mine dust, filler, whiting agent, fluxstone, soil conditioner, manufacture of lime, and many other chemical uses.

Much of the United States' marble production is from Tennessee, but Vermont, Georgia, Alabama, Colorado, Maryland, and North Carolina are also significant producers.

Sand and gravel. Sand ranges in size from 0.05 to 2 mm, and gravel is material coarser than 2 mm. Sand and gravel differ from each other in size as well as in composition. Most sands are made up predominantly of quartz grains. Gravels, however, are nonuniform in composition, being composed chiefly of various types of rock fragments. Sand and gravel are commonly mixed, but sand quite often occurs alone.

Sand and gravel are deposited by streams flowing at high volume and high velocity. Because sand and gravel are unconsolidated, they may be upgraded by screening, washing, and sizing. For general uses, sand and gravel must be clean, have good resistance to abrasion, be sound, and have several size ranges for different uses. Being a low-priced product, sand and gravel must be very close to its market because it cannot pay its own transportation for any distance.

In the United States, over 3000 sand and gravel plants are active along present day or somewhat older streams in many areas. The largest production comes from California, but other areas of significant production include deposits along the Colorado River, North Platte River, Missouri River, and the Tennessee River. Substantial amounts of sand and gravel are also produced from the Atlantic and Gulf Coastal Plains. In addition to California, other leading states are Michigan, Ohio, Illinois, Wisconsin, Texas, New York, Minnesota, Indiana, and Utah.

Sand and gravel are used chiefly in the construction and paving industry as concrete aggregate, for dams, piers, foundations, airport runways, and highways. Loose sand and gravel are used for highway subgrade material.

Sandstone. Many types of sedimentary sandstone have been used commercially, but the most important type is the so-called "pure" sandstone, containing generally 90 percent or more quartz grains. Sandstones often are poorly cemented and can be easily crumbled into sand. However, strong, durable sandstone, with tan and gray colors, is used as dimension sandstone for exterior facing and trim for buildings, in houses, as curbstones, in bridge abutments, and in retaining walls of various types. Firm sandstones are crushed for use as concrete aggregate, railroad ballast, and riprap. However, many commercial sandstones are weakly cemented and thus are crumbled and used for molding sand and glass sand.

Sand with good cohesiveness and refractoriness is used for making molds of special shapes into which molten metal is poured in the manufacture of metal pieces for many applications; glass sand is used as the main ingredient in glass. Other types include abrasive sand, engine sand (on wet rails), furnace sand in foundries, and filter sand.

Good commercial sandstones are quite numerous in the United States. The Berea sandstone in northern Ohio, the Oriskany sandstone in the New York-Virginia area, and the St. Peter sandstone in the Wisconsin, Minnesota, Iowa, Illinois, Missouri area are three of the nation's important ones. There are many other good commercial sandstones being utilized.

Clay. Commercial uses of clays are numerous and varied. Clays themselves are of varied types and of varied compositions. Clay consists of hydrous aluminum silicates (clay minerals), usually of colloidal or near-colloidal sizes, and is material which is more or less plastic when wet. The clay minerals include the kaolinites, the illites, the montmorillonites, the mixed-layer clays, and attapulgite. Nearly all clays are formed by weathering of other rock, and then remain as residual clay, or they may be moved and deposited as transported clay. Some clay is also of hydrothermal origin.

Clay-mineral composition controls the utility to which it can be applied. About two-thirds of all clay (and shale) is used in ceramic products such as pottery, chinaware, stoneware, sanitary ware, tile, and porcelain, and in structural clay products such as brick, drain tile, and sewer pipe. Clay is plastic and thus can be molded into special shapes before firing in a kiln. Much clay is used in refractories (molded and fired clay products which withstand high temperatures), as liners in steel furnaces, cement kilns, lime kilns, etc. Clay (nonfired) can be used as fillers and coatings in paper, as drilling mud, as absorbents for oils of many types, in cement, as lightweight aggregate, and many other applications.

Commercial clays are very numerous in the United States. Important ones are in North Carolina, in northeastern Wyoming, Arizona, Maryland, Missouri, Georgia, South Carolina, Tennessee, Kentucky, California, Ohio, Pennsylvania, and other parts of the country.

Limestone and dolomite. Limestone and dolomite (dolostone) are sedimentary rocks made up of 50 percent or more of carbonate minerals, calcite and dolomite. If calcite mineral dominates over dolomite mineral, the rock is limestone; if dolomite mineral is more abundant than calcite, it is dolomite rock (dolostone). Intermediate types of limestones and dolomites exist in which calcite and dolomite minerals are in nearly equal amounts.

Limestone and dolomite have an extremely large number of uses, and hence their physical properties are very varied. Half of the production is crushed stone, for uses as concrete aggregate, road metal, railroad ballast, filter beds, poultry grit, stucco, coal mine dust, filler, and whiting. A considerable amount of crushed limestone is used as flux stone in steel manufacture, also as soil conditioner, for manufacture of lime, and chemical raw material in glass making. A substantial quantity of limestone and some dolomite is used as dimension stone in the "marble" trade, being used in exteriors and interiors of buildings, steps, columns, trim, floors, or sills. Limestone, but not dolomite, finds additional large consumption as raw material for the manufacture of portland cement, which has a great many uses in the construction field.

For most uses as crushed stone, limestone and dolomite must be uniform in texture and composition, nonporous, and be fairly clean of chert, pyrite, and organic debris. For dimension stone, the rock must be uniform in color and texture. Good limestones (for different utilities) are located in many states: Pennsylvania, Ohio, Indiana, Great Lakes region, Virginia, Tennessee, Michigan, Florida, Texas, and California, are the leaders; there are still others.

Rock gypsum. The mineral gypsum, hydrated calcium sulfate, $CaSO_4 \cdot 2H_2O$, occurs chiefly as sedimentary evaporites from marine brines. Three varieties exist: selenite, as a cleavable form; satin spar, as a fibrous form; and alabaster, as a massive and finely crystalline form. Commercial gypsum, however, is usually finely crystalline or granular *rock gypsum*, having white, gray, pink, or yellow colors. During evaporation of saline marine waters, calcium sulfate is usually deposited first as anhydrite, $CaSO_4$. At a later time, during weathering, anhydrite hydrates and is secondarily converted to gypsum.

When heated to around 175°C, gypsum loses 75 percent of its water, producing plaster of paris. When it cools, it can be mixed with water and then spread or molded before it hardens. Practically all of the gypsum produced is used in this way, as plaster, wallboard, and similar construction materials. Small amounts of raw gypsum are added to portland cement to retard setting time; otherwise, portland cement would set before it could be properly formed. Raw gypsum, in powdered form, is used as fertilizer, as insecticide carrier, as filler, and in yeast growth.

Gypsum rock deposits are fairly widespread, and only readily accessible deposits are exploited commercially. About two-thirds of the United States' production comes from California, Michigan, Iowa, Texas, and Nevada. But other production is from New York and Oklahoma. About 70 percent of the world production comes from the United States, Canada, France, United Kingdom, and Russia.

Rock salt. Commercial rock salt is made up almost exclusively of the mineral halite, NaCl, occurring as solid beds of evaporites. Halite itself is usually colorless or white but is often tinted gray, blue, pink, or yellow. Common impurities in bedded rock salt are anhydrite, calcium and magnesium chlorides, and sodium and magnesium sulfates. Rock salt beds range from a few feet to several hundred feet thick. The high solubility of rock salt prevents its outcropping at the surface; consequently, commercial deposits are exploited by subsurface mining methods. Salt mines produce large quantities of the solid material, while minor amounts of salt brine (liquid) are obtained by pumping water down a drill hole, thus dissolving the salt, and returning the brine to the surface. Sodium chloride (halite) also is obtained from sea water, salt lakes, dried up salt lakes, and brine springs. Nearly all of the commercial salt is obtained from sedimentary rock salt beds.

Salt has innumerable uses in the chemical industry. In nearly all cases, it is separated into sodium and chlorine. Sodium goes for the making of soda ash, with all of its ramifications (see Soda), and the chlorine is applied in the manufacture of chlorates, hydrochloric acid, etc. Salt is also used in soaps, dyes, textile processing, ice control, and extensively in the food industry. Michigan leads the United States in production, followed by Texas, New York, Louisiana, and Ohio.

Mineral Fuels

The mineral fuels are those natural rock and mineral materials which find their greatest value as sources of heat and power. In other words, they are sources of usable energy. They are generally well-known materials: coal, petroleum and natural gas, wood, water power, the sun, and radioactive materials. The three most important of these are coal, petroleum and natural gas, and uranium for atomic energy. Coal is still the most important source of energy units, followed strongly by petroleum and natural gas. In fact, petroleum and natural gas have given very strong competition to coal and are preferred for mobile power units. Atomic energy, a relatively new type of energy, has not yet reached its peak as an important power. However, great strides are being made, and atomic power will undoubtedly progressively advance as a source of unit energy.

Coal

Coal is divided into four main groups: anthracite, bituminous, lignite, and cannel coal. Cannel coal is a special type, but the other three are ranked according to their fuel ratio, i.e., fixed carbon/volatile matter. In anthracite, fuel ratio is high; in lignite, fuel ratio is low. Both the fixed carbon and volatile components burn, the volatiles giving a long smoky flame, while the fixed carbon gives a short, hot, smokeless flame and is a lasting source of heat. Thus, anthracite offers the best heat supply. Coal usually contains a small amount of sulfur in the minerals marcasite and pyrite. Any noncombustible material such as clay or silt is referred to as ash.

Coal is used principally as a source of heat and power. About 20 percent (bituminous type) is used as coking coal. With its higher carbon content after special coking treatment, the resultant coke is used to smelt iron ore in the steel industry. Most coal (more than 50 percent) is used for power generation in electric utilities. Coal is also used to generate steam for railway transportation, for domestic heating, and other industrial purposes. Coal is derived from luxurient vegetation accumulating in lakes and swamps. The largest reserves of coal in the world, about one-third, are in the United States, but China and Russia both have quite large reserves, too. Germany and the United States are the biggest producers. Major United States' production is from West Virginia, Pennsylvania, Illinois, Kentucky, Ohio, and other states.

Petroleum

Petroleum has become extremely valuable to modern civilization, and because of this, it is often called "black gold." Many different products are made from petroleum, giving petroleum its great value. The total value of the products of petroleum is greater than from any other mineral material, and transportation, modern industry, and even warfare are greatly dependent on these various products. Since petroleum is a liquid which is rather inexpensively removed from the ground (as crude oil), it can also be shipped easily.

Petroleum is a fairly complex mixture of chemical compounds called hydrocarbons, which may be present as gas, liquid, or solid portions of the petroleum. Hydrocarbons are substances composed principally of hydrogen (H) and carbon (C) in various combinations. Other compounds containing oxygen (O), nitrogen (N), and sulfur (S) are usually present in crude oil, and water and inorganic matter are commonly present in very small amounts. Consequently, various crude oils have somewhat different physical and chemical properties. The chemical compositions of crude oils are summarized in Table 7-3.

Petroleum varies in color from nearly colorless, light yellow, red, green, brown, to nearly black. The heavier crude oils are amber or green in color, and the lighter ones are brown to black. Crude oils vary in odor also according to the composition. Odors are generally agreeable, but as the hydrogen sulfide and sulfur content increases, the odor becomes more disagreeable.

All crude oils are lighter than water and hence their specific gravities are less than 1.0. Specific gravities in crude oils range from about 0.97 to 0.61.

Table 7-3: CHEMICAL COMPOSITION OF CRUDE OILS

Carbon	82.2—87.1 percent
Hydrogen	11.7—14.7 percent
Sulfur	0.1— 5.5 percent
Nitrogen	0.1— 2.4 percent
Oxygen	0.1— 4.5 percent
Mineral matter	0.1— 1.2 percent

(summarized from several sources)

The specific gravity is an index to the value of the oil; the lighter oils are more valuable because they have greater gasoline content and other valuable components.

The various hydrocarbons which make up crude oil are grouped into two general types: the aliphatics (paraffins or methanes), and the carbocyclics (benzenes). In the aliphatics the carbon atoms are linked in straight chains as shown by butane, C_4H_{10}:

There are many members of this group, such as methane, CH_4, ethane, C_2H_6, propane, C_3H_8, pentane, C_5H_{12}, butylene, C_4H_8, and hexylene, C_6H_{12}. The carbocyclics contain carbon atoms arranged in closed chains (or rings), as shown by benzene, C_6H_6:

Other members of this group include toluene, $C_6H_5CH_3$, cyclopentane, C_5H_{10}, cyclohexane, C_6H_{12}, and naphthalene, $C_{10}H_8$.

Sulfur (S) and sulfur compounds such as hydrogen sulfide, H_2S, thiophene, C_4H_4S, alkyl sulfides, and others are usually present. Nitrogen compounds such as $C_{12}H_{17}N$ and $C_{17}H_{21}N$ are also usually present. Oxygen is commonly present in small amounts in naphthenic acids, phenolic compounds, and other compounds. Inorganic constituents, usually in very small amounts reported in ash of crude oil, may include iron, aluminum, phosphorus, magnesium, copper, vanadium, nickel, and sodium.

Petroleum must be refined before it can be used commercially. The refining process actually separates the petroleum into its component parts, each having special properties for a particular utilization. Refining usually requires one of three processes. The most important process is *fractional distillation*, which is based on condensation temperatures of the vapors of the various components. The oil is heated and its vapors rise in the fractionation tower. Vapors which condense at high temperatures form on plates or in pans located low in the tower. Vapors which condense at low temperatures form on plates or in pans high in the tower. The various fractions, such as grease, lubricating oils, fuel oils, gasoline, and naphtha are removed through drains.

Refining also may be accomplished by a *cracking* process which often yields up to 60 percent gasoline, the most valuable product. The heavier oils contain large complex molecules and these are cracked or split into smaller gasoline molecules. A third process, called *polymerization*, involves small, lighter molecules. They are made to combine at low pressure into larger molecules which yield more gasoline or certain oils and greases.

About 2500 products are made from petroleum, and another 3000 various petrochemicals are processed. A barrel of crude oil is refined into useful major products in the approximate following proportions:

Gasoline	45 percent
Fuel oil	36 percent
Kerosene	5 percent
Asphalt and road oil	2 percent
Lubricants	4 percent
Others	8 percent

(data from American Petroleum Institute, 1971)

Petroleum thus provides mankind with many useful fuel products, the most important of which is gasoline. Gasoline makes up about 45 percent of all refined products and is used in automobiles, buses, trucks, and tractors. Some 60 billion gallons are used annually in the United States. Airplanes use another 2 billion gallons each year.

The second most important petroleum product is fuel oil. Various kinds of fuel oils provide heating and power in homes, buildings, factories, smelters, trucks, railroads, and ships. About 55 billion gallons of fuel oil are used yearly in the United States. Other products such as lubricating oils and greases are used to lubricate moving parts of machinery and machines, thereby preventing friction which would cause rapid wear. Domestic use requires some 2 billion gallons of lubricants per year. Jet fuel for modern aircraft is a mixture of gasoline, kerosene, and certain oils. At the present time, nearly 8 billion gallons of jet fuel are used in the United States annually, but consumption is increasing.

Asphalt and various road oils are used for road surfacing and to cover roofs. Over 90 percent of the streets and roads and over 70 percent of the airport runways in the United States are covered with asphalt. Some 20 million tons of asphalt are produced annually.

Various petrochemicals are made from petroleum. These various products find application in antiseptics, synthetic rubber, plastics, cleaning fluids, detergents, paint, cosmetics, printing ink, anesthetics, fertilizers, drugs, nylon, explosives, rust preventives, insecticides, and fungicides. Wax is also obtained from petroleum, and it is used in candles, paper coatings, and various other household items.

It is fairly well agreed among scientists that oil (and gas) are of organic origin. Evidence is overwhelming in favor of low forms of marine or brackish-water life as the source of oil. However, no such agreement exists as to details of conversion to petroleum. Microscopic plants and animals are thought to accumulate in shallow marine muds and become buried. After burial, such factors as pressure, moderate temperatures, bacteria, and time are important in causing the organisms to yield natural hydrocarbons. These same types of microscopic organisms, both plant (diatoms and algae) and animal (one-celled) exist today in oceanic waters.

The source rocks in which most oil originates appear to be organic shales, but in some instances limestones, siltstones, and lake sediments serve as source rocks. Once formed, oil tends to migrate naturally to more porous rocks, as a result of compaction of the source rock by overlying sediments, buoyancy due to the fact that oil is lighter than salt water in rocks, or because of artesian water pressures in the rocks. Eventually the oil may migrate into rocks porous and permeable enough to contain sizable reservoirs. Porous rocks such as sandstones and conglomerates are common as reservoir rocks. Oil often accumulates in limestones and dolomites such as coral reefs because such structures have good porosity and permeability.

Migration of oil continues as long as the porosity, permeability, and buoyancy effects are suitable. However, in many cases, migration is halted by impermeable cap rock which holds the oil in a reservoir rock below, forming an oil trap. Oil traps are of two general types: structural traps and stratigraphic traps (Fig. 7-2). Structural traps are produced by folding, faulting, intrusion, or various other geological activity. Stratigraphic traps are formed as a result of vertical and lateral changes in permeability caused by original depositional factors of the reservoir rock.

About 80 percent of the oil produced is obtained from structural traps called *anticlines* or domes. Oil migrates upward in the rocks and accumulates at the crest below a caprock. *Fault traps*, which yield about 1 percent of the production, are formed where tilted reservoir rocks become sealed against impermeable rock; oil subsequently collects at the top of the porous areas. *Salt domes* produced by forceful injection of salt upward into sedimentary rocks produce various structures where sealed upper ends of porous beds create traps.

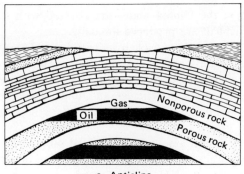

a. Anticline
STRUCTURAL

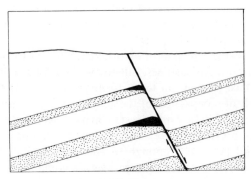

b. Fault
STRUCTURAL

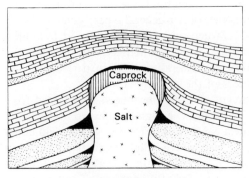

c. Salt dome
STRUCTURAL

Fig. 7-2: Types of oil traps.

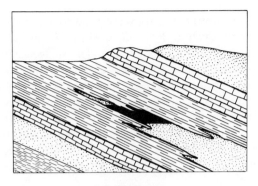

d. Pinch out
STRATIGRAPHIC

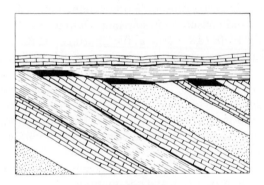

e. Unconformity
STRATIGRAPHIC

f. Buried reef
STRATIGRAPHIC

Fig. 7-2: Continued.

Several types of stratigraphic traps are good sources of oil, yielding about 18 percent of the production. Rock units which "wedge out" laterally often become enclosed by rock of another type. The wedged-out ends of porous sandstones, if enclosed by impermeable rock, commonly make good *pinch-out* traps. Where angular *unconformities* exist, good stratigraphic traps may be formed, especially if a good impermeable rock such as shale acts as the impermeable layer above. About 3 percent of the production comes from various unconformities. A third type of stratigraphic trap is the *buried reef*. Reefs are quite porous and oil can migrate into them and accumulate. Reefs yield about 3 percent of the production. Several other types of stratigraphic traps are also known.

Sandstone reservoir rocks yield about 59 percent, limestone about 40 percent, and other types of rock yield about 1 percent of the production. Oil production is measured in barrels, and one barrel is equivalent to 42 U. S. gallons or to 35 imperial gallons. World production of oil is about 6 billion barrels a year, and consumption is rising. Despite the discovery of new oil fields, the early 1970s saw a critical fuel shortage. At the present time, some 230 billion barrels are estimated to lie beneath the surface as reserves. The greatest amount of reserves is in Asia, but about 97 percent is centered around the Middle East (Iraq, Iran, Kuwait, and Saudi Arabia). North America has about 35 billion barrels of crude oil reserves, furnishing a little over half of the world consumption, more than that of all the other continents combined. South American reserves are estimated at about 18 billion barrels, centered almost exclusively in Venezuela.

The United States, holding about 85 percent of North America's reserves, leads the world in production and consumption. United States' production is over 3 billion barrels per year, followed by Russia (1.1 billion barrels) and Venezuela (1.0 billion barrels). The leading states in production are Texas, California, Louisiana, and Oklahoma. In addition, Kansas, Wyoming, New Mexico, and Illinois are significant producers. Recent new discoveries in Alaska are substantial.

Natural gas is universally associated with oil, being found in all oil-producing fields, but varying in amount and composition from field to field. It consists of the more highly volatile components (gaseous hydrocarbons) of petroleum at ordinary temperatures and pressures. These consist principally of methane, CH_4 and ethane, C_2H_6, but various amounts of propane, C_3H_8 and butane, C_4H_{10} may be present. Gaseous impurities are nitrogen, carbon dioxide, and hydrogen sulfide. Others in extremely small amounts are hydrogen, oxygen, and carbon monoxide. Helium is present in certain gases in Texas and Utah. Over 13 trillion cubic feet of gas are produced in the United States yearly, some 9 trillion coming from the Texas and Louisiana area.

Natural gas is used chiefly as a fuel, but it is also very important as a source of raw material for the chemical industry. Valuable components of natural gas, suitable for the chemical industry, are removed before the gas is sent to consumers. These components find application in inks, synthetic rubber, carbon black, explosives, drugs, and many other products. Liquid petroleum gas (bottled gas), used in heating and cooking, consists of butane or propane (or both). As it comes from the ground, natural gas often contains impurities which give it an unpleasant odor and makes the gas unpopular as a fuel. Researches were able to remove the impurities which produced the odor and also discovered many valuable components of gas at the same time. Clean, dry gas has no odor; so a chemical odorizer is added before it is sent to consumers. This odorizer produces a smell which is strong enough to be detected before the gas becomes dangerous. Natural gas, as well as manu-factured gas, is used in both factories and homes, serving mankind for heat, cooking, air conditioning, laundry drying, water heating, refrigeration, and burning of various wastes.

Natural asphalts and bitumens often accumulate where asphalt-base or paraffin-base crude oils seep out at the surface. These natural asphalts and bitumens are used chiefly for road materials, but they are also used in paints, pipe coatings, waterproofing, roofing, and floor tile.

Shales containing substantial amounts of bituminous material are called oil shale. The bituminous material, called kerogen, can be removed only by distillation (heating) processes and yields between 4 and 150 gallons of oil per ton of rock. However, the yield usually is between 18 and 40 gallons. These oil shales may be original source rock of petroleum from which no migration has occurred. Oil shales do not appear oily, and they are usually reddish-brown to black in color. Oil shales in Colorado, Utah, and Wyoming yield between 10 and 70 gallons of oil per ton, but at high costs which do not threaten the petroleum industry. However, large reserves are present.

Uranium

Uranium is a silvery-colored metal and is softer than iron. The metal and its compounds are poisonous, and chips of the metal may ignite spontane-ously in air. It melts at 1132°C (2065°F). The metal is both ductile and malle-able. At the present time, uranium is a minor metal used in weaponry and in alloys of steel, copper, and nickel. Uranium salts are used to give yellow and brown colors to glasses and glazes. However, the future outlook indicates that uranium will be extremely important as fuel in power-generating nuclear react-ors, in seawater desalinating nuclear reactors, and also as power supplies for ships, aircraft, and space vehicles. Its future uses all depend upon the fact that uranium is radioactive and is fissionable, causing the release of large amounts of energy.

The richer uranium deposits are of hydrothermal origin, being veins con-taining chiefly pitchblende, UO_2. Pure pitchblende is 100 percent UO_2, but it is seldom pure. Uranium also is found as irregular disseminations in sandstones and limestones, occurring chiefly as carnotite, $K_2(UO_2)_2(VO_4)_2 \cdot 3H_2O$, carrying about 63 percent UO_2. Uranium is also found as autunite, $Ca(UO_2)_2(PO_4)_2 \cdot 10H_2O$, meta-torbernite, $Cu(UO_2)(PO_4)_2 \cdot 8H_2O$, and tyuyamunite, $Ca(UO_2)_2(VO_4)_2 \cdot nH_2O$. Large world reserves are in the Blind River area, the Great Bear Lake area, and the Lake Athabasca area of Canada, and in the Witwatersrand area of South Africa. Fairly substantial reserves are located in the United States, as well as in other countries. Major United States' production comes from New Mexico, Utah, Colorado, Wyoming, South Dakota, and Arizona.

Water

Our most important and valuable natural resource is water. Water plays a critical role in the development of our society and civilization. One important role of water is that it determines the location of cities. It plays an important part in the selection of areas for new subdivisions, recreation sites, country homes, vacation resorts, and it even partly regulates fire insurance rates on various real estate assets. And, of course, it is essential in cooking and meal preparation.

Much of the Earth's water is *surface water*, located in such places as freshwater lakes, saline lakes, inland seas, stream channels, polar ice caps, and oceans. About 97 percent of the surface water is in oceanic areas, but oceanic water is too salty for general utility. The large polar ice caps are too far from major civilization and are not usable. Some of the Earth's water lies in rocks below the surface, and this is called *subsurface water*, located partly in shallow zones at depths of one-half mile or so and partly in deeper zones below a depth of one mile. Problems of water usually do not have to do with abundance, but rather with distribution and rates of supply. The abundance factor varies from locality to locality, and water's availability determines urban growth and population expansion. Subsurface water is much more important as a source of water than is surface water.

Subsurface water occurs in extremely large quantities. However, several problems exist regarding its recovery. In many places, rock porosity is so low and permeability so poor that flow of water is so slow that it cannot be obtained at a reasonable rate. Also, replenishment by rainfall often is slow because of excessive river run-off. Another factor deals with quality of water because, as water moves through various subsurface rocks, it dissolves various salts, often making it unfit for human consumption in many localities.

Uses of Water

Subsurface water is used in a great many ways, but these can be grouped into four main categories. These may be listed, along with the approximate amounts consumed daily in the United States, as follows:

Uses	Billion Gallons Per Day Approximate	Percent of Total Approximate
Rural (stock and domestic)	3.0	5.0
Municipal	8.0	13.1
Industrial	8.0	13.1
Irrigation	42.0	68.8

About half the people in the United States use groundwater in their homes, including those in rural areas and those in urban areas with public waterworks. Subsurface water is generally more feasible for use in small communities than is surface water. But large cities, with few exceptions, often obtain upland surface waters from great distances, or they use large lakes or rivers nearby. Subsurface water is generally readily available, and associated waterworks are inexpensive to install because wells are pretty cheaply constructed. On the other hand, production of surface waters usually means expensive reservoirs, with aqueducts and even offshore intake installations and tunnels to pumping stations, usually afforded only by large cities.

In addition, a great deal of subsurface water is recovered for industrial uses. However, slightly over two-thirds of the recovered subsurface water is used for irrigation purposes. With continued increase in population and industrialization, demand for water is growing rapidly, and search for new sources is very active. Groundwater usually is recovered through wells which may be drilled, bored, dug, or driven.

Dug wells are commonly shallow, and care must be taken that they are dug on up-gradient sides of possible sources of contamination. Bored wells are completed by hand or by powerdriven augers, ranging from 6 to 30 inches in diameter. Casing is placed in the holes, and the lower end is perforated for water inflow.

Driven wells are completed by forcibly driving pointed steel pipe into unconsolidated sediments. The lower end of the pipe is perforated for water inflow.

Drilled wells are usually fairly deep. They are drilled by cable tool or rotary methods. These wells are commonly cased, with perforations at the lower end. The water usually must be pumped, or in many cases, it may be artesian, flowing under its own pressure.

Subsurface water is continually being replenished as a result of surface water (precipitation) seeping downward. As long as withdrawal rates do not exceed replenishment rates, then continued use of subsurface water can be assured. But if utilization rates exceed natural return rates, then water-budgeting programs will need to be implemented.

In the United States, subsurface ground water is contained in a variety of rock masses known as aquifers. The quality and the amount of water available at any particular place are dependent upon geographic, geologic, and climatic conditions. Groundwater provinces in the United States differ somewhat according to supply and availability, but there are four major regions. These are the Coastal Plain region (New York to Texas), the East Central region (Maine to Minnesota to Oklahoma), the Great Plains region (Montana to West Texas), and the Western Mountain region (Washington to Montana to Southern California to New Mexico.

There seems to be a constant increased demand for water as technology advances; consequently, water is becoming more and more important as a factor in determining the location of manufacturing. Industries which require large quantities of water, such as steel, pulp and paper, food processing, and various chemical industries, will make sure of water quality and supply before constructing a new plant in any area. In fact, in many areas, the quantity of good water is a major problem as a result of contamination by sewage and chemicals.

Water needed for hydroelectric purposes is surface running water, with a good steady flow throughout the year, and with a slope or gradient sufficient to provide a satisfactory "head" or "fall." This requires that hydroelectric plants are constructed at the site of waterpower, which may not be close to market areas, thus possibly causing various economic problems.

Quality of Water

The use to which water is put determines how pure it should be. A chemical analysis is usually required to determine the quality, an analysis which determines the quantities and kinds of ions dissolved in the water. Bacteriological analyses also are often made to determine if drinking water is contaminated. Other tests to determine temperature, color, odor, and taste can also be made.

Standards for drinking water are set by the United States Health Service and adopted by the American Waterworks Association. Mandatory maximum limits for dangerous elements are as follows:

Lead	0.10 ppm (parts per million)
Fluorides	1.50 ppm
Arsenic	0.05 ppm
Selenium	0.05 ppm
Chromium	0.05 ppm
Total solids	1000.00 ppm

Standards for many less-dangerous elements are also set, as well as standards for physical characteristics and bacteria. Temporary "hardness" of water, which usually refers to the amount of calcium carbonate and magnesium carbonate in solution, is often of concern for laundering purposes, since these chemicals prevent dissolving of soap and reduce lathering. Permanent "hardness," caused by sulfates, can be removed only by complicated chemical processes.

Many industries, especially food industries, require extremely pure water or water with low concentrations of certain ions. However, such specifications vary considerably from industry to industry. Waters used for agricultural purposes must meet certain standards also, because some plants cannot tolerate certain salts in the irrigation waters. In areas where irrigation is desired, waters often contain such dissolved salts which often create severe problems of plant intolerance. In addition, pumping water to the surface over a period of time causes more salts to be returned to the groundwater supply, as the water slowly finds its way back downward. The ultimate result can be that contamination increases to the point of making the water supply unsuitable.

Quality of subsurface water often is decreased by several other factors, such as the following:

(1) inadequate treatment of sewage waters;

(2) disposal of sewage into streams;

(3) disposal of foods, oil, chemicals, etc., into surface streams or lakes;

(4) natural seepage of salt waters into subsurface supplies; and

(5) movement of poor-quality underground water into a supply of high quality.

The development of groundwater supplies continues to be important in areas where the annual precipitation rate is less than the annual withdrawal rate, such as in the arid and semiarid areas of southwestern United States. Long-range planning for any area must be based on the premise that supplies can always be greater than quantities used, even if good-quality water must be transported over fairly great distances.

Bibliography

AMERICAN PETROLEUM INSTITUTE, *Petroleum Facts and Figures*. Washington, D.C.: Amer. Petr. Inst., 1971.

AZAROFF, L. V., AND M. J. BUERGER, *The Power Method in X-ray Crystallography*. New York: McGraw-Hill Book Co., 1958.

BATEMAN, ALAN M., *Economic Mineral Deposits*. New York: John Wiley & Sons, Inc., 1950.

BATES, ROBERT L., *Geology of the Industrial Rocks and Minerals*. New York: Harper & Row, Publishers, 1960.

BLATT, HARVEY, G. MIDDLETON, AND R. MURRAY, *Origin of Sedimentary Rocks*. Englewood Cliffs, N. J.: Prentice-Hall, Inc., 1972.

BOWEN, N. L., *Evolution of the Igneous Rocks*. New York: Dover Publications, Inc., 1956.

BRAGG, L., G. F. CLARINGBULL, AND W. H. TAYLOR, *Crystal Structures of Minerals*. Ithaca: Cornell University Press, 1965.

CAMERON, E. N., R. H. JAHNS, A. M. McNAIR, AND L. R. PAGE, "Internal Structure of Granitic Pegmatites." *Econ. Geol.*, Mono. No. 2. Urbana, Ill.: Economic Geology Publishing Co., 1949.

COMMITTEE ON RESOURCES AND MAN, NAT. ACAD. SCI.–NAT. RES. COUNCIL, *Resources and Man*. San Francisco: W. H. Freeman & Co., 1969.

DANA, EDWARD S., (William E. Ford, *ed*), *A Textbook of Mineralogy*, 4th Ed., New York: John Wiley & Sons, Inc., 1948.

DUNBAR, CARL O., AND JOHN RODGERS, *Principles of Stratigraphy*. New York: John Wiley & Sons, Inc., 1957.

—————————, *Encyclopaedia Britannica*. Chicago: Encyclopaedia Britannica, Inc., 1970.

ERNST, W. G., *Earth Materials*. Englewood Cliffs, N. J.: Prentice-Hall, Inc., 1969.

EVANS, R. C., *Crystal Chemistry*. Cambridge: Cambridge University Press, 1966.

FENTON, CARROLL, AND MILDRED A. FENTON, *The Rock Book*. Garden City, N.Y.: Doubleday & Co., Inc., 1940.

FLAWN, PETER T., *Mineral Resources*. Chicago: Rand McNally & Co., 1966.

FRONDEL, CLIFFORD, *Dana's System of Mineralogy*, 7th ed., Vol. III–Silica Minerals. New York: John Wiley & Sons, Inc., 1962.

GREEN, JACK, "Geochemical Table of the Elements for 1959." *Geol. Soc. Amer. Bull*. Vol. 70 (1959), pp. 1127–1184.

HARKER, ALFRED, *Metamorphism*. New York: E. P. Dutton Co., Inc., 1939.

HURLBUT, CORNELIUS S., JR., *Dana's Manual of Mineralogy*. New York: John Wiley & Sons, Inc., 1971.

JAHNS, R. H., "The Study of Pegmatites." *Econ. Geol*. 50th Ann. Vol., Urbana, Ill.: Econ. Geol. Publishing Co., 1952, pp. 1025–1130.

KEMP, JAMES FURMAN, (Frank F. Grout, *ed*). *A Handbook of Rocks*, 6th Ed., New York: D. Van Nostrand Co., Inc., 1940.

LAMEY, CARL A., *Metallic and Industrial Mineral Deposits*. New York: McGraw-Hill Book Co., 1966.

LANDSBERG, H. H., *Natural Resources for U.S. Growth: A Look Ahead to the Year 2000*. Baltimore: Johns Hopkins Press, 1964.

LEET, L. DON, AND SHELDON JUDSON, *Physical Geology*. Englewood Cliffs, N.J.: Prentice-Hall, Inc., 1971.

MACDONALD, GORDON A., *Volcanoes*. Englewood Cliffs, N.J.: Prentice-Hall, Inc., 1972.

MACFALL, RUSSELL P., *Collecting Rocks, Minerals, Gems, and Fossils*. New York: Hawthorne Books, Inc., 1963.

MCALESTER, A. L., *The History of Life*. Englewood Cliffs, N.J.: Prentice-Hall, Inc., 1968.

MASON, BRIAN, "Trap Rock Minerals of New Jersey." *New Jersey Geol. Surv. Bull. 64*, 1960.

MASON, BRIAN, AND L. G. BERRY, *Elements of Mineralogy*. San Francisco: W. H. Freeman & Co., 1968.

MCDIVITT, J. F., *Minerals and Men*. Johns Hopkins Press, 1965.

MCGUINNESS, C. L., *The Role of Ground Water in the National Water Situation*. *U.S. Geol. Surv., Water-Supply Paper 1800*, 1963.

MEGGERS, WILLIAM F., *Key to the Welch Periodic Chart of the Atoms*. Chicago: W. M. Welch Manufacturing Co., 1959.

MERO, J. L., *The Mineral Resources of the Sea*. New York: Elsevier, 1965.

NOBLE, J. A., "The Classification of Ore Deposits." *Econ. Geol.*, 50th Ann. Vol., Part I. Urbana, Ill.: Econ. Geol. Publishing Co., 1955.

PAGE, L. R., et al., "Pegmatite Investigation, 1942–1945, Black Hills, South Dakota." *U.S. Geol. Surv. Prof. Paper 247*, 1953.

PALACHE, C., "The Minerals of Franklin and Sterling Hill, Sussex County, New Jersey." *U.S. Geol. Surv. Prof. Paper 180*, 1935.

PALACHE, C., et al., *Dana's System of Mineralogy.* New York: John Wiley & Sons, Inc., Vol. I, 1944; Vol. II, 1951.

PARK, CHARLES F., AND ROY A. MACDIARMID, *Ore Deposits.* San Francisco: W. H. Freeman & Co., 1970.

PEARL, RICHARD M., *Rocks and Minerals.* New York: Barnes & Noble, Inc., 1956.

PETTIJOHN, F. J., *Sedimentary Rocks.* New York: Harper & Row, 1957.

PHILLIPS, F. C., *An Introduction to Crystallography.* New York: John Wiley & Sons, 1968.

POUGH, FREDERICK H., *A Field Guide to Rocks and Minerals.* Boston: Houghton-Mifflin Co., 1951.

RILEY, CHARLES M., *Our Mineral Resources.* New York: John Wiley & Sons, Inc., 1959.

SCHURR, S. H., AND B. C. NETSCHERT, *Energy in the American Economy, 1850–1975.* Baltimore: Johns Hopkins Press, 1960.

SIMPSON, BRIAN, *Rocks and Minerals.* New York: Pergamon Press, 1966.

SINKANKAS, JOHN, *Mineralogy: A First Course.* Princeton, N.J.: D. Van Nostrand Co., Inc., 1966.

SKINNER, BRIAN J., *Earth Resources.* Englewood Cliffs, N.J.: Prentice-Hall, Inc., 1969.

STOKES, W. L., AND SHELDON JUDSON, *Introduction to Geology: Physical and Historical.* Englewood Cliffs, N.J.: Prentice-Hall, Inc., 1968.

SWINNERTON, H. H., *Outlines of Paleontology.* London: Edward Arnold & Co., 1947.

TRAVIS, RUSSELL B., "Classification of Rocks." *Quarterly of the Colorado School of Mines*, Vol. 50, No. 1 (Jan., 1955).

TURNER, FRANCIS J., AND JOHN VERHOOGEN, *Igneous and Metamorphic Petrology.* New York: McGraw-Hill Book Co., Inc., 1960.

TYRRELL, G. W., *Principles of Petrology.* London: Methuen & Co., Ltd., 1929.

U.S. BUREAU OF MINES, *Minerals Yearbook.* U.S. Gov. Print. Office, 1972.

VANDERS, IRIS, AND PAUL F. KERR, *Mineral Recognition.* New York: John Wiley & Sons, Inc., 1967.

WILLIAMS, H., F. J. TURNER, AND C. M. GILBERT, *Petrography.* San Francisco: W. H. Freeman, 1950.

Appendix

Table 1: METALS PRODUCTION, U.S. 1970*

Ferroalloy:		Short Tons
Iron ore		95,893,600
Manganese ore		373,039
Molybdenum (concentrates)		55,109
Nickel (in ore and concentrates)		15,933
Tungsten (in ore and concentrates)		8,194
Vanadium (in ore and concentrates)		5,319
Nonferrous:		Short Tons
Aluminum ore		2,290,200
Copper (in ores)		1,719,657
Lead (in ores)		571,767
Zinc (in ores)		534,136
Precious:	Troy Ounces	Short Tons
Gold	(1,743,322) (equivalent)	54
Silver	(45,005,000) (equivalent)	1,406
Minor:		Short Tons
Titanium (concentrates)		920,964
Mercury (29,360 76-lb flasks) (equivalent)		1,037

*After U.S. Bureau Mines *Minerals Yearbook* for 1970, 1972

Table 2: NONMETALLIC MINERAL PRODUCTION, U.S. 1970*

	Short Tons
Feldspar	713,104
Mica	118,843
Fluorite	269,221
Barite	854,000
Asbestos	125,314
Talc (incl. soapstone and pyrophyllite)	1,027,929
Vermiculite	285,000
Diatomite	597,636
Potash	2,729,000
Soda	2,688,000
Borates	1,041,000
Sulfur	7,060,900
Phosphates	38,739,000
Gypsum	9,436,000
Salt	45,804,000

*After U.S. Bureau Mines *Minerals Yearbook* for 1970, 1972

Table 3: INDUSTRIAL ROCK PRODUCTION, U.S. 1970*

	Short Tons
Stone: includes granite, basalt, slate, marble, sandstone, limestone	874,512,000
Pumice	3,134,000
Perlite	456,134
Sand-gravel	943,941,000
Clay	54,853,000

*After U.S. Bureau Mines *Minerals Yearbook* for 1970, 1972

Table 4: FUELS PRODUCTION, U.S. 1970*

Coal	612,661,000 Short tons
Petroleum (42-gallon barrels)	3,517,450,000 Bbl
Uranium (recoverable as U_3O_8)	12,341 Short tons

*After U.S. Bureau Mines *Minerals Yearbook* for 1970, 1972

Index

NOTE: Entries and page numbers in **boldface** indicate principal minerals described in considerable detail in Chapter 3 and the description locations.

Entries in *italics* are principal rocks described in some detail in the text.

Deposits *(cont.)*
  nature, 229
  nature of solutions, 229
Descartes, Rene, 350
Descloizite, 372
Diabase, 256, 393
Diagenesis *(see* Lithification)
Diamond:
  described, **100**
  geologic setting, 389
  nature, 388-389
  polymorph of carbon, 48
  sources, 389
  synthesized, 43
  uses, 388-389
Diaphaneity of minerals, 59-60
Diaspore, 376
Diatomaceous earth, 389
Diatomite:
  geologic setting, 389
  nature, 298-299, 389
  sources, 390
  uses, 389
Diatoms, 299
Diffractogram, 40-42
Diffractometer, 39-41
Dihexagonal dipyramid, 30-31
Dihexagonal prism, 30-31
Dike, 234-235
Diopside, 177, 188, 204
Diorite:
  colors, 253
  composition, 252-253
  texture, 252-253
  varieties, 252-253
Diorite porphyry:
  composition, 252
  texture, 252
Diploid, 28-29
Dipyramid, 30-35
Disseminated deposits, 353-354
Dissolved load, 270
Distorted crystals, 18, 21
Ditetragonal dipyramid, 32-33
Ditetragonal prism, 32-33
Dodecahedral cleavage, 72
Dodecahedral habit, 61
Dodecahedron, 28-29
Dogtooth calcite (Dogtooth spar), 141
Dolomite, 146
Dolomite rock:
  bedding, 306
  composition, 306, 380
  texture, 306
  uses, 398
Dolostone, 398
Double-chain silicates, 93

Dravite, 182
Dull luster, 58, 64
Dunite, 258

Early search for gems, 348
Early uses of minerals, 347-349
Earth materials:
  classification, 362-363
  definition, 1
Earthy luster *(see* Dull luster)
Effervescence, 141
Electrical properties, 76-77
Electrolytic refining, 366
Electromagnetic forces, 8
Electron(s), 6-13
Electrum, 96, 377
Elements:
  atomic radii, 6
  common, 5
  in common minerals, 81-83
  definition, 5
  symbols, 9, 81-83
  table, 9
  transition group, 52
Embryo crystals, 217
Emerald, 180, 384
Emery, 126
Enargite, 121, 373
Enstatite-hypersthene, 185
Environments of sedimentation:
  marginal, 271
  marine, 272
  terrestrial, 271
Epidote, 178
Eruptions (volcanic), 230
Evaporates:
  anhydrite, 296
  nitrates, 296
  rock gypsum, 296
  rock salt, 296-297
  soda salts, 296
Exfoliation, 266
Expanded perlite, 394-395
Exploration and discovery, 364
External structures in sedimentary rocks, 311
Extrusive igneous rocks *(see* Igneous rocks)

Fabric *(see* Structures of metamorphic rocks)
Faces on crystals:
  development, 20-23
  Miller indices, 22-23
  parameters, 21-23
  perfection, 20-21
  relation to atomic structure, 20-23
  relation to axes, 21-23